GRANULAR PATTERNS

Granular Patterns

Igor S. Aranson
Materials Science Division, Argonne National Laboratory, 9700 South Cass Avenue, Argonne, Illinois 60439, USA

Lev S. Tsimring
Institute for Nonlinear Sciences, University of California, San Diego, 9500 Gilman Drive, La Jolla, California 92093, USA

OXFORD
UNIVERSITY PRESS

Great Clarendon Street, Oxford OX2 6DP

Oxford University Press is a department of the University of Oxford.
It furthers the University's objective of excellence in research, scholarship,
and education by publishing worldwide in

Oxford New York

Auckland Cape Town Dar es Salaam Hong Kong Karachi
Kuala Lumpur Madrid Melbourne Mexico City Nairobi
New Delhi Shanghai Taipei Toronto

With offices in

Argentina Austria Brazil Chile Czech Republic France Greece
Guatemala Hungary Italy Japan Poland Portugal Singapore
South Korea Switzerland Thailand Turkey Ukraine Vietnam

Oxford is a registered trade mark of Oxford University Press
in the UK and in certain other countries

Published in the United States
by Oxford University Press Inc., New York

© I. Aranson & L. Tsimring, 2009

The moral rights of the authors have been asserted
Database right Oxford University Press (maker)

First Published 2009

All rights reserved. No part of this publication may be reproduced,
stored in a retrieval system, or transmitted, in any form or by any means,
without the prior permission in writing of Oxford University Press,
or as expressly permitted by law, or under terms agreed with the appropriate
reprographics rights organization. Enquiries concerning reproduction
outside the scope of the above should be sent to the Rights Department,
Oxford University Press, at the address above

You must not circulate this book in any other binding or cover
and you must impose the same condition on any acquirer

British Library Cataloguing in Publication Data

Data available

Library of Congress Cataloging in Publication Data

Data available

Typeset by
Printed in Great Britain
on acid-free paper by
the MPG Books Group

ISBN 978–0–19–953441–8 (Hbk)

10 9 8 7 6 5 4 3 2 1

To our parents

Preface

Granular materials are ubiquitous in our daily lives. Because of their technological importance, they have been a subject of intensive engineering research for centuries. Yet in the last two decades granular matter attracted significant attention of physicists interested in non-equilibrium phenomena and pattern formation. Granular materials proved to be a fertile ground for a crop of novel spatiotemporal patterns and mechanisms operating at vastly different scales, from bio-molecules to sand dunes. The main difficulty we had to face while writing this book was the lack of common approach in the theoretical description of emerging patterns. A number of competing conceptually different theoretical models have been proposed for describing the onset of collective behavior and pattern formation in granular matter. This books attempts to give a snapshot of the flurry of activity and bring a semblance of order and logic in this young and rapidly developing field.

This book is written primarily for physicists and engineers entering this field. However, we hope that experimentalists and theorists already working in the field will also find the book useful for its comprehensive review of the state-of-the-art in experimental and theoretical studies of granular patterns. It can also serve as a supplementary reading material for advanced graduate courses on granular physics, self-organization in non-equilibrium systems, and, to a small degree, biophysics. The book is accompanied by a CD with a collection of movies illustrating various aspects of pattern formation in granular systems.

We would like to thank our friends and colleagues Guenter Ahlers, Bob Behringer, Eshel Ben Jacob, Eli Ben-Naim, Daniel Blair, Jean-Philippe Bouchaud, Paul Chaikin, Hugues Chaté, Philippe Claudin, Eric Clement, Sue Coppersmith, George Crabtree, Adrian Daerr, Pierre-Gilles de Gennes, James Dufty, Jacques Duran, Douglas Durian, Robert Ecke, Denis Ertas, Chay Goldenberg, Isaac Goldhirsch, Ray Goldstein, Jerry Gollub, Gary Grest, Thomas Halsey, Jeff Hasty, Haye Hinrichsen, Jacob Israelachvili, Henrich Jaeger, James Jenkins, Leo Kadanoff, Devang Khakhar, Evelyne Kolb, Lorenz Kramer, Arshad Kudrolli, Wai Kwok, James Langer, Anaël Lemaître, Jie Li, Wolfgang Losert, Stefan Luding, Baruch Meerson, Francisco Melo, Stephen Morris, Sid Nagel, Jeffrey Olafsen, Julio Ottino, Len Pismen, Thorsten Pöschel, Olivier Pouliquen, Jacques Prost, Sriram Ramaswamy, Maksim Saposhnikov, Peter Schiffer, Leo Silbert, Alexey Snezhko, Harry Swinney, Julian Talbot, Paul Umbanhowar, Jeffrey Urbach, Alexandre Valance, Martin van Hecke, Wim van Saarloos, Valerii Vinokur, Dmitri Volfson, Thomas Witten, Dietrich Wolf, Falko Ziebert, and many others from the granular physics community for many useful discussions and suggestions in the course of preparation of this book.

Contents

1	**Introduction**	1
2	**Experimental overview of patterns in granular matter**	6
	2.1 Pattern formation in vibrated layers	6
	2.2 Gravity-driven granular flows	10
	2.3 Flows in rotating cylinders	13
	2.4 Granular impacts and craters	18
	2.5 Grains with complex interactions	19
	2.6 Granular patterns in biology	23
3	**Main theoretical concepts and tools**	30
	3.1 Fundamental microscopic properties of granular matter	30
	3.2 Multiparticle effects	34
	3.3 Kinetic theory and granular hydrodynamics	39
	3.4 Statistical mechanics of dense granular matter	45
	3.5 Continuum mechanics of granular materials	50
	3.6 Phenomenological models	54
	3.7 Molecular dynamics simulations	56
4	**Phase transitions, clustering, and coarsening in granular gases**	59
	4.1 Clustering in freely cooling granular gases	59
	4.2 Patterns in granular gases heated by the side walls	68
	4.3 Granular thermoconvection	78
	4.4 Clustering of grains on a vibrated substrate	84
5	**Surface waves and patterns in periodically vibrated granular layers**	91
	5.1 Heaping in vertically vibrated layers	91
	5.2 Standing wave patterns	95
	5.3 Simulations of vibrated granular layers	98
	5.4 Phenomenological continuum theories	101
	5.5 Hydrodynamic description of pattern formation in vibrated layers	110
6	**Patterns in gravity-driven granular flows**	117
	6.1 Rheological properties of dense granular flows	118
	6.2 Continuum models of gravity-driven granular flows	125
	6.3 Avalanches in thin granular layers	142
	6.4 Longitudinal instability of granular flow along a rough inclined plane	156
	6.5 Pattern-forming instabilities in rotating cylinders	159
	6.6 Self-organized criticality and statistics of granular avalanches	164

7 Patterns in granular segregation — 167
- 7.1 Basic mechanisms of granular segregation — 167
- 7.2 Granular stratification — 179
- 7.3 Radial segregation patterns in a rotating drum — 184
- 7.4 Axial segregation in rotating drums — 190
- 7.5 Other examples of granular segregation — 200

8 Granular materials with complex interactions — 203
- 8.1 Vortices in vibrated layers of granular rods — 203
- 8.2 Swirling motion in monolayers of vibrated elongated particles — 212
- 8.3 Global rotation in the system of chiral particles — 223
- 8.4 Patterns in solid–fluid mixtures — 225
- 8.5 Electrically driven granular media — 241
- 8.6 Magnetic particles — 253

9 Granular physics of biological objects — 259
- 9.1 Nematic ordering of growing colonies of non-motile bacteria — 259
- 9.2 Self-organization of microtubules interacting via molecular motors — 267
- 9.3 Collective dynamics of self-propelled particles — 285
- 9.4 Collective swimming of motile bacteria in thin fluid films — 294

References — 311

Notations — 339

Index — 341

1
Introduction

Granular materials are collections of many macroscopic solid grains with a typical size large enough that thermal fluctuations are negligible. Despite this seeming simplicity, bulk properties of granular materials are often different from conventional solids, liquids, and gases due to the dissipative nature of forces acting on interacting grains, such as inelasticity of collisions, dry friction, or viscous drag. For granular systems to remain active they have to gain energy from external sources. In addition to shear and vibrations, external volume forces such as gravity, electric and magnetic fields, flows of interstitial fluids may also activate grains. When subjected to a large enough driving force, a granular system may exhibit a transition from a granular solid-like state to a liquid or flowing state, and various ordered patterns of grains may develop. These patterns are the subject of this book.

Understanding the fundamentals of granular materials draws upon and gives insights into many fields at the frontier of modern physics: plasticity of solids, fracture and friction, complex systems such as colloids, foams, suspensions, and a variety of biological multicellular systems. Particle-laden flows are widespread in geophysics, for example, they are involved in dune formation and migration, erosion/deposition processes, landslides, formation of coastal geomorphology etc. Particulate flows are essential to many industrial applications including chemical, pharmaceutical, food, metallurgical, agricultural, and construction.

From a theoretical point of view, it is tempting to invoke an analogy between granular materials and ordinary condensed matter systems and to regard the grains as the equivalent of classical atoms. However, as we pointed out above, dissipative interactions among grains leads to essentially different phenomenology that cannot be captured by a standard equilibrium condensed matter theory. In particular, dissipation is responsible for the fact that most static regimes of granular matter are metastable and the system does not typically evolve towards certain "free energy" minimum. The macroscopic size of the grains renders thermal fluctuations negligible and most standard thermodynamic concepts, such as energy equipartition, generally inapplicable. This situation calls for the formulation of new theories explicitly taking into account the non-equilibrium character of granular dynamics. For dilute granular systems such a theory can be developed along the lines of the kinetic theory of dilute gases on the basis of the Boltzmann equation with the collision integral adapted to take into account inelasticity of interparticle collisions. From the corrected Boltzmann equation, granular hydrodynamic theory can then be derived following a standard Chapman–Enskog expansion. While this theory can satisfactorily explain the behavior of dilute granular systems (granular gases), a similar theory of dense granular assemblies is far less developed. In the subsequent chapters of the book we discuss several alternative approaches to this difficult and still unsolved problem.

The scientific studies of granular materials began more than two centuries ago and are associated with such illustrious names as Leonard Euler, Michael Faraday, Osborne Reynolds, and Charles Coulomb; however during most of the twentieth century it was predominantly the domain of applied engineering research. But in the last two decades this field has experienced a renaissance of sorts. The renewed interest among physicists in granular materials was probably spurred by the seminal idea of self-organized criticality introduced by Per Bak *et al.* (1987), which was originally associated with the behavior of sandpiles. While it was later realized that real sandpiles do not exhibit self-organized criticality in the strict sense, the physics community "rediscovered" a fascinating world of granular physics, and a new generation of granular studies using modern experimental techniques and theoretical concepts ensued. Another reason behind the resurgence of granular physics renaissance is that many physicists interested in pattern formation in "ordinary" fluids (gases, liquids) turned their attention to the new and exciting domain of granular media. From a theoretical perspective, granular physics lies at the crossroads of fluid dynamics, non-equilibrium statistical mechanics, and pattern formation. The "avalanche" of papers dealing with all the aspects of granular physics began in the beginning of the 1990s and continues to grow to this day. Annual meetings of the American Physical Society in the last several years held several "granular" sessions every day, and conferences and schools on granular physics proliferate at a remarkable rate.

This rapid progress necessitates periodic reviews and reflections. Even though such reviews usually become obsolete the moment they go into press, their usefulness far outweighs the lack of completeness. There have been several reviews written on the subject of granular physics, starting from the influential Behringer, Jaeger, and Nagel 1996 review "Granular gases, liquids, and solids", followed by several others (de Gennes (1999); Gollub and Langer (1999); Kadanoff (1999); Rajchenbach (2000), Kudrolli (2004), Aranson and Tsimring (2006*b*). Several monographs have been published as well, such as "Pattern formation in Granular Materials" (2001) by G. Ristow, "Sands, Powders, and Grains" (1997) by J. Duran, "Kinetic Theory of Granular Gases" (2004) by N. Brilliantov and T. Pöschel. The closest to the theme of this monograph is G. Ristow's book, however, it is already more than six years old and it is mostly devoted to experiments rather than to theoretical aspects of granular dynamics and pattern formation. That is why we believe that the time for publication of a monograph with the emphasis on the latest theoretical developments and modelling in the field of granular physics is ripe.

We believe that the lack of comprehensive reviews or monographs focused on theoretical developments in granular physics is not coincidental. The difficulty in writing such a book lies in the fact that granular physics is still a mixture of many different concepts, modelling tools, and phenomenological theories. Different experimental results are routinely described by different theories not having a common denominator in the form of a fundamental set of equations such as Navier–Stokes equations for fluids or Maxwell equations for plasmas. In our recent review (Aranson and Tsimring, 2006*b*) we took the extensive approach describing this multitude of different theoretical concepts to a broad variety of pattern-formation problems in granular media. While slightly reducing the scope of material, in this book we develop a more logical and didactic approach to the theory of granular patterns. Of course, we had to reflect the variety of sometimes incompatible approaches and methods being developed by many theoretical groups for description of different (or sometimes the same) experimental problems, but we tried to emphasize common threads and ideas that are employed in the theoretical modelling. While having a unified description of all different types of granular materials may indeed be an unreachable goal, it should be possible to develop such a theory for at least some canonical model systems, e.g. monodisperse hard frictionless inelastic spheres. In fact, a unified description exists for a dilute system of rapid grains in the form of the kinetic theory of dissipative gases. However, for the densities approaching the close-packing density (and this situation represents the vast majority of practically relevant cases), no universal description exists to date. We believe (admitting that this belief may be affected by a personal bias) that some combination of the fluid dynamics with phase-field modelling based on an order-parameter description of the state of the granular material, can be one possible candidate for the unified theory.

The scope of granular physics has become so broad that we chose to limit ourselves to a subfield of granular pattern formation leaving out many interesting and rapidly developing areas. We loosely define pattern formation as a dynamical process leading to the spontaneous emergence of non-trivial spatially non-uniform and possibly time-dependent structures that are weakly dependent on initial and boundary conditions. Consequently, we have left out detailed description of such interesting and important topics as anisotropic stress propagation in quasistatic granular materials, jamming, acoustic phenomena, and many others. The unique feature of our book is that we make a strong effort to connect concepts and ideas developed in granular physics with new emergent fields, especially in biology, such as cytoskeleton dynamics, molecular motors transport, organization of active (self-propelled) particles and dynamic self-assembly.

The structure of the book is the following. For the sake of simplicity and transparency of our presentation, through the entire span of the book we discuss the granular patterns in the context of phenomena rather than the methods. However, we make a consistent and systematic effort to emphasize the similarity of various techniques and approaches to seemingly different phenomena occurring in granular patterns when this is appropriate.

The book contains nine chapters focused on various classes of granular patterns. We start with a brief experimental overview of key granular patterns and phenomena (Chapter 2). There we introduce the primary experimental systems, such as vertically and horizontally vibrated granular layers, rotating drums, flow down an inclined plane, a variety of experiments with "complex" grains (charged, magnetic, and "biological" particles), and present the most relevant patterns that will be discussed in detail in the following chapters.

In Chapter 3 we introduce the main theoretical approaches and models employed in the physics of granular media, such as the kinetic theory of diluted granular gases, various methods of molecular dynamics simulations (event driven, soft particles, contact dynamics), order-parameter-based phase-field models, depth-averaged and two-fluid models of dense flows, and a variety of other phenomenological models.

In Chapter 4 we focus on clustering and coarsening phenomena in (initially) dilute granular systems, such as shock formation and dynamics in freely cooling granular gases, phase separation, van der Waals-like instability in driven granular gases, clustering, solid–gas transitions, and coarsening of multiple clusters in vertically vibrated granular submonolayers. A variety of different methods are applied, from granular hydrodynamics to molecular dynamics and phenomenological theories, including the Burgers, Cahn–Hilliard and van der Waals normal form equations. The theoretical approaches are discussed in the context of relevant experiments with vertically vibrated granular sub-monolayers. In this chapter we also discuss granular "thermoconvection" occurring in granular layers for very high frequency and amplitude of vibration when the vibrating bottom is equivalent to the "hot wall". In this situation the phenomenon of convection can be captured in the framework of granular hydrodynamics for not too dense gases.

Standing-wave resonant patterns, such as localized oscillons and periodic squares, stripes and hexagons, emerging in periodically vibrated granular layers are discussed in Chapter 5 in the framework of various phenomenological theories based on the Ginzburg–Landau-type amplitude equations or coupled maps lattices. Also, specific effects of noise on pattern formation in granular systems are considered.

In Chapter 6 we focus on fundamental properties and models of patterns in dense shear granular flows energized by gravity, such as avalanches and other surface flows down an inclined plane. In the beginning of this chapter we summarize primary experimental facts on the rheology of dense granular flows. Then, we describe the phenomenon of granular avalanche formation and evolution in the framework of several complimentary theoretical approaches: granular hydrodynamics, partial fluidization theory for dense shear flows, depth-averaged (Saint-Venant) equations, and two-phase models. The applicability and limitations of the theories are discussed in detail in connection with experimental results for avalanches and surface flows.

Granular segregation phenomena in polydisperse granular systems are treated in Chapter 7. There we focus on the theoretical models for axial and radial segregation in rotating drums, granular stratification in granular flows down an inclined plane (chute) and stripe formation in horizontally vibrated thin layers. While some aspects of segregation phenomena can be understood in the context of kinetic theory for bidisperse gases, the majority of segregation pattern occur in the regime of dense granular flows where the kinetic theory is not applicable. Thus, phenomenological approaches are used to obtain insights into the observed behavior.

Patterns found in granular assemblies with complex interactions are discussed in Chapter 8. There we consider non-trivial phenomena occurring in particle–fluid mixtures such as sand dunes and ripples, patterns formed in ensembles of electrically and magnetically interacting or driven grains, and dynamic structures such as vortices and swirls, emerging in driven layers of anisotropic grains. Since in these situations rigorous treatment of the problem is usually impossible or prohibitively difficult, the progress is achieved mostly by a combination of numerical modelling and phenomenological approaches.

Finally, in Chapter 9 we introduce the rapidly growing field of pattern formation in multiunit biological systems that may fall into the category of granular systems with non-trivial interaction rules. In particular, we discuss ordering and crystallization of growing bacteria colonies, formation of large-scale patterns due to interaction of microtubules and molecular motors, and organization of self-propelled micro-organisms, such as swimming bacteria or other motile cells. In this chapter we make a consistent effort to illustrate connections between the emergent collective behavior in concentrated populations of bioparticles and more traditional areas of the granular physics.

2
Experimental overview of patterns in granular matter

In this chapter we give a short overview of the key experiments and phenomena occurring in granular media that will be discussed in greater depth in the following chapters. We classify experiments according to the way energy is injected into the system: vibration, gravity, or shear. Despite tremendous diversity of granular systems and patterns, one nevertheless may find some degree of universality irrespective of the way energy is supplied to the system: the onset of large-scale collective motion in sufficiently dense systems, hydrodynamic-like instabilities, clustering, segregation.

2.1 Pattern formation in vibrated layers

A mechanically vibrated thin layer of grains is probably the most studied system exhibiting remarkably regular large-scale patterns. In a typical experiment a thin layer of grains is energized by precise vertical vibrations produced by an electromechanical shaker, see Fig. 2.1(a). In order to reduce the effects of ambient air experiments are sometimes performed in evacuated containers; evacuation becomes especially important for the experiments with very small (under 100 micrometers) grains. The evolution of granular patterns is monitored by a high-speed digital video camera suspended above the container.

There are several important dimensionless parameters affecting the dynamics of vibrated layers: thickness of the layer (in number of particle diameters d) h, aspect ratio, i.e. the ratio of the horizontal system size to the thickness of the layer, dimensionless acceleration Γ, i.e the amplitude of the vibration acceleration of the bottom plate normalized by the gravity acceleration g: $\Gamma = 4\pi^2 f^2 A_0/g$, where A_0, f are the displacement amplitude and the frequency of vibration.

For sufficiently high magnitude of vibrations $\Gamma > 2$, a submonolayer of grains, i.e. an assembly with less than 100% coverage by particles of the bottom plate, behaves as a quasi-two-dimensional gas with nearly uniform spatial density. However, upon reducing Γ below a certain critical value Γ_c that depends on the area fraction, a surprising transition to a bimodal regime occurs, which is characterized by a single dense cluster of closely packed almost immobile grains surrounded by a gas of agitated particles, (Olafsen and Urbach, 1998), Fig. 2.1(b). A similar clustering transition occurs in a non-driven (freely cooling) gas of inelastic particles (Goldhirsch and Zanetti, 1993). In the freely cooling gas, hot (fast) particles collide with cold (slow) particles in the clusters and lose their energy due to inelasticity, cooling the system down. Clusters gradually absorb more and more particles and eventually the relative motion of all particles stops, i.e. the system freezes.[1] In vibrated submonolayers, the energy is constantly supplied to particles, and a dynamical equilibrium is established eventually in which the energy supply to hot particles is balanced by the energy losses due to collisions of particles with each other, the bottom plate, and the cold cluster.

The vibrated submonolayer experiment is sometimes considered a "benchmark" experiment in granular physics because the positions and velocities of all particles can be precisely measured, and thus a detailed comparison with theoretical models is feasible. A detailed consideration of clustering phenomena in submonolayers of grains is given in Chapter 4.

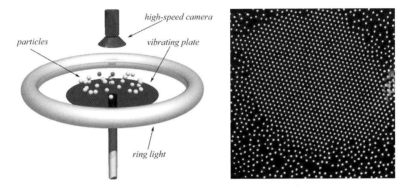

Fig. 2.1 Left panel: a typical experiment with vibrated granular layers. Particles are energized through collisions with a vibrating plate. The ring light uniformly illuminates particles for high-speed imaging using a video camera suspended above the plate. Right panel: top view of a submonolayer of particles (1-mm steel spheres) on a vibrated plate. Here, a dense immobile cluster with nearly crystalline order coexists with dilute granular gas, courtesy of Jeffrey Olafsen and Jeffrey Urbach.

Vibrated multilayers of granular materials can exhibit spectacular pattern formation. In fact, the first observations of patterns in vibrated layers were made more than two centuries ago by Chladni (1787) and Faraday (1831) who observed that thin powder is accumulated along certain lines on a membrane excited by a violin bow, see Fig. 2.2. Faraday realized that those lines correspond to nodal lines of a particular mode of oscillations of the membrane excited by the bow. He was also the first who pointed out the important role of interstitial air that affected heaping of thin powders. One puzzling result by Chladni was that a very thin powder would collect at the antinodal regions where the amplitude of vibrations is maximal. As Faraday demonstrated by evacuating the container, this phenomenon is caused by air permeating the grains in motion. A vibrating membrane generates convective flows in the air above it due to so-called "acoustic streaming", and these flows transport powder towards the antinodal regions. Evidently, the interstitial gas becomes important when the terminal

---

[1] However, a rigid–body rotation of cold (frozen) clusters is possible due to conservation of total mechanical momentum.

velocity of a free-falling particle $v_t = \nu g d^2/18\eta$ (η is viscosity of air, ν density of the particle) becomes comparable with the plate velocity, and this condition is fulfilled for $10 - 20 \mu$m particles on a plate vibrating with frequency 50 Hz and acceleration amplitude g.

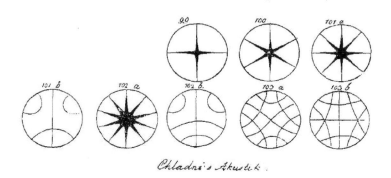

Fig. 2.2 Reproduction of some original Chladni figures, hand-drawings by Chladni (1787). A more complete set of reproductions can be found in Waller (1938)

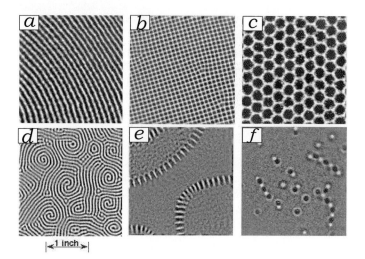

Fig. 2.3 Patterns in a vertically vibrated granular layer for various values of frequency and amplitude of the vibration: stripes, squares, hexagons, spirals, interfaces, and localized oscillons. Panels show snapshots of the layer surface under low-angle oblique lighting: troughs appear dark and crests are light. Courtesy of Paul Umbanhowar.

Interest in patterns in vibrated granular layers was renewed at the end of the 1980s, in particular due to the development of new experimental techniques and data-acquisition methods. In a "typical" modern experiment, a layer of grains with thickness

Fig. 2.4 An oscillon in a vertically vibrated granular layer, courtesy of Paul Umbanhowar.

h of about 10–30 particle diameters is energized by a precise vertical vibration produced by an electromagnetic shaker. Experiments have been performed with various particle types (smooth spherical glass beads, copper shot, rough sand particles, plant seeds, etc.), in vacuum or under atmospheric pressure, with systematically varying amplitude and frequency of vibration. Depending on experimental conditions, a plethora of patterns has been observed, from stripes and squares to hexagons and interfaces; see Figure 2.3. The resulting patterns are typically *subharmonic*: two cycles of external driving are necessary to repeat the original pattern waveform. In a narrow range of parameters at the threshold of the primary pattern-forming instability, remarkable localized objects, dubbed *oscillons*, were discovered by Umbanhowar *et al.* (1996), see Fig. 2.4. Detailed consideration of these phenomena and their theoretical description will be given in Chapter 5.

For very high frequency of vibration, when for a fixed value of Γ, the plate displacement becomes smaller than the particles diameter, a new interesting phenomenon of granular convection emerges, see Fig. 2.5. The particle collisions with the rapidly vibrating bottom plate can be approximately described as the interaction with a "hot wall". Because of inelastic collisions, the effective granular temperature decays off the wall, and so a density inversion occurs. When the effective temperature of the wall is sufficiently high, this density inversion becomes unstable, and a set of slowly rotating convective rolls emerges. This granular convection is reminiscent of the Raleigh–Bénard convection in conventional fluids heated from below. While convection rolls reported by Wildman *et al.* (2001) could possibly be caused by the friction with sidewalls, more recent experiments by Eshuis *et al.* (2007) provide strong evidence of the density inversion and the thermal mechanism of granular convection.

Vertical vibration is not the only way to inject the energy in a granular layer. Interesting patterns are also observed when a thin granular layer is shaken horizontally (Ristow, 1997; Liffman *et al.*, 1997; Tennakoon *et al.*, 1998; Medved, 2002). While there are certain common features between resulting patterns, such as subharmonic regimes and instabilities, horizontally vibrated systems have not shown the richness of behavior typical of the vertically vibrated systems, and non-trivial flow regimes, such as convection rolls, are typically localized near the side walls, see Fig. 2.6.

When the granular matter is polydisperse, vertical or horizontal shaking often leads to segregation. The most well-known manifestation of this segregation is the so-called "Brazil-nut" effect when under vertical shaking large particles rise to the surface of a thick granular layer (Rosato *et al.*, 1987; Breu *et al.*, 2003; Shinbrot, 2004), see Fig. 2.7. The direction of motion of larger particles depends sensitively on the mass density ratio between large and small particles and the parameters of vibration; a reverse Brazil-nut effect (when large grains sink to the bottom) is possible for certain driving conditions. Horizontal shaking can also produce interesting segregation band patterns oriented orthogonally to the direction of shaking (Mullin, 2000; Mullin, 2002), see Fig. 2.8.

2.2 Gravity-driven granular flows

Gravity-driven granular systems such as flows down inclined planes (chutes) and sandpiles often exhibit non-trivial patterns and spatiotemporal structures. One of the most familiar examples of a gravity-driven granular flow is an *avalanche*. Obviously, the dynamics of natural avalanches is "contaminated" by many factors such as topography, inhomogeneity of granular matter (from rocks to snow and dust), weather, vegetation, etc. Fundamental aspects of the avalanche dynamics are usually studied in controlled laboratory experiments with dry or wet granular piles. Granular slopes can be characterized by two *angles of repose* – the *static* angle of repose φ_s, which is the maximum angle at which the granular slope can remain static, and the *dynamic* angle of repose φ_d, or a minimum angle at which the grains can still flow down the slope. Typically, in dry granular media the difference between the static and dynamic angles of repose is relatively small, about $2° - 5°$, e.g. for smooth glass beads $\varphi_s \approx 25°$, $\varphi_d \approx 23°$. Avalanches typically occur in the bistable regime when the slope angle $\varphi_d < \varphi < \varphi_s$. The bistability is explained by the need to *dilate* the granular material for it to enter the flowing regime (*Reynolds dilatancy*, see Section 3.2.5). An avalanche can be initiated by a small localized perturbation from which the fluidized region expands downhill and sometimes also uphill, while the sand of course always slides downhill. An avalanche in a deep sandpile usually involves a narrow layer near the surface, see Fig. 2.9. Interesting spatiotemporal dynamics of avalanches has also been observed

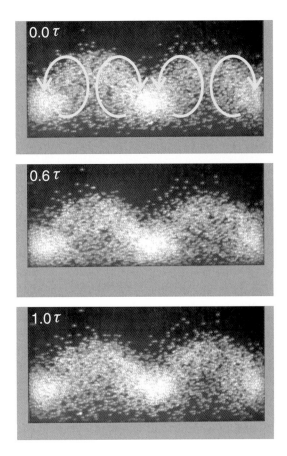

Fig. 2.5 Four rolls of granular convection in the layer of 1-mm glass beads, layer depth 8.1 particles diameters, frequency of vibration $f = 73$ Hz, dimensionless acceleration $\Gamma = 64$ at three different moments of time during one period of external vibration $\tau = 2\pi/f$. From Eshuis et al. (2007)

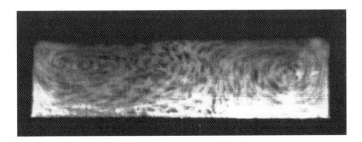

Fig. 2.6 Two symmetrically positioned convection rolls localized near sidewalls in a horizontally vibrated container. The streaks is due to the motion of dark tracer particles, from Medved (2002)

12 Experimental overview of patterns in granular matter

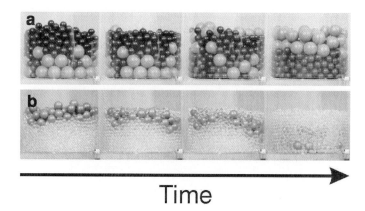

Fig. 2.7 Illustrations of the normal (a) and reverse (b) Brazil-nut effects, from Breu *et al.* (2003). In the upper row of images light large particles (15-mm polypropylene spheres) rise to the top of the layer of 8-mm glass beads (classic Brazil nut effect) upon vertical shaking. In the lower set of images the heavier 10-mm bronze spheres sink to the bottom of the layer of lighter 4-mm glass beads.

in thin granular layers on inclined planes (Daerr and Douady, 1999; Daerr, 2001). A typical laboratory-size setup for studies of granular avalanches in chutes is shown in Fig. 2.10. To avoid slippage of grains at the bottom of the chute, the latter is often made rough, either by covering it by a soft cloth or by gluing a layer of particles. The inclination angle can be adjusted by a computer-controlled winch. Patterns of flowing sand can be made visible by a sidewise illumination with a laser light sheet: elevated areas appear brighter than troughs. Using relatively long exposures or by subtracting an image of the unperturbed granular layer from snapshots of developing avalanches, flowing sand within the avalanche can be visualized. The two-dimensional structure of a developing avalanche depends on the thickness of the granular layer and the slope angle. For thin layers and small angles, wedge-shaped avalanches are formed similar to the loose snow avalanches (Fig. 2.11(a)). In thicker layers and at higher inclination angles, avalanches have a balloon-type shape and expand both down- and uphill (Fig. 2.11(b)).

Gravity-driven granular flows are prone to a variety of secondary instabilities arising under certain flow conditions: fingering of avalanche front, see Fig. 2.12 (Pouliquen *et al.*, 1997), modulation waves and longitudinal vortices in rapid chute flows (Forterre and Pouliquen, 2001; Forterre and Pouliquen, 2003), see Fig. 2.13, and others.

A rich variety of patterns and instabilities also exists in flows of granular matter either completely or partially immersed in liquid. Underwater avalanches exhibit transverse instability of avalanche fronts, fingering, pattern formation in the sediments behind the avalanche, etc. (Daerr *et al.*, 2003; Malloggi *et al.*, 2006). These patterns are generally similar to those in "dry" granular layers, however, long-range hydrodynamic forces (Chapter 8) are responsible for certain qualitative differences in the observed phenomenology.

Interesting regular channelization patterns are often observed on beaches or sand

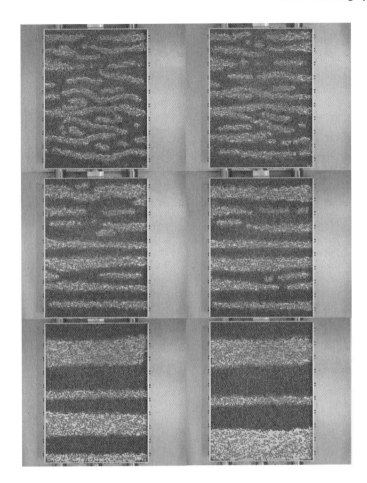

Fig. 2.8 Snapshots demonstrating segregation of a layer of copper balls/poppy seeds mixture in a horizontally shaken cavity (frequency 12.5 Hz, amplitude 2 mm) at times 5 min, 10 min, 15 min, 30 min, 1 h, and 6 h after the start of the experiment, from Mullin (2000).

dunes after heavy rains, see Fig. 2.14. These erosion patterns are formed as water seeps through an inclined surface of a granular layer. Detailed laboratory experiments and numerical simulations (Daerr et al., 2003; Schorghofer et al., 2004) revealed that the small channels are typically initiated (incised) by the transversal instability of the thin surface water runoff, but then these channels are enlarged and coarsen due to the secondary Wentworth instability and the "competition" for seeping groundwater. The characteristic spatial scale of the pattern (the separation between major channels) increases with time and the distance from the beginning of the channel.

2.3 Flows in rotating cylinders

Energy can also be supplied into a granular system through a shear imparted by moving boundaries through friction. A typical experimental realization of this type of

14 *Experimental overview of patterns in granular matter*

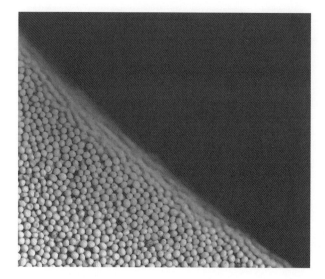

Fig. 2.9 Surface flow in a pile of mustard grains. Only seeds within the top few layers participate in the flow, as indicated by the blurred streaks in this exposure, from Jaeger *et al.* (1996).

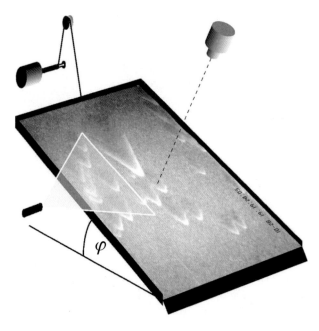

Fig. 2.10 Sketch of a typical avalanche experiment. The grains are spread on an inclined plane covered by velvet cloth. Camera fixed above the plane records patterns illuminated by a laser light sheet at a small incident angle, from Daerr (2001).

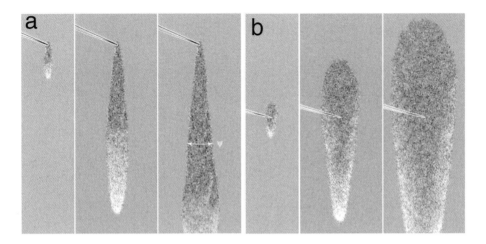

Fig. 2.11 Evolution of avalanches in granular layers on inclined chute for two different layer thicknesses: (a) development of a wedge avalanche in a thin layer; (b) evolution of a balloon avalanche in a thicker layer. The images ere obtained by subtracting the photo of the unperturbed layer from the photos of the layer with developing avalanches; from Daerr and Douady (1999).

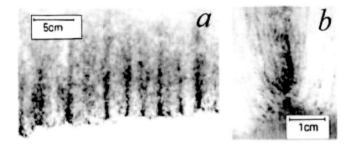

Fig. 2.12 Fingering instability in a chute flow. Images taken from the bottom (a) and front (b) of the layer illustrate accumulation of coarse particles between advancing fingers, from Pouliquen et al. (1997).

energy supply is the Taylor–Couette flow of grains in an annulus (a narrow gap between two counter-rotating cylinders with the axis aligned with the direction of gravity (Losert et al., 2000; Mueth et al., 2000). While in experiments with conventional fluids at high enough rotation rates an array of so-called Taylor vortices stacked along the cylinder axis forms, no non-trivial dynamic patterns have been observed in monodisperse granular Taylor–Couette experiment even for very high shear rates. However, an analog of Taylor vortices has been observed in bidisperse granular mixtures (Conway et al., 2004).

Another commonly used system for studies of granular flows and mixing is the

16 *Experimental overview of patterns in granular matter*

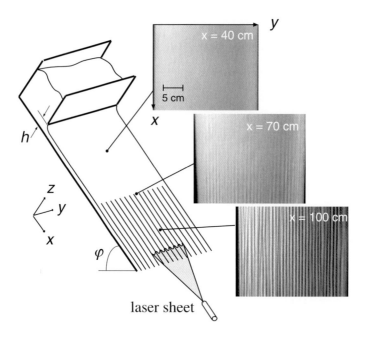

Fig. 2.13 Longitudinal vortices in a rapid granular flow down a rough incline. Three photographs are taken from above the flowing layer at different distances from the inlet, from Forterre and Pouliquen (2001).

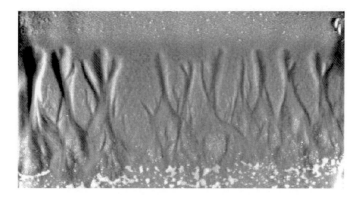

Fig. 2.14 Examples of multiple drainage channels in a laboratory experiment on seepage erosion of granular layer, from Schorghofer *et al.* (2004).

so-called *rotating drum*, a partially filled horizontal cylinder rotating around its axis. Rotating drums and tumblers (cylinders with non-circular cross-sections) are often used in chemical engineering for mixing or separation of particles. For not too high rotating rates the flowing grains are localized near the free surface, while the bulk exhibits an almost solid-body rotation. In slowly rotating drums the near-surface flow is intermittent, with quiescent periods of the gradual increase of the free-surface angle towards the static angle of repose and subsequent fast relaxation to a lower dynamic angle of repose via an avalanche, after which the process repeats. At higher rotation rates, as the time interval between avalanches becomes comparable with the avalanche duration, a transition to the steady flow regime occurs (Rajchenbach, 1990). Scaling of various flow parameters with the rotation speed (e.g. the width of the fluidized layer etc.) and the development of correlations in "dry" and "wet" granular matter was recently studied by Tegzes et al. (2002), Tegzes et al. (2003). For example, in the case of "wet avalanches", the increase of moisture content in a certain range significantly changes the avalanche morphology and roughness due to attractive interactions between wet grains. As a result, large correlated regions of grains begin moving together down the free surface.

Rotating drums are often used to study size segregation in mixtures of granular materials. Two types of segregation are often distinguished: radial and axial. Radial segregation is a relatively fast process and occurs after a few revolutions of the drum. In the case of a mixture of particles of different sizes, radial segregation leads to the expulsion of larger particles to the periphery, and a core of smaller particles is formed near the axis of rotation (Metcalfe et al., 1995; Khakhar et al., 1997b; Metcalfe and Shattuck, 1996; Ottino and Khakhar, 2000); see Fig. 2.15. For very low rotation rates (in the regime of discrete avalanching) the circular core loses its stability and spectacular multipetal structures of segregated particles develop, see Fig. 2.16.

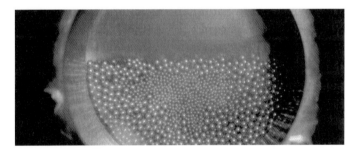

Fig. 2.15 Radial size segregation in a rotating drum: a core of smaller particles is formed in the bulk of the drum, while larger particles are expelled to the periphery, courtesy of Wolfgang Losert.

Axial segregation of binary mixtures in the long drums occurs on a much longer time scale (hundreds of revolutions). As a result of axial segregation, bands of segregated materials are formed along the drum axis (Zik et al., 1994; Hill and Kakalios, 1994; Hill and Kakalios, 1995); see Fig. 2.17. The segregated bands exhibit very slow coarsening behavior. Under certain experimental conditions axial segregation patterns show oscillatory behavior and travelling waves (Choo et al., 1997; Fiedor and Ottino, 2003; Arndt et al., 2005; Charles et al., 2006). The mechanisms leading to segregation of polydisperse granular mixtures will be discussed in Chapter 7.

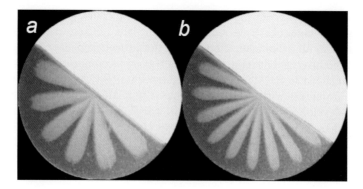

Fig. 2.16 Multipetal structures developed in mixtures of 0.71-mm (dark) and 0.12-mm (white) glass beads in a thin drum rotating with the angular speed $\omega = 0.13$ (a); $\omega = 0.09$ (b) rad/s, from Zuriguel et al. (2006).

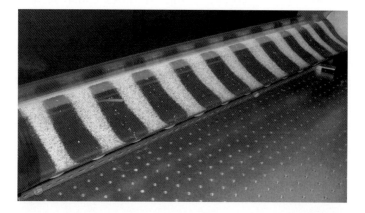

Fig. 2.17 Long rotating drum partially filled with a mixture of large black glass spheres and small transparent glass spheres showing axial size segregation after many revolutions, courtesy of Christopher Charles, Zeina Khan, and Stephen Morris.

2.4 Granular impacts and craters

When a large rapidly moving object falls into a deep granular layer, it produces a splash, and after all motion stops, a static crater. Many such craters exist on planetary surfaces. These craters keep a fossil record of meteorite impacts. However, it is difficult to reconstruct the dynamics of the impact just by studying the impact craters because the shape of the crater is determined by many factors such as the shape, velocity, and direction of the projectile, as well as the structure of the surface. In the laboratory, a number of model experiments has been performed, however, the possibility of scaling the results of laboratory experiments to geological craters remains an open question.

In recent experiments by Walsh et al. (2003) and Ambroso et al. (2005) the relation between the impact energy of a projectile and the morphology of the crater has been studied. Indeed, a progression of crater morphologies analogous to that seen in craters on the Moon was observed by Walsh *et al.* (2003) as the energy of a ball falling into a deep layer of sand was increased. Furthermore, the scaling of the crater diameter D and penetration depth H with impact energy E agreed remarkably well with the empirical law known in geophysics, $D, H \propto E^{1/4}$. Interestingly, for low energy of impact, when the ball is not completely submerged by sand after the impact, the penetration depth of the projectile scales differently with the impact energy, $D \propto E^{1/3}$ (Ambroso *et al.*, 2005).

In a very loose sand,[2] granular impact can lead to the formation of a very strong jet of sand emitted from the impact crater right after a heavy projectile enters the granular layer (Thoroddsen and Shen, 2001). Experiments by Lohse *et al.* (2004) revealed a complicated sequence of the impact events in this system, see Fig. 2.18. Upon impact, grains are blown away in all directions in a crown-like splash, and immediately a deep cavity is formed. Then this cavity collapses, and a primary granular jet emerges. A secondary jet is emitted down into the air bubble formed during the penetration of the projectile, thus pushing surface material deep inside. The air bubble rises slowly towards the surface, causing a granular eruption. Remarkably, the height of the primary (cumulative) jet created as a result of cavity collapse may exceed the release height of the ball.

Since high-speed low-density granular jets are produced in the course of the collapse of the cavity, interstitial air potentially may play an important role. Royer *et al.* (2005) conducted systematic studies of the granular jet formation over a wide range of air pressures. The experiments revealed a sensitive dependence of the jet height on the air pressure. In particular, jets practically disappear in vacuum.

2.5 Grains with complex interactions

Traditionally, most studies in granular physics have been focused on the dynamics of spherical or nearly spherical grains with purely repulsive contact forces. However, a wealth of interesting phenomena emerge beyond this domain. Novel collective behaviors emerge when the interaction between the grains is modified by additional effects such as non-spherical shape, adhesion, interstitial fluid, magnetization, polarization, etc. In these situations short-range collisions, the hallmark of "traditional" granular systems, are augmented by long-range and, typically, anisotropic forces. The interplay of these forces can lead to new phenomena and structures. Studies of granular systems with complex interactions serve as a natural bridge to different classes of systems such as foams, dense colloids, dusty plasmas, ferrofluids. The detailed discussion of various fascinating phenomena emerging in ensembles of grains with complex interactions will be presented in Chapter 8.

Remarkably regular and dynamic patterns emerge in layers of vibrated rods. At sufficiently strong vertical acceleration, rods spontaneously assume an almost vertical orientation. In fact, rods jump on their ends slightly tilted and slowly drift in the direction of their tilt. While the rods themselves are symmetric, the tilt breaks the symmetry, turning the rod into a *polar self-propelled object*. Travelling rods slowly

---

[2]Very loose initial configuration is obtained via fluidizing sand by blowing air through it from below and slowly tapering down the airflow

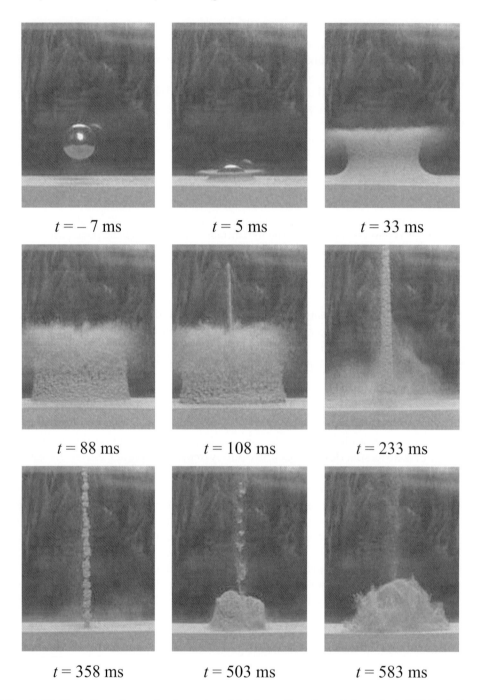

Fig. 2.18 Evolution of granular impact and jet formation, from Lohse et al. (2004). A steel ball of radius 1.25 cm touches very loose sand at time $t = 0$. Frames 2–4 illustrate consecutive phases of splash, frames 5 and 6 jet formation, and frames 8 and 9 granular eruption.

self-organize in large vortices, see Fig. 2.19. Similarly, a symmetric dimer vibrated on a plate can break the symmetry and start bouncing on one end and roll on the other. Because of frictional interaction with the surface, these dimers drift in a direction that is selected by the symmetry breaking in the bouncing mode (Dorbolo et al., 2005). The inherent anisotropy of non-symmetric particles also can induce directed ratchet-like motion on a vibrated substrate (Kudrolli et al., 2008). These objects collectively may form complex spatiotemporal patterns.

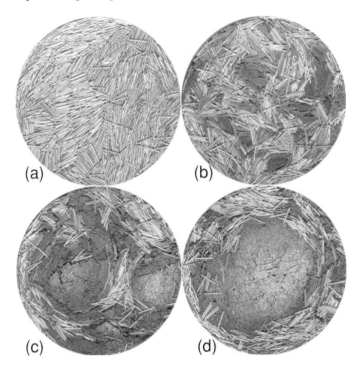

Fig. 2.19 Patterns observed in a system of vertically vibrated rods at frequency $f = 50$ Hz as the magnitude of vibrations is increased (snapshots taken from above) and for different filling fractions ϕ: (a) nematic-like gas phase, acceleration $\Gamma = 2.42$, $\phi = 0.152$; (b) moving domains of nearly vertical rods, $\Gamma = 3.38$, $\phi = 0.344$; (c) multiple rotating vortices, $\Gamma = 3.29$, $\phi = 0.551$; (d) single vortex, $\Gamma = 3$, $\phi = 0.535$, from Blair et al. (2003).

The details of a pattern depend sensitively on the shape and physical properties of grains. Vibrated rice particles in a certain range of frequencies and magnitudes of vibration spontaneously form a nematic crystalline-like state similar to those in liquid crystals (Narayan et al., 2006). Similar nematic and sometimes smectic patterns are also formed by metallic particles with tapered ends. However, cylindrical particles with flat ends demonstrate tetratic order, see Fig. 2.20. Nematic states often rotate as a whole in a circular container or exhibit smaller-scale swirls. These swirls are fundamentally different from the vortices of tilted rods in the experiment by Blair et al. (2003): here the particles on average are parallel to the plate, the direction of their motion is not related to the tilt or particle anisotropy (rice particles are presumably) but rather to an instability of the slow circular drift caused by small chirality of external excitation (see Chapter 8 for details).

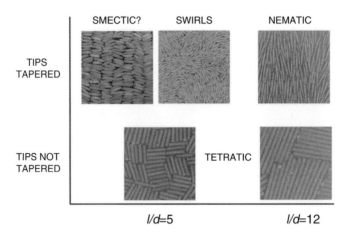

Fig. 2.20 Patterns in monolayers of vibrated elongated particles for different aspect ratios (l and d are particle length and diameter, respectively) and for different particles shapes, from Narayan et al. (2006).

In addition to the intrinsic particle anisotropy, the long-range electric or magnetic fields may introduce strong anisotropic effects leading to a non-trivial collective behavior. For example, mechanically or electrostatically driven magnetized grains exhibit formation of chains, rings, or interconnecting networks (Blair and Kudrolli, 2003; Snezhko et al., 2005), see Fig. 2.21. These patterns are formed as a result of a tail-to-head alignment of magnetic moments of particles due to an interplay of the long-range magnetic dipole–dipole interaction with usual hard-core collisions.

Ordered clusters and non-trivial dynamic states can be caused by long-range hydrodynamic interaction of grains immersed in liquids. Such patterns can be observed even in very small systems of particles (Voth et al., 2002; Thomas and Gollub, 2004), see Fig. 2.22. Fluid-mediated interaction between particles in a vibrating cavity leads to both long-range hydrodynamic attraction and short-range repulsion. A plethora of non-trivial patterns including rotating vortices, pulsating rings, chains, hexagons emerge in the system of conducting particles in direct (dc) electric field immersed in poor electrolyte (Sapozhnikov et al., 2003b), see Fig. 2.23. The non-trivial competition between electrostatic forces and self-induced electrohydrodynamic flows determines the structure of emerging patterns.

Magnetized grains can also be energized by an alternating (ac) magnetic field. Snezhko et al. (2006) conducted experiments with small (90 μm) magnetic spheres suspended at the water/air interface and energized by the vertical ac magnetic field. In a certain range of amplitudes and frequencies of the magnetic field and for certain

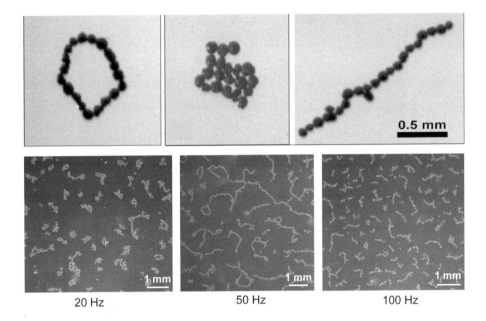

Fig. 2.21 Top row: structures formed in a submonolayer of magnetic microparticles subjected to an alternating magnetic field: rings, compact clusters, and chains. Bottom row: changes in the pattern morphology with the increase of the magnetic field frequency, from Snezhko et al. (2005).

concentrations of particles highly non-trivial dynamic objects, *magnetic snakes*, were discovered, see Fig. 2.24. The origin of these snakes is also related to the long-range hydrodynamic interaction, only in this case it is mediated by surface waves in the liquid excited by the collective response of magnetic microparticles to the alternating magnetic field. Segments of the snakes exhibit long-range antiferromagnetic ordering, while each segment is composed of ferromagnetically aligned chains of microparticles. The snakes' ends generate large-scale hydrodynamic vortices.

2.6 Granular patterns in biology

Most living organisms, from the simplest unicellular species such as bacteria, amoeba, algae, to higher animals exhibit highly organized collective behavior. From a physicist's point of view, a biological population sometimes can be considered a granular medium in which "particles" interact according to some rules that can range from relatively simple to extremely complicated. These interactions can sometimes lead to spectacular pattern formation. For example, many types of bacteria and amoeba form fractal-like aggregation patterns, fish aggregates in schools of hundreds or even thousands, birds organize in large flocks, locust swarm *en masse*, see, e.g. Ben-Jacob et al. (2000) and Buhl et al. (2006). Patterns also emerge on the subcellular level, for example, in the course of cytoskeleton formation from rigid (microtubules) and soft (actin) filaments

24 *Experimental overview of patterns in granular matter*

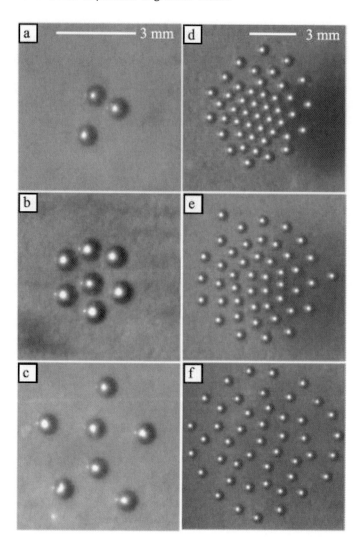

Fig. 2.22 Arrangements of particles near the bottom of a vibrated water-filled container when both attraction and repulsion are important. All images are taken for the vibration frequency 20 Hz and for different values of the dimensionless vibration acceleration Γ or for different initial conditions: (a) & (b) $\Gamma = 3$; (c) & (d) $\Gamma = 3.7$: (e) $\Gamma = 3.9$ and (f) $\Gamma = 3.5$, from Voth *et al.* (2002).

mediated by molecular motors and cross-links (Smith *et al.*, 2007; Nédélec *et al.*, 1997; Surrey *et al.*, 2001).

Pattern formation in biological systems is an enormously rich and complicated field. Biological patterns are forged by a variety of mechanisms, from general physical, such as steric repulsion in dense colonies or hydrodynamic entrainment in aqueous solutions, to highly specific biological, such as chemotaxis (drift in the direction of chemical

Granular patterns in biology 25

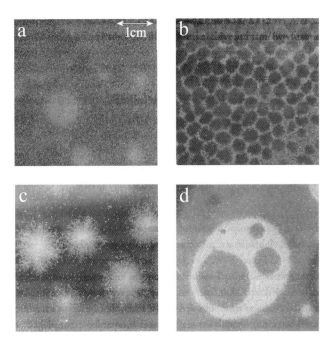

Fig. 2.23 Structures of electrostatically driven granular media in a weakly conducting non-polar fluid (toluene–ethanol mixture). Representative patterns for different values of applied field and concentration of ethanol: (a) static clusters, (b) honeycombs, (c) dynamic vortices, and (d) pulsating rings, from Sapozhnikov et al. (2003b).

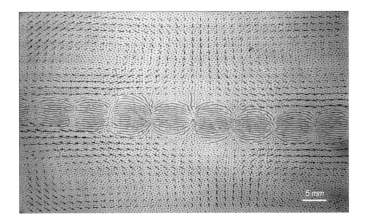

Fig. 2.24 Magnetic "snakes" formed by 90-μm nickel spheres suspended at the water/air interface and energized by a vertical ac magnetic field. Arrows show particle trajectories obtained by particle-image velocimetry, courtesy of Alexey Snezhko.

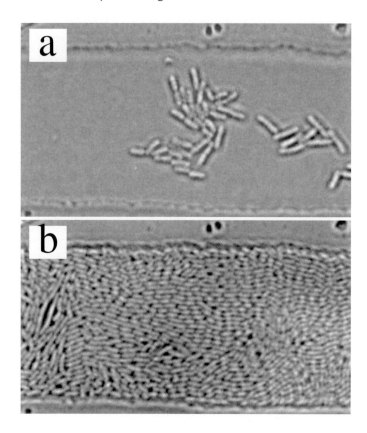

Fig. 2.25 Ordering and crystallization of a colony of bacteria *E. coli* in a microfluidic chamber (Volfson *et al.*, 2008). Top panel shows initial arrangement of cells and the bottom panel shows the colony 6 hours later.

gradients) or "social" interactions due to visual, audio, or chemical communication between the species. The comprehensive overview of patterns in biological systems is certainly beyond the scope of this book. Here we only focus on the biological patterns caused by roughly the same physical mechanisms that are at work in ordinary granular systems, however, even in this limited context "live grains" exhibit interesting and unique features.

One of the fundamental problems in the morphogenesis of multicellular colonies and organisms is understanding the nature of highly ordered (sometimes almost crystalline) states in which individual cells are arranged. It is believed that mechanical interaction among cells plays a role of a feedback loop that controls growth and arrangement of cells in tissues and colonies (Shraiman, 2005). If cells have a rod-like structure (as many bacteria, for example), their growth and expansion can itself lead to their ordering due to the steric repulsion between neighboring cells. To study this process *in vitro*, Volfson *et al.* (2008) grew a flagella-less strain of bacteria *E. coli* in a narrow open microfluidic channel (1 µm depth, 30 µm width). In the course of growth and expansion, the colony

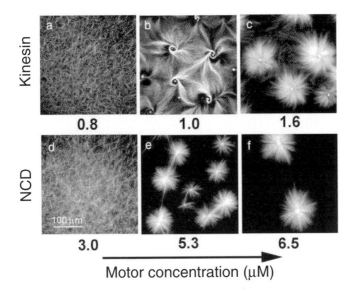

Fig. 2.26 Vortices and asters observed experimentally in microtubules/molecular motors mixtures for various concentration of motors and for two different types of motors (kinesin and NCD), from Surrey *et al.* (2001).

acquired local nematic order (see Fig. 2.25) with cells oriented along the direction of expansion. Unlike molecules forming a liquid crystal, non-motile cells constrained by the microfluidic device do not exhibit a significant Brownian motion, and only move when pressed by other bacteria in the course of colony expansion. Thus, the mechanism of cell ordering and colony crystallization differs from the classical Onsager mechanism for stiff rods (Onsager, 1949) based on the free-energy minimization. We discuss the order-parameter theory of cell ordering and crystallization in Chapter 9.

As we mentioned above, one of the main functions of molecular motors is to assemble microtubules into a highly organized cytoskeleton that stabilizes the cell's shape and provides "tracks" for all kinds of transport processes in a living cell. The most active phase of motor–tubule interaction is during mitosis and meiosis, when two spindles are formed that separate chromosomes in two halves of the mother cell that then become daughter cells (Howard, 2001). In order to understand the details of this complex self-organization process, a number of *in-vitro* experiments were performed (Nédélec *et al.*, 1997; Surrey *et al.*, 2001). In these experiments, the interaction of molecular motors and microtubules was studied in isolation from other biophysical processes simultaneously occurring *in vivo* in a living cell. At low concentration of motors and/or tubules the mixture of motors and microtubules remains disordered and isotropic. At sufficiently large concentration of molecular motors and microtubules, the latter organize into ray-like *asters* or rotating *vortices* depending on the type and the concentration of molecular motors, see Fig. 2.26. For kinesin-type motors, vortices give way to asters with an increase of the motor concentration. For NCD-type molecular motors (gluththione-S-transferase-nonclaret disjunctional fusion protein) no vortices

28 *Experimental overview of patterns in granular matter*

were observed for any concentration: asters emerge immediately after the breakdown of the isotropic state. We will show in Chapter 9 that under a certain approximation the problem of self-organization of microtubules and motors can be directly mapped on a granular problem of inelastic collisions of polar rods.

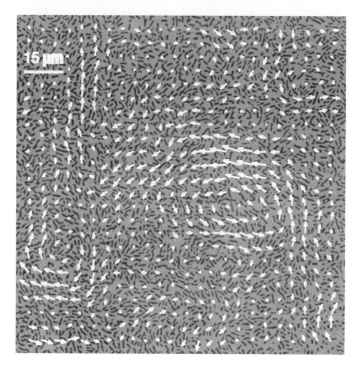

Fig. 2.27 Patterns of collective swimming in the colony of *Bacillus subtilis* confined in a thin free-suspended fluid film. Bacteria are seen as black short stripes on the gray background of the image. The instantaneous velocity field was obtained by particle-image velocimetry, and is shown by arrows (the longest arrow represents a velocity of 100 μm/s). Courtesy of Andrey Sokolov.

There has been growing interest among physicists in the dynamical properties of interacting, self-propelled organisms (Toner and Tu, 1998; Grégoire and Chaté, 2004; Czirók and Vicsek, 2000; Toner *et al.*, 2005). This interest was motivated by self-organization phenomena in biology, such as flocking, fish schools, etc. While the mechanisms of interaction among organisms can be drastically different, the result of interaction is often a synchronized motion of the whole colony in a certain direction.

Swimming bacteria may serve as a model system for this type of self-organization. Bacteria are possibly the simplest self-propelled "bioparticles". One of the most studied examples of patterns of swimming micro-organisms is so-called *bioconvection* that occurs due to cells swimming upwards to the liquid/air interface and thus creating an unstable density inversion (Pedley and Kessler, 1992). However, even in a confined geometry of thin-film samples where the oxygen gradient is not relevant, swim-

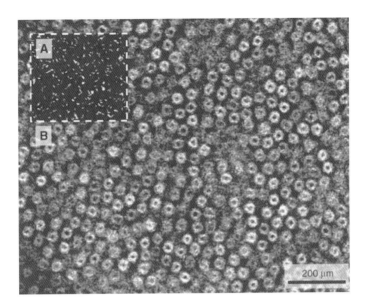

Fig. 2.28 Two-dimensional hexagonal arrays of rings formed by circulating sea urchin spermatozoa cells in a thin film. (a) Dark-field contrast image showing heads of the sea urchin spermatozoa. (b) The average intensity of 25 consecutive frames showing the arrangement of rings, each containing 9 spermatozoa swimming clockwise, from Riedel *et al.* (2005).

ming bacteria in concentrated solutions can form intricate patterns of swirls and jets (Mendelson *et al.*, 1999). These swirls lead to greatly enhanced diffusion and even superdiffusion of passive tracers (Wu and Libchaber, 2000).

Experiments performed with dense populations of swimming bacteria *Bacillus subtilis* (Mendelson *et al.*, 1999; Dombrowski *et al.*, 2004; Sokolov *et al.*, 2007) and sea urchin sperm cells (Riedel *et al.*, 2005) have demonstrated the onset of highly correlated collective motion of cells (Figs. 2.27 and 2.28). The correlation length of the resulting patterns can exceed the size of individual cells by more than an order of magnitude. Furthermore, the velocity of the collective flow may significantly exceed the swimming velocity of individual bacteria (about 20 μm/s). In contrast, dilute suspensions show no collective flow, and the correlation length is comparable to the size of a single micro-organism. Indeed, the advective motion on the scale of individual cells is negligible compared to diffusion due to the smallness of the corresponding Peclet number.[3] These experiments provide strong indication that hydrodynamic interaction among rapidly moving bacteria in liquid media may serve as an important mechanism of collective motion along with more traditional chemotactic interaction usually occurring among slowly moving cells in aggregation patterns of amoebae *Dictyostelium Discoideum* or bacteria *E. coli* (Keller and Segel, 1971; Budrene and Berg, 1991). Theoretical models of flocking and swimming will also be presented in Chapter 9.

---

[3]Peclet number is defined as $Pe = UL/D$ where U is flow velocity, L is characteristic dimension of the problem and D is the diffusion coefficient

3
Main theoretical concepts and tools

The physics of granular media is a diverse and eclectic field incorporating many different concepts and ideas, from hydrodynamics to the theory of glasses and phase transitions. Consequently, many fundamentally different theoretical approaches have been proposed to address observed phenomena. They can roughly be divided into three major classes: (i) microscopic models and molecular dynamics simulations, (ii) statistical mechanics and kinetic theories (iii) continuum and phenomenological models. In this chapter we will consider all of these, however we will begin with a brief overview of the fundamental properties of granular materials that differentiates them from ordinary gases, liquids and solids.

3.1 Fundamental microscopic properties of granular matter

3.1.1 Inelasticity

Probably the most fundamental common property of granular materials is irreversible energy dissipation in the course of interactions (collisions) between the particles. For the case of so-called dry granular materials, i.e. when the interaction with an interstitial fluid such as air or water is negligible, the encounter between grains results in dissipation of energy while their total mechanical momentum is conserved. The amount of irreversibly dissipated kinetic energy, or inelasticity of an individual collision is often characterized by the *restitution coefficient* e,

$$e = \frac{(\mathbf{v}_1' - \mathbf{v}_2') \cdot \mathbf{n}_{12}}{(\mathbf{v}_2 - \mathbf{v}_1) \cdot \mathbf{n}_{12}} \tag{3.1}$$

where $\mathbf{v}_{1,2}, \mathbf{v}_{1,2}'$ are the velocities of particles 1 and 2 before and after collision, respectively, and $\mathbf{n}_{12}$ is the vector connecting the centers-of-mass of the particles at the time of collision (which is assumed instantaneous here). For a fully elastic collision $e = 1$ while for a fully inelastic collision, $e = 0$. For $0 < e < 1$ the total energy loss is of the form

$$\Delta E = -\frac{1-e^2}{4}|(\mathbf{v}_1 - \mathbf{v}_2) \cdot \mathbf{n}_{12}|^2.$$

The amount of energy dissipation in a single collision depends on many factors, most importantly, of the material(s) that the two grains are made of, see Table 3.1. Indeed, the energy dissipation during collision is cased by deformations in the bulk of the two grains, and thus is dependent on the elastic and dissipative parameters of the materials.

Table 3.1 Coefficients of restitution e, static friction μ_s and dynamic friction μ_d for some materials. The friction and restitution coefficients values are provided for the same materials (steel on steel, etc.).

Material	e	μ_s	μ_d
Steel	0.6	0.78	0.42
Glass	0.9	0.9–1	0.4
Copper	0.5	1	0.3
Rubber	0.8	1–2	-
Teflon	0.72	0.04	-
Nylon	0.88	0.15–0.25	-

Other important factors determining restitution are the impact velocity $v = |(\mathbf{v}_1 - \mathbf{v}_2) \cdot n_{12}|$ itself (Brilliantov and Pöschel, 2004), the shape of the grains and the point of contact (Goldsmith, 1964). For metallic particles, at large collision velocities (say, $v > 5$ m/s), the dissipation comes mainly from the plastic deformation, and the restitution coefficient $e \propto v^{-1/4}$ (Johnson, 1989). In the limit of a very large impact velocity, the restitution coefficient approaches zero, as two particles stick to each other. For small impact velocities ($v < 0.1$ m/s), the shape of the grains after collisions is restored, and dissipation mostly stems from viscoelastic interaction, i.e. the relation between a dissipative stress tensor and the deformation rate tensor (Landau and Lifshitz, 1986). Brilliantov et al. (1996), Ramirez et al. (1999) calculated the restitution coefficient for two viscoelastic spheres in the limit of small impact velocity and found $1 - e \propto v^{1/5}$. According to this expression, the restitution coefficient approaches 1 (elastic limit) as $v \to 0$. Figure 3.1 demonstrates that these asymptotics indeed approximate well the experimental data from Lifshitz and Kolsky (1964).

These factors are often ignored in theoretical modelling. Description of collisions between particles by a fixed restitution coefficient is very simple and intuitive; however this approximation may lead to unphysical results. For example, cooling of a gas of hard inelastic particles in event-driven numerical simulations (Section 3.7.2) with fixed e leads to so-called *inelastic collapse* (McNamara and Young, 1996), the divergence of the number of collisions in a finite time. Of course, in reality the number of interparticle collisions remains finite, because eventually the interval between collisions becomes comparable with the collision time, and so the assumption of "hard particles" breaks down.

3.1.2 Friction

Tangential friction forces play an important role in the dynamics of granular matter, especially in dense systems. The friction, arising from complicated intermolecular and surface forces between the particles, is hysteretic (the contact between two grains can be either stuck or sliding depending on the history of interaction). Tribology, the branch of science studying the origins of friction, lubrication, and wear between the surfaces in contact, is a subject of active research (Israelachvili, 1985; Persson, 2000). Strongly non-linear behavior makes the analysis of frictional granular materials extremely difficult. In the majority of numerical studies, the simplest Coulomb law is

32 *Main theoretical concepts and tools*

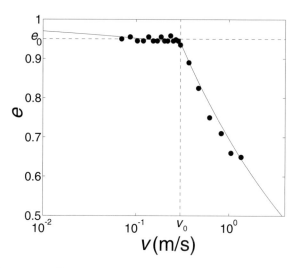

Fig. 3.1 The restitution coefficient e as a function of the impact velocity v from the experiment of Lifshitz and Kolsky (1964) (symbols). Solid line indicates matching of two asymptotic laws for plastic deformations at large v and viscoelastic deformations at small v, from McNamara and Falcon (2005)

adopted: two grains maintain a static contact if the ratio of tangential force to the normal force acting at the contact point is less than a *static friction coefficient* μ_s, otherwise they set into motion at which this ratio remains constant (dynamic friction coefficient μ_d) independent of the relative sliding velocity, see Table 3.1. Usually, a static friction coefficient is slightly larger than dynamic one, thereby giving rise to *stick–slip motion*: spontaneous jerking motion that can occur while two objects in contact are sliding past each other. While incorporating sliding friction into computational algorithms is relatively straightforward, accurate representation of static frictional forces is notoriously difficult (Walton, 1993). It is well known that frictional contact forces among solid particles exhibit indeterminacy in the case of multiple contacts per particle because there are fewer force-balance constraints than stress components, see, e.g. McNamara and Herrmann (2004) and Unger et al. (2005). To resolve this indeterminacy in simulations, various approximate algorithms have been proposed, see Section 3.7.

In reality, contact forces between two grains can be significantly more complex than the simple Coulomb approximation suggests. Indeed, two particles entering the contact may have rotational degrees of freedom, and thus contact may begin as sliding but then (as the sliding velocity becomes zero due to sliding friction, enter a static contact phase. Furthermore, for non-spherical grains, even more complex slide reversal collisions are possible (Stronge, 1990).

Frictional effects also significantly complicate non-frontal collisions of individual particles and collisions of rotating particles, even if the particles are purely spherical. The accurate treatment of these problems turns out to be extremely complex, however in simulations the problem is often simplified by introduction of the *tangential* restitution coefficient $-1 < e_\tau < 1$ similar to that of the normal restitution coefficient eqn. 3.1:

$$e_\tau = \frac{v'_{12\tau}}{v_{12\tau}} \qquad (3.2)$$

where $v'_{12\tau}$ and $v_{12\tau}$ are the projections of relative velocities in the tangential direction after and before the collision. For the frictionless case $e_\tau = 1$, and $e_\tau = -1$ for the no-slip (large friction) case.

3.1.3 Adhesion and other interparticle forces in granular media

Standard Hertz theory of particle–particle interaction assumes that grains are non-adhesive, so the interaction force turns zero when tensile stress is applied. The interaction law between two adjacent elastic spheres (known as the Hertz law after German physicist Heinrich Hertz who started the field of contact mechanics in 1881) is obtained analytically in the linear elasticity theory (Landau and Lifshitz, 1986). It predicts that the relation between the force F_0 exerted on the particle and the particles overlap distance Δ (the difference of the distance between their centers and the sum of their radii) is $F_0 \sim \Delta^{3/2}$ for spheres (three dimensions) and $F_0 \sim \Delta$ for disks (two dimensions).

However, any two particles exhibit at least small adhesion when the distance between them becomes comparable with the lengthscale of van der Waals forces, typically of the order of 1 micrometer. Attard and Parker (1992) extended the Hertz theory by approximating adhesion with the Lennard–Jones potential, and obtained the adhesive force between two surfaces in the form

$$F_a = \frac{H}{6\pi\delta^3}\left[\frac{\delta_0^6}{\delta^6} - 1\right], \qquad (3.3)$$

where δ is the distance between the surfaces, δ_0 is the equilibrium distance, and H is the Hamaker constant characterizing the van der Waals attraction between particles in vacuum. Several other theories have been proposed to phenomenologically describe molecular adhesion between small particles, see Brilliantov and Pöschel (2004) for a review. Generally, adhesion forces are only relevant for very fine powders (so-called group C powders by the Geldart (1973) classification with mean particle size $d < 10$ μm). However, adhesion forces can become very significant for much larger particles if even a small amount of interstitial fluid is present. The reason is that fluids can form so-called capillary liquid bridges between particles that provide attraction that is absent in dry granular media. This attraction is responsible for a non-zero tensile stress and a very different overall rheology of wet granular matter. Hornbaker et al. (1997) demonstrated that a very small amount of fluids lubricating grains (less than 50 nm on millimeter-size glass beads) can dramatically increase the repose angle of a sandpile (Fig. 3.2), the effect responsible for stability of almost dry sandcastles. This behavior was studied in recent laboratory experiments by Nowak et al. (2005) in a rotating drum. They showed that a theoretical description of the maximum angle of stability based on a frictionless liquid-bridge model, combining surface stability considerations with some failure criterion for liquid bridges within the bulk of the material, fit well the experimental data for different diameter drums, grain sizes and interstitial fluids. Moreover, Nowak et al. (2005) demonstrated that past discrepancies in experimental

34   *Main theoretical concepts and tools*

reports can be attributed to the sidewall effects that can significantly increase the stability of granular material.

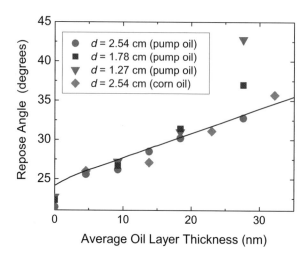

Fig. 3.2 The angle of repose of a sandpile as a function of the thickness of liquid on spherical beads, from Hornbaker et al. (1997)

When a granular medium is completely immersed in a moving fluid, it can experience a collective viscous drag force. In turn, a granular mass provides a porous medium that has to be taken into consideration in modeling the fluid flow. Furthermore, particles can have a long-range interaction due to viscous drag forces. These forces also can strongly affect the dynamics of granular materials, see Fig. 2.22. Gas-driven particulate flows is an active research area in the engineering community; see e.g. Jackson (2000). Fluid–particle interactions are also involved in many geophysical processes, e.g. dune formation (Bagnold, 1954). Whereas interaction of small individual particles with fluid is well understood in terms of the Stokes law, collective interaction and mechanical momentum transfer between multiple particles and fluid remains an open problem. Various phenomenological constitutive relations are used in the engineering community to model fluid–particulate flows; see e.g. Duru et al. (2002).

Finally, small particles can acquire an electric charge or a magnetic moment. In this situation fascinating collective behavior emerges due to competition between short-range collisions and long-range electromagnetic forces; see, e.g. (Aranson et al., 2000; Blair and Kudrolli, 2003; Sapozhnikov et al., 2003b). Effects of complex interparticle interactions on pattern formation in granular systems will be discussed in detail in Chapter 8.

3.2 Multiparticle effects

3.2.1 Indeterminacy

The fundamental problem of *granular statics* is easy to formulate: given the position of all grains (packing), determine the structure of the stress transmitted through it. However, the exact solution to this problem meets an unsurmountable difficulty because of the inherent non-linearity of the interparticle contact forces, both normal (elastic) and tangential (friction). Furthermore, sufficiently large shear stress may cause contacts to fail, which provides a major additional source of non-linearity. In general, it can be shown that the stress distribution cannot be found from the force balance conditions alone. This infamous *indeterminacy* of forces among solid grains stems from the well-known fact that for a pair of particles in a static contact with a specified normal force between them, the Coulomb friction law only specifies the maximum tangential force. The stress indeterminacy problem can already be seen in a very simple system of a single spherical particle in a groove (Halsey and Ertaş, 1999), see Fig. 3.3. The forces acting on the particle from the sidewalls of the groove cannot be determined simply by geometry (the particle diameter and the angle of the groove), but depend on the history of contact loading. The same is true in general for granular statics: in order to determine the stress distribution one has to augment the force-balance conditions by the constitutive relations that encode the construction history of a particular granular packing. The formulation of such constitutive relations is a subject of active current research (Wittmer et al., 1997; Edwards and Grinev, 1999; Ball and Blumenfeld, 2002; Goldenberg and Goldhirsch, 2002).

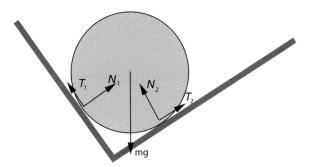

Fig. 3.3 A ball in a groove. Individual normal $N_{1,2}$ and tangential $T_{1,2}$ forces cannot be determined from the geometry and the mass of the particle.

In a large set of grains, the task of finding all interparticle forces is principally solvable only under the assumption of isostaticity, i.e. that the coordination number z (the number of contacts per particle) for *each* particle is $z_i = D+1$ for frictional contacts ($z_i = 2D$ for frictionless contacts) where D is the spatial dimension. Indeed, the number of unknown force components $zD/2$ per particle is then equal to the number of constraints (force and torque balance) $D + D(D-1)/2$ at $z = D+1$. Most real granular assemblies under external compression are *hyperstatic*, with the number of contacts greater than z_i. Furthermore, while for isostatic systems the forces are uniquely determined, the determination of coarse-grained stress distributions remains a non-trivial task. Indeed, in three dimensions the coarse-grained force and torque balance conditions

$$\nabla : \boldsymbol{\sigma} + \mathbf{F}_{ext} = 0, \quad \boldsymbol{\sigma} = \boldsymbol{\sigma}^T, \qquad (3.4)$$

(where $\boldsymbol{\sigma}$ is the stress tensor, $\mathbf{F}_{\text{ext}}$ is the external force such as gravity, and superscript T denotes transpose) only give three constraints for 6 independent components of $\boldsymbol{\sigma}$, and have to be augmented by one constitutive relation (in two dimensions one has, correspondingly, only 2 constraints for 3 independent components of $\boldsymbol{\sigma}$). In regular elastic solids, this constitutive relation relates the stress with the strain tensor (Landau and Lifshitz, 1986), but in frictional granular packing this relation does not apply. Instead, a scalar relation between the components of the stress tensor

$$\sum_{ij} q^{ij}\sigma_{ij} = 0 \tag{3.5}$$

is usually postulated, where q^{ij} are components of a tensor that depends on local geometry. Ball and Blumenfeld (2002) related this tensor to the fabric tensor that is determined for each individual particle. Subsequently, Blumenfeld (2004a) suggested a proper coarse-graining procedure allowing a macroscopic form of the constitutive relation (3.5) to be derived from the first principles. However, the derivation of these constitutive relations is based on the assumption that granular packing is close to the isostatic configuration, so its validity to a real granular system that are typically hyperstatic remains unclear.

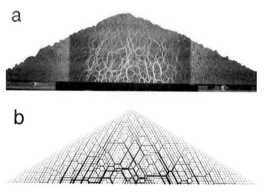

Fig. 3.4 (a) Network of force chains observed in a two-dimensional granular pile (grains are birefrigent plastic disks), from Vanel et al. (1999); (b) contact network inside a sandpile, from numerical simulations, from Luding (1997). The line thickness indicates the magnitude of the contact force.

3.2.2 Force chains

When a dense granular system is loaded, the forces among the grains are distributed very non-uniformly. Figure 3.4(a) illustrates the stress distribution in a jammed two-dimensional granular assembly. The force is visualized using photorefractive disks that change light polarization under mechanical stress. Force non-uniformity is also apparent in numerical simulations (see Fig. 3.4(b) where plots of contact forces are shown by lines whose thickness is proportional to the force magnitude). Since a strong force

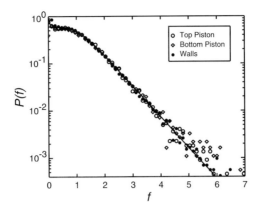

Fig. 3.5 The distribution $P(f)$ of normalized forces f in a static granular packing under uniaxial compression onto the interior surfaces of a confining vessel. The symbols show data for the top piston (open circles), the bottom piston (diamonds), and the walls (solid circles). The solid curve shows the fit to the equation $P(f) \sim (1 - b^2 f^2) \exp(-\beta f)$, from Mueth et al. (1998).

acting on a particle causes it to exert a similarly large force on other neighbors, forces form rather long chains extending several particle diameters. Much recent work has been devoted to the quantitative description of force distributions in granular packing (Liu et al., 1995; Radjai et al., 1996; Mueth et al., 1998; Cates et al., 1998; Howell et al., 1999; Longhi et al., 2002; Blair et al., 2001). In particular, the high force tail of the force distribution (which corresponds to the strong force chains) typically exhibits an exponential behavior, $P(f) \propto \exp(-f/f_0)$, as illustrated in Fig. 3.5. This behavior appears to be independent of the particle properties (rigidity, friction) and the degree of packing (Blair and Kudrolli, 2001; Erikson et al., 2002). It has also been observed in a number of molecular dynamics simulations (Nguyen and Coppersmith, 1999; Silbert et al., 2002b). The exponential statistics of the force distributions were first explained by the so-called q-model (Coppersmith et al., 1996) and then reproduced by other more realistic models (Socolar, 1998; Claudin and Bouchaud, 1998; Snoeijer et al., 2004). Furthermore, it was shown that a disordered packing is not required to obtain the exponential tail of the force distribution: even lattice systems exhibit similar scaling (Goldenberg and Goldhirsch, 2004). Recent work has focused on the non-universal behavior of force distributions at *small* forces. In fact, it was shown that this behavior contains some additional non-trivial information about the state of the system: in a jammed state, $P(f)$ exhibits a peak at some small value of f that disappears with the increase of applied shear stress (O'Hern et al., 2001; Silbert et al., 2002b; Snoeijer et al., 2004).

3.2.3 Macroscopic elastic behavior of granular packings

It has been argued that the existence of force chains makes jammed states fundamentally inelastic even at arbitrary small deformations, and therefore an entirely different

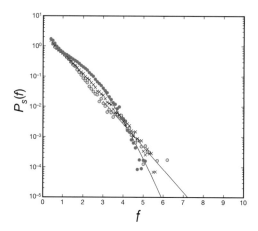

Fig. 3.6 Azimuthally averaged force distributions, $P_s(f)$, evaluated inside annuli at a distance s from the rotation center for filling height $h = 20$ grain diameters. The data from three annuli are shown: $0 < s < 3.5$ (open circles), $10.5 < s < 14$ (crosses), and $18.5 < s < 21$ (filled circles). In the non-sheared regions near the center, $P(f)$ exhibits a nearly exponential shape as for a static pack. In the shearing regions, $P(f)$ acquires the shape of an equilibrium distribution given by eqn 3.6, from Corwin *et al.* (2005)

"elasticity theory" of disordered granular materials is necessary (Bouchaud *et al.*, 1995; Wittmer *et al.*, 1996; Wittmer *et al.*, 1997). On the other hand, based on the exact constitutive relations valid for two-dimensional isostatic granular medium, it can be shown that the resulting stress equations are hyperbolic, and their solutions also give rise to force chains (Ball and Blumenfeld, 2002; Blumenfeld, 2004a; Blumenfeld, 2004b). In contrast, Goldenberg and Goldhirsch (2004) argue that in a macroscopic granular packing force chains do not play an important role in the stress transmission, except near the boundaries. It was demonstrated using two- and three-dimensional models with interparticle harmonic interactions that quasistatic granular materials exhibit departures from elasticity even at small loadings and small scales (below 100 particle diameters), at which continuum elasticity theory is invalid. However, the departures from elasticity vanish at large scales, and far away from the boundaries stress propagation in macroscopic granular packing can be described by the classic elasticity theory. Goldenberg and Goldhirsch (2004) obtained force chains on small scales, and the force and stress distributions that agree with experimental findings.

3.2.4 Jamming and force distribution

The contact force f distribution function $P(f)$ may also carry sensitive information on salient properties of granular packing under shear close to the onset of motion. As we mentioned above, for static, frictional, jammed granular systems, the function $P(f)$ decays exponentially above the average force, $\langle f \rangle$, and has a plateau or a small peak at force magnitudes below f_0, see Fig. 3.5. However, the function $P(f)$ changes when the shear is applied. In order to investigate this problem, Corwin *et al.* (2005) conducted experiments with cylindrical columns of grains of different height sheared by rotation of the upper confining lid. The force distribution function $P(f)$ was measured with a photoelastic plate at the bottom of the pack. As illustrated by Fig. 3.6, the function $P(f)$ exhibits a qualitative change at the onset of jamming. In the central annulus where the shear strain rate is minimal, the force distribution has an almost exponential shape, as expected for static system (compare with Fig. 3.5). However, with the increase in the shear strain rate (close to the outer wall of the annulus) a significant deviation from the exponential shape is observed. There, the shape of $P(f)$ is well approximated by the following (faster than exponential) distribution derived for three-dimensional elastic spheres interacting via Hertzian contacts

$$P(f) \sim \left[1 + f^{2/3} \frac{\langle \Delta \rangle}{d}\right]^2 \frac{1}{f^{1/3}} \exp\left[-\frac{\beta}{\beta_0} f^{5/3}\right], \tag{3.6}$$

where d is the grain diameter, $\langle \Delta \rangle$ is the mean grain–grain overlap (see Section 3.1.3), and parameter $\beta = 1/k_\mathrm{B} T_\mathrm{eff}$, T_eff is the effective temperature and k_B is Boltzmann's constant and $1/\beta_0$ is the temperature scale set by the average force per bead and the bead elastic modulus. The experiments showed that the effective temperature obtained by fitting the force distribution to eqn 3.6 was independent of the height of the column and the shear rate. This effective temperature can be interpreted as the temperature at which an amorphous granular glass (or granular solid) becomes a liquid. These measurements provide a way to introduce an effective temperature for the jamming/unjamming transition in granular materials.

3.2.5 Dilatancy

Flow of granular materials is a very complex phenomenon in which static and sliding contacts among grains are constantly created and eliminated. The important property of granular materials that was first described by Reynolds (1885), is *dilatancy*, or an increase of voidage of granular material in response to a deformation, as illustrated by Fig. 3.7. It can be readily observed on the beach, when water-saturated sand becomes dryer around a foot pressing on it, because water sinks into enlarged voids between the grains. There have been many attempts to incorporate dilatancy in the constitutive relations for granular flows, and, in particular, to relate it to the normal stresses and the granular packing stability (Nixon and Chandler, 1999; Goddard and Alam, 1999; Massoudi and Mehrabadi, 2001). On the basis of the simple model of granular friction (a slider block on frictional plane) Lacombe et al. (2000) argued that the dilatancy plays an important role in the dynamics by the modification of the granular packing, and therefore the friction force, especially for the stick–slip instability at a low velocity of sliding. In this aspect the dilatancy is related to the order parameter that we will consider later in the context of the phase-field description of dense granular materials, see Section 3.6.2.

3.3 Kinetic theory and granular hydrodynamics

3.3.1 Statistical properties of dilute granular gases

Superficially, a dilute gas of randomly colliding nearly elastic particles resembles ordinary molecular gases. This makes it tempting to apply the vast and well-developed

40 *Main theoretical concepts and tools*

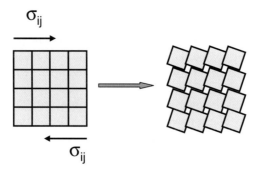

Fig. 3.7 Dilatancy: increase of voidage fraction in granular packing under shear.

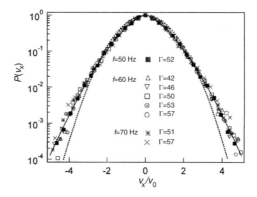

Fig. 3.8 Non-Maxwellian velocity distribution in driven inelastic gases. The horizontal velocity distribution function $P(v_x)$ (normalized to be 1 at the maximum) vs scaled velocity v_x for areal (number) density $\phi = 0.15$ and for different values of the vibration frequency f and normalized acceleration Γ in a vertical rectangular cell. The solid line is a stretched–exponential fit $P \sim \exp[-\beta(v_x/v_0)^\zeta]$, $\zeta \approx 1.52$, $\beta \approx 0.78$. The dotted line is a Gaussian: $\exp[-(v_x/v_0)^2/2]$, where $v_0 = \sqrt{\langle v_x^2 \rangle}$ is the root mean square velocity. The Gaussian curve overestimates small velocities and underestimates large velocities, from Rouyer and Menon (2000).

toolkit of classical statistical mechanics to the description of granular gases. Adopting fundamental principles of equipartition and equiprobability, one could introduce a canonical ensemble and arrive at the Maxwell velocity distribution and thermodynamic laws, as well as fluctuation–dissipation relationships. However, a deeper analysis shows that even for arbitrary small inelasticity, the statistical properties of granular gases are profoundly different from those of molecular gases. Since the energy is lost at every collision, the granular gas is a manifestly non-equilibrium system, whether it is in a free cooling regime with decaying kinetic energy or a driven state where energy loss is balanced by an external input. It has been shown in a number of laboratory and numerical experiments that the equipartition principle does not hold in both these situations even for very dilute granular gases (Luding et al., 1995; Wildman and Parkar, 2002; Feitosa and Menon, 2002). For example, Luding et al. (1995) showed numerically that in a monodisperse gas of spherical particles, the mean kinetic energy of particles differs from the mean rotational energy. A similar effect was observed numerically in a freely cooling gas of long needles (Huthmann et al., 1999). Furthermore, the velocity distributions of granular gases are profoundly non-Maxwellian and have overpopulated tails characterized by stretched-exponential distributions $P(v) \sim \exp[-\text{const} \times (v/v_0)^\zeta]$ with ζ close to $3/2$ (Losert et al., 1999a; Olafsen and Urbach, 1999; Rouyer and Menon, 2000; Aranson and Olafsen, 2002), see Fig. 3.8. Thus, the concept of temperature in granular systems does not have the same strict thermodynamic meaning as in equilibrium systems. However, these complications notwithstanding, one can successively develop an approximate statistical description of granular gases using the assumption of *molecular chaos*, introducing a pseudo-Liouville operator for inelastic particle collisions and the corresponding Boltzmann–Enskog equation (Jenkins and Richman, 1985).

3.3.2 Boltzmann–Enskog equation

Kinetic theory deals with the probability distributions functions describing the state of a granular gas. The corresponding equations, similar to the Boltzmann equations for rarefied gases, can be rigorously derived for the dilute gas of inelastically colliding particles with fixed restitution coefficient (Goldshtein and Shapiro, 1995). Kinetic theory is formulated in terms of the Boltzmann–Enskog equation for the probability distribution function $f(\mathbf{v}, \mathbf{r}, t)$ to find the particles with the velocity $\mathbf{v}$ at point $\mathbf{r}$ at time t. In the simplest case of identical frictionless spherical particles of radius d with fixed restitution coefficient e it assumes the following form

$$(\partial_t + \mathbf{v}_1 \cdot \nabla) f(\mathbf{v}_1, \mathbf{r}_1, t) = I[f], \tag{3.7}$$

with the binary collision integral $I[f]$ in the form

$$I = d^2 \int d\mathbf{v}_2 \int d\mathbf{n}_{12} \Theta((\mathbf{v}_2 - \mathbf{v}_1) \cdot \mathbf{n}_{12}) |(\mathbf{v}_2 - \mathbf{v}_1) \cdot \mathbf{n}_{12}|$$
$$i \times [\chi f(\mathbf{v_1}'', \mathbf{r}_1, t) f(\mathbf{v_2}'', \mathbf{r}_1 - d\mathbf{n}_{12}, t) - f(\mathbf{v}_1, \mathbf{r}_1, t) f(\mathbf{v}_2, \mathbf{r}_1 + d\mathbf{n}_{12}, t)], \tag{3.8}$$

where $\chi = 1/e^2$, Θ is the theta function ($\Theta(x) = 0, x < 0$, $\Theta(x) = 1, x \geq 0$), and postcollision velocities $\mathbf{v}_{1,2}$ and pre-collision velocities $\mathbf{v}_{1,2}''$ are related as follows

$$\mathbf{v}_{1,2}'' = \mathbf{v}_{1,2} \mp \frac{1+e}{2e}[\mathbf{n}_{12}(\mathbf{v}_1 - \mathbf{v}_2)]\mathbf{n}_{12}, \tag{3.9}$$

(cf. eqn 3.1). Equation 3.7 is derived with the usual "molecular chaos" approximation that implies that all correlations between colliding particles are neglected. One should keep in mind, however, that in dense granular systems this approximation can be rather poor due to excluded-volume effects and inelasticity of collisions introducing velocity correlations among particles, see, for example, (van Noije et al., 1998; Brito and Ernst, 1998; Soto and Mareschal, 2002; Brilliantov and Pöschel, 2004).

3.3.3 Granular hydrodynamics

Hydrodynamic equations are obtained by truncating the hierarchy of moment equations obtained from the Boltzmann–Enskog equation (3.7) via an appropriately modified Chapman–Enskog procedure (a detailed derivation can be found in the original papers (Jenkins and Richman, 1985; Goldshtein and Shapiro, 1995; Brey et al., 1998; Garzó and Dufty, 1999) and the book by Brilliantov and Pöschel (2004)). As a result, a set of continuity equations for mass, momentum and fluctuation kinetic energy (or "granular temperature") is obtained. The hydrodynamic variables, viz. the density ν or filling fraction $\phi = \nu/\nu_0$, (ν_0 is the density of grains),[1] the coarse-grained velocity $\mathbf{v}$ and granular temperature T, are the corresponding moments of distribution function f:

$$\nu(\mathbf{r},t) = \int d\tilde{\mathbf{v}} f(\tilde{\mathbf{v}},\mathbf{r},t)$$

$$\mathbf{v}(\mathbf{r},t) = \frac{1}{\nu(\mathbf{r},t)} \int \tilde{\mathbf{v}} d\tilde{\mathbf{v}} f(\tilde{\mathbf{v}},\mathbf{r},t) \qquad (3.10)$$

$$T(\mathbf{r},t) = \frac{1}{2\nu(\mathbf{r},t)} \int (\mathbf{v}-\tilde{\mathbf{v}})^2 d\tilde{\mathbf{v}} f(\tilde{\mathbf{v}},\mathbf{r},t).$$

The mass, momentum and energy conservation equations in granular hydrodynamics have the form

$$\frac{D\nu}{Dt} = -\nu \nabla \cdot \mathbf{v}, \qquad (3.11)$$

$$\nu \frac{D\mathbf{v}}{Dt} = -\nabla \cdot \boldsymbol{\sigma} + \nu \mathbf{g}, \qquad (3.12)$$

$$\nu \frac{DT}{Dt} = -\boldsymbol{\sigma} : \dot{\boldsymbol{\gamma}} - \nabla \cdot \mathbf{q} - \varepsilon, \qquad (3.13)$$

where $D/Dt = \partial_t + \mathbf{v} \cdot \nabla$ is the material derivative, $\mathbf{g}$ is the gravity acceleration, $\boldsymbol{\sigma}$ is the stress tensor, $\mathbf{q}$ is the energy flux vector, $\dot{\boldsymbol{\gamma}}$ is the strain rate tensor with components $\dot{\gamma}_{\alpha\beta} = (\partial_\alpha v_\beta + \partial_\beta v_\alpha)/2$, and ε is the energy dissipation rate. Equations 3.11–3.13 are structurally similar to the Navier–Stokes equations for conventional fluids except for the last term ε in eqn 3.13 for the granular temperature that accounts for the energy loss due to inelastic collisions.

Equations 3.11-3.13 have to be supplemented by the constitutive relations for the stress tensor $\boldsymbol{\sigma}$, energy flux $\mathbf{q}$, and the energy dissipation rate ε. For dilute systems, linear relations between tensors of stress $\boldsymbol{\sigma}$ and strain rate $\dot{\boldsymbol{\gamma}}$ and vectors of the "heat flux" $\mathbf{q}$ and the temperature gradient ∇T are obtained,

$$\sigma_{\alpha\beta} = [p + (\eta - \eta_b)\text{Tr}\dot{\boldsymbol{\gamma}}]\delta_{\alpha\beta} - \eta \dot{\gamma}_{\alpha\beta}, \qquad (3.14)$$

---

[1]The distinction between the filling fraction ϕ and the density ν becomes important for heterogeneous granular materials.

$$\mathbf{q} = -\kappa \nabla T, \tag{3.15}$$

where p is the isotropic pressure, η and η_b are the shear and bulk viscosities, and κ is the thermal conductivity of the granular material. In the kinetic theory of a two-dimensional gas of slightly inelastic hard disks by Jenkins and Richman (1985), eqns 3.11–3.15 are closed with the following equation of state

$$p = \frac{4\nu T}{\pi d^2}[1 + (1+e)G(\phi)], \tag{3.16}$$

and the expressions for the shear and bulk viscosities

$$\eta = \frac{\nu T^{1/2}}{2\pi^{1/2} d G(\phi)}\left[1 + 2G(\phi) + \left(1 + \frac{8}{\pi}\right)G(\phi)^2\right], \tag{3.17}$$

$$\eta_b = \frac{8\nu G(\phi) T^{1/2}}{\pi^{3/2} d}, \tag{3.18}$$

the thermal conductivity

$$\kappa = \frac{2\nu T^{1/2}}{\pi^{1/2} d G(\phi)}\left[1 + 3G(\phi) + \left(\frac{9}{4} + \frac{4}{\pi}\right)G(\phi)^2\right], \tag{3.19}$$

and the energy dissipation rate

$$\varepsilon = \frac{16\nu G(\phi) T^{3/2}}{\pi^{3/2} d^3}(1 - e^2). \tag{3.20}$$

Function $G(\phi)$ in these expressions is the radial pair distribution function due to particle–particle correlations in a finite-density granular gas. For a relatively dilute two-dimensional gas of elastic hard disks $G(\phi)$ can be approximated by the two-dimensional analog of the Carnahan–Sterling formula (Carnahan and Starling, 1969; Song et al., 1989):

$$G_{\text{CS}}(\phi) = \frac{\phi(1 - 7\phi/16)}{(1 - \phi)^2}. \tag{3.21}$$

This formula is expected to work for densities roughly below 0.7. An even more accurate approximation is sometimes used,

$$G_4(\phi) = \frac{\phi(1 - 7\phi/16)}{(1 - \phi)^2} - \frac{\phi^3}{128(1 - \phi)^4}. \tag{3.22}$$

For high-density granular gases, this function has been calculated using free-volume theory by Buehler et al. (1951), which in two dimensions is of the form:

$$G_{\text{FV}} = \frac{1}{(1+e)\left[(\phi_c/\phi)^{1/2} - 1\right]}\left(1 + O(\phi - \phi_c)\right). \tag{3.23}$$

where $\phi_c \approx 0.82$ is the random close packing density for disks.[2] Luding et al. (2001) proposed a global fit

---

[2] In order to avoid "crystallization" in two dimensions the disks should be slightly polydisperse

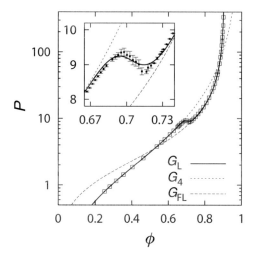

Fig. 3.9 Dependence of dimensionless "excess" pressure $P = 2G(\phi)$ vs. filling fraction ϕ. The short dashed line shows the dependence of P on ϕ using expression 3.22 in the equation of state (eqn 3.16) valid for low density, and the long dashed line corresponds to eqn 3.23 valid for high density. The solid line corresponds to the modified function G_L, eqn 3.24, symbols indicate results of simulations with 576 monodisperse hard disks. The inset shows the blowup of the region near the critical filling fraction $\phi_c \approx 0.7$ for the crystallization transition of hard disks, from Luding et al. (2001).

$$G_L = G_4 + (1 + \exp(-(\phi - \phi_0)/m_0))^{-1})(G_{FV} - G_4), \qquad (3.24)$$

with empirically fitted parameters $\phi_0 \approx 0.7$ and $m_0 \approx 10^{-2}$. The dependence of the scaled pressure P on ϕ is shown in Fig. 3.9. The expression (3.24) indeed works reasonably well in a wide range of densities, from dilute to almost close-packed density. Moreover, eqn 3.24 succeeds in capturing the pressure drop in the vicinity of the crystallization transition known for identical hard disks in two dimensions near the critical filling fraction $\phi_c \approx 0.7$, see inset to Fig. 3.9.

In the context of the crystallization transition, Khain and Meerson (2006) suggested to extend the hydrodynamics beyond the granular density corresponding to the melting point of the equilibrium phase diagram of elastic hard spheres. This is achieved by employing conventional granular hydrodynamics with constitutive relations all of which (except for the shear viscosity) diverge at the crystal-packing density, while the shear viscosity diverges at a smaller density. However, even with this extension, the continuum theory comprised of eqns 3.11–3.20 apparently cannot describe the effects of force chains that transmit stress via persistent contacts remaining in dense granular flows, as well as the hysteretic transition from solid to static regimes and coexisting solid and fluid phases.

Granular hydrodynamics is probably the most universal (however, not always the most appropriate) tool for modelling large-scale collective behavior in driven granular matter. Granular hydrodynamics equations 3.11–3.13 and their modifications are widely used in the engineering community to describe a variety of large-scale granular flows, especially for design of gas-fluidized bed reactors (Gidaspow, 1994). In the physics community granular hydrodynamics is used to understand various instabilities and structures in relatively dilute flows, such as flows past obstacles (Rericha et al., 2001), granular convection (Khain and Meerson, 2003), floating clusters (Livne et al., 2002a; Livne et al., 2002b; Meerson et al., 2003), longitudinal rolls (Forterre and Pouliquen, 2002; Forterre and Pouliquen, 2003), patterns in vibrated layers (Bougie et al., 2005), and others. However, eqns 3.11–3.13 are often used far beyond their applicability limits. In contrast to conventional hydrodynamics, the applicability of granular hydrodynamics is often suspect because typically there is no clear separation of scales between microscopic and macroscopic motions, except in the case of almost elastic particles with the restitution coefficient $e \to 1$, see, e.g. Tan and Goldhirsch (1998).

The formal absence of the scale separation for shear flows follows from the following simple argument. According to Tan and Goldhirsch (1998), in planar shear flow with the shear rate $\dot{\gamma}$ the granular temperature is of the form

$$T = C \frac{\dot{\gamma}^2 l_0^2}{1 - e^2}, \qquad (3.25)$$

where l_0 is the mean free path of the particle and C is the pre-factor of the order one. The "microscopic" time scale τ can be associated with the mean free time $\tau = l_0/T^{1/2}$, and the "macroscopic" time scale is given by $\dot{\gamma}^{-1}$. Using eqn 3.25 we obtain their ratio $\tau \dot{\gamma} = \sqrt{(1 - e^2)/C}$ that is the $O(1)$ quantity unless the particles are nearly elastic (e.g. for $e = 0.9$, $\sqrt{1 - e^2} \approx 0.44$). Consequently, certain parameters and constitutive relations need to be adjusted heuristically in order to accommodate the observed behavior. For example, Bougie et al. (2002) had to introduce artificial non-zero viscosity in eqn 3.12 for $\nu \to 0$ in order to avoid the spurious blowup of the solution. Similarly, Losert et al. (2000) introduced the viscosity diverging when the density approaches the close-packed limit as $(\nu - \nu_c)^\beta$ with the fitting parameter $\beta \approx 1.75$ in order to describe the structure of dense shear granular flows.

3.4 Statistical mechanics of dense granular matter

3.4.1 Fluctuation–dissipation relationships in granular systems

In equilibrium statistical mechanics, there is the rigorous fluctuation–dissipation theorem that relates the transport and diffusion properties of the fluid. For example, the famous Einstein relation connects the diffusion coefficient of a tracer particle D_{tr} with the mobility ξ of a particle driven by a constant force F via the surrounding liquid at the temperature T,

$$D_{\text{tr}} = \xi T. \qquad (3.26)$$

As we already mentioned in the beginning of this chapter, statistical mechanics of inelastically colliding grains is fundamentally different from that of a molecular gas. In particular, since there is no equipartition, there is no unique temperature characterizing the state of the system. So one cannot a priori expect fluctuation–dissipation relations like eqn 3.26 to hold. However, these relationships and their generalizations have been successfully used in other out-of-equilibrium systems such as glasses (Mezard et al., 1987). It was shown that a generalized fluctuation–dissipation theorem holds for glass systems, but the corresponding effective temperature is different from the bath temperature (Cugliandolo et al., 1997). There were also studies of a similar relation for granular materials as well. Makse and Kurchan (2002) conducted a direct numerical and experimental study of the fluctuation–dissipation relation in a granular flow. They performed time-resolved measurements of particle displacements $x(t) - x(0)$ in a sheared cell with polydisperse glass beads and a force-induced displacement $x_t(t) - x_t(0)$ of a tracer particles of different sizes, and obtained that the Einstein-like relation

$$\langle |x(t) - x(0)|^2 \rangle = 2T_{\text{eff}} \frac{\langle x_t(t) - x_t(0) \rangle}{F} \qquad (3.27)$$

holds with the well-defined effective temperature $T_{\text{eff}} \approx 10^{-4}$ which as expected is much larger than kT at room temperature. Furthermore, Makse and Kurchan (2002) concluded that this effective temperature within computational error coincides with the compactivity (see below) X measured by averaging over an ensemble of jammed configurations at a given deformation energy (which in turn is related to the volume fraction). One has to keep in mind, however, that the effective temperature determined via the Einstein relation is different from the granular temperature defined as the variance of the velocity distribution function.

The existence of simple Einstein-type relations for dense granular flows has been a controversial matter. More recent experiments by Utter and Behringer (2004) show that the diffusion of tracers in granular shear flows is anisotropic, and the drift of tracers may appear either sub- or superdiffusive. Moreover, for small shear rates the drag force shows a rather non-trivial logarithmic dependence on the velocity (Geng and Behringer, 2005).

3.4.2 Edwards hypothesis

In the foundation of classical statistical mechanics lies the notion of a statistical ensemble. It comprises all possible configurations of the system, and in isolated systems, the energy is conserved, and all configurations with the same energy are assumed to be equally likely. In thermal systems, persistent stochastic motion of particles lets the system sooner or later explore all available phase-space configurations. As we have seen in the previous section, for dilute granular systems an approximate statistical description can be developed similar to that of regular gases and fluids with an important difference that the energy is lost at particle collisions. While there are some fundamental difficulties in this approach, it works rather well in dilute gases of weakly inelastic particles.

For dense systems, the situation is significantly more complicated. Jamming of granular packing precludes the system from exploring the available phase space configurations, and therefore the principle of equiprobability does not hold even approximately. In athermal systems such as a jammed granular system composed of rigid grains, the analog of the phase space is the space of all possible jammed configurations characterized by coordinates of all particles ζ. While one can introduce the probability distribution for all possible jammed configurations, this distribution density cannot be expressed in terms of the total energy of the system since mechanical energy has no role in the description of the system state. Edwards and Grinev (1999) suggested that the volume of the system rather than its energy is the key macroscopic quantity governing the behavior of the granular matter. Thus, the *volume function* $W(\zeta)$ rather than the Hamiltonian $H(p,q)$ is assumed to determine the statistics of the granular packing. The microcanonical ensemble is characterized by the density of states Σ with a given volume V

$$\Sigma(V) = \int \delta(V - W(\zeta))\theta(\zeta)d\zeta \qquad (3.28)$$

Here, function $\theta(\zeta)$ imposes a constraint that summation is only performed over reversible jammed configurations and not all static configurations. The key (and rather counterintuitive) assumption of the theory is that all microstates with a given volume are equiprobable, which leads to the entropy-like quantity

$$S(V) = \lambda \ln \Sigma(V), \qquad (3.29)$$

where λ plays the role of the Boltzmann constant. The derivative of "entropy" S with respect to volume V yields a temperature-like quantity

$$X = \left[\frac{\partial S(V)}{\partial V}\right]^{-1}, \qquad (3.30)$$

which is called *compactivity*. The compactivity characterizes "fluffiness" of a granular system. As the system approaches the dense limit, the compactivity of the system approaches zero, since the number of states with a given volume diminishes. In the close-packed limit, there is only one configuration remaining (e.g. FCC periodic lattice for identical spherical particles) and therefore $S = 0$ and $X = 0$. Conversely, for $X = \infty$ all mechanically stable configurations for a given macroscopic volume V including the most "fluffy" are equiprobable. The behavior of entropy and compactivity as a function of volume fraction ϕ is qualitatively shown in Fig. 3.10.

Continuing the analogy with the classical canonical ensemble, we can consider configurations of N identical grains. We can partition this configuration in Voronoi cells around each grain, and then the total volume of the system V will be a sum of volumes of elementary Voronoi cells,

$$V = \sum_{i=1}^{N} v_i. \qquad (3.31)$$

Since compactivity plays the role of temperature, and the volume the role of energy, the probability for the occurrence of a state with volume V should be proportional to $\exp(-V/X)$. Assuming that the volumes of elementary Voronoi cells are uniformly distributed between certain minimum and maximum values v_{min} and v_{max} (a mean-field approximation), we can write down the partition function

$$Z = \int_{v_{min}}^{v_{max}} ... \int_{v_{min}}^{v_{max}} e^{-\sum_i v_i/X} dv_1...dv_N = \left(\int_{v_{min}}^{v_{max}} e^{-v/X} dv\right)^N$$

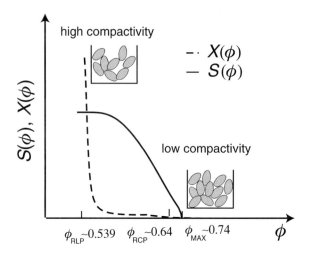

Fig. 3.10 Illustration of dependence of compactivity X (dashed line) and entropy S (solid line) on packing fraction ϕ. Compactivity is high at random loose-packing fraction ϕ_{RLP}, low at random close-packing fraction ϕ_{RCP}, and zero at maximum (e.g. FCC for spheres) packing fraction ϕ_{MAX}, from Makse et al. (2004).

$$\mathcal{Z} = \left[2Xe^{-v_{\text{mid}}/X} \sinh\left(\frac{\Delta v}{X}\right)\right]^N \tag{3.32}$$

where $v_{\text{mid}} = (v_{\text{min}} + v_{\text{max}})/2$ and $\Delta v = (v_{\text{min}} - v_{\text{max}})/2$. Using the partition function we can immediately find the average volume per particle (Edwards and Oakeshott, 1989; Srebro and Levine, 2003)

$$\langle v \rangle = \frac{\langle V \rangle}{N} = \frac{1}{N}\frac{\partial \ln Z}{\partial X} = v_{\text{mid}} + X - \Delta v \coth\left(\frac{\Delta v}{X}\right). \tag{3.33}$$

As compactivity $X \to 0$, the system approaches the "ground state" in which $\langle v \rangle = v_{\text{min}}$, whereas at $X \to \infty$, $\langle v \rangle = v_{\text{mid}}$, the mean volume of all possible states.

In order to use expression (3.33) one must know the values of v_{mid} and Δv, or v_{min} and v_{max}. The value of v_{min} can be found from purely geometrical considerations, as the minimum volume per particle is achieved for the hexagonal packing in two dimensions (2D) and the corresponding FCC packing in three dimension (3D). Accordingly, in 2D, $v_{\text{min}} = \sqrt{12}r^2$ and in 3D, $v_{\text{min}} = \sqrt{32}r^3$ where r is particle radius. The corresponding maximum packing fractions are $\phi_{\text{max}}^{\text{2D}} = \pi r^2/v_{\text{min}}^{\text{2D}} = \pi/\sqrt{12} \approx 0.91$ and $\phi_{\text{max}}^{\text{3D}} = (4\pi/3)r^3/v_{\text{min}}^{\text{3D}} = \pi/\sqrt{18} \approx 0.74$.

The maximal volume v_{max} cannot be determined solely by geometry and must depend on the frictional properties of grains. Rough particles may form arches and thus increase the volume of the Voronoi cells. Srebro and Levine (2003) estimated the value of v_{max} using minimal "toy" systems of just a few grains (three in two dimensions and four in three dimensions) and extrapolating the results toward large granular assemblies. As a result of this calculation the volume fraction ϕ can be found as a function of the static friction coefficient at different values of compactivity (see Fig. 3.11). This dependence agrees with the qualitative picture of Fig. 3.10 that packing fraction decreases with compactivity, but it also determines how the packing fraction depends on the friction coefficient when the compactivity is kept fixed.

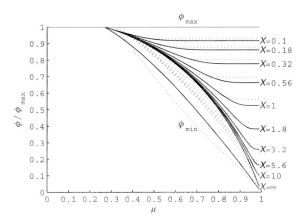

Fig. 3.11 Normalized packing fraction ϕ as a function of friction coefficient μ for several values of compactivity X for two dimensions (solid lines) and three dimensions (dotted lines), from Srebro and Levine (2003).

Edwards and collaborators (Edwards and Oakeshott, 1989; Edwards and Grinev, 1999; Makse et al., 2004) have developed a complete kinetic theory of jammed systems using the concepts of entropy and compactivity outlined above. In particular, they proposed a "granular Boltzmann equation" for the probability distribution of a certain packing configuration evolving under a sequence of small "taps" and showed that the Second Law of thermodynamics $dS/dt \geq 0$ follows from this theory, and that the equal sign (equilibrium) is achieved on the Gibbs distribution $f(W) \sim \exp(-W/\lambda X)$. Some predictions of this theory appear to be in agreement with numerical simulations as well as with recent experimental studies of jammed configurations using confocal microscopy of emulsions with matched refractive index of solid particles and interstitial liquid (Makse et al., 2004).

One major difficulty of this theory is that it is based on the notion of compactivity which is difficult to directly measure in experiment. However, compactivity can be inferred from the analog of a fluctuation–dissipation theorem which in ordinary thermodynamics allows one to write the specific heat in two different ways,

$$C_V = \partial E_0/\partial T|_V = \langle (E - E_0)^2 \rangle / k_B T^2 \tag{3.34}$$

where E is the instantaneous energy of the system at equilibrium, E_0 is the mean energy, k_B is the Boltzmann constant, T is the temperature, and brackets denote time average. In Edwards' theory, the analogous relation is postulated (Nowak et al., 1998),

$$C = dV_{ss}/dX = \langle (V - V_{ss})^2 \rangle / \lambda X^2, \tag{3.35}$$

where V_{ss} is the steady-state volume. This relation allows the compactivity to be expressed via the variance of the volume fluctuation around the steady state,

$$X = \left[\lambda \int_{V_{RLP}}^{V} \frac{dV_{ss}}{\langle (V - V_{ss})^2 \rangle} \right]^{-1}, \qquad (3.36)$$

where V_{RLP} is the volume of random loose-packing state at which the compactivity should diverge. Nowak et al. (1998) and Schröter et al. (2005) used this relation to estimate (up to the unknown constant λ) the compactivity of granular systems subjected to periodic excitations, tapping in Nowak et al. (1998) and periodic pulses of gas fluidization in Schröter et al. (2005). The latter approach is more robust as it avoids a slow compaction and creates a history-independent sequence of mechanically stable configurations. The mean packing fraction ϕ of grains after each fluidization pulse could be controlled by the gas flow rate during the pulse. Figure 3.4.2 demonstrates the dependence of the compactivity on the packing fraction ϕ obtained by fitting the experimental data to relation 3.36 taking into account that $\phi = V^{-1}$. By construction, the compactivity diverges at ϕ_{RLP} and decays with the volume fraction, however, it appears to level off at 10^{-2}. Presumably, the compactivity should approach zero as the packing fraction approaches the random close-packing limit $\phi_{RCP} \approx 0.64$, however, this limit could not be accessed in the periodic fluidization experiments by Schröter et al. (2005).

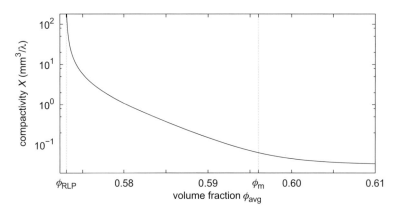

Fig. 3.12 Compactivity as a function of the average volume fraction. $\phi_{RLP} \approx 0.573$ is the packing fraction of the random loose state, and ϕ_m is the packing fraction at which the system has a minimum in the variance of the volume fluctuations (which corresponds to a minimum in the number of particles within a statistically independent region), from Schröter et al. (2005).

3.5 Continuum mechanics of granular materials

Continuum description of dense granular systems rests on the same fundamental conservation laws of mass and momentum and energy as hydrodynamics of dilute granular flows described above in Section 3.3. The major difference lies in the constitutive relations. In granular hydrodynamics they are derived from the kinetic theory via perturbative schemes like the Chapman–Enskog approximation. For dense granular flows, such derivations are no longer valid in any asymptotic sense, but still constitutive relations are kept in the same functional form, but with excluded volume corrections. However this approach neglects the fundamental property of granular systems, namely, that granular assemblies behave elastically at very small stresses, and yield plastically at larger stresses.

3.5.1 Stress–strain relations

The elastic response of granular systems is often described similarly to the ordinary solids, via linear stress–strain constitutive relations. In classical linear elasticity theory (Landau and Lifshitz, 1986), the stress tensor is a linear function of the strain tensor, $\sigma_{xy} = C_{xy\alpha\beta}\gamma_{\alpha\beta}$, where $\gamma_{\alpha\beta} = \frac{1}{2}\left(\partial_{x_\alpha} u_\beta + \partial_{x_\beta} u_\alpha\right)$ is the linear strain tensor, $\mathbf{u}(\mathbf{r},t)$ is the displacement field, and $C_{\alpha\beta\gamma\delta}$ is the tensor of elastic constants. Goldhirsch and Goldenberg (2002) generalized classical derivations of the linear elasticity theory to disordered solids that can serve as a simplified (dissipationless) approximation of dense granular materials. Using simplified two- and three-dimensional models with harmonic interparticle interactions, Goldhirsch and Goldenberg (2002) demonstrated that continuum elasticity is valid on large scales (exceeding $O(100)$ particle diameters), and on smaller scales deviations from continuum elasticity (e.g. force chains) occur even for very small loads.

3.5.2 Mohr–Coulomb criterion

For sufficiently large shear stresses, granular material sets in motion. In the engineering literature, this process is called *yield* and is characterized by the yield function $Y(\sigma_{\alpha\beta})$, which is negative in the stable static regime and positive where the static regime is unstable. Thus, the critical stresses at which a granular system only begins to flow (state of *incipient failure*), correspond to a locus of points where Y vanishes in the stress space $\sigma_{\alpha\beta}$. This description is widely used in soil mechanics (Nedderman, 1992). It originates in the metal plasticity (Mase, 1970) but uses more general forms of yield functions.

The simplest yield function corresponds to the cohesionless Coulomb material,

$$Y = |\tau| - \mu\sigma, \tag{3.37}$$

where n and s are coordinates normal and parallel to the slip plane, σ and τ are correspondingly normal and shear components of the stress on this surface, and μ is the friction coefficient. The yield function can only be negative or zero. When it is negative, the granular material is stable and motionless. Condition $Y = 0$ gives the locus of incipient yield, which for a cohesionless Coulomb material is a cone $|\tau| = \mu\sigma$. For cohesive Coulomb materials, the yield function can be generalized $Y = |\tau| - \sigma\mu + c$, where the cohesion constant c determines the shear yield stress at zero normal stress. The corresponding yield condition is

$$|\tau| = \mu\sigma + c. \tag{3.38}$$

To determine whether a particular static stressed granular configuration is stable or not one needs to check the yield criterion for all possible orientations of the slip plane. However, in two dimensions there exists an elegant graphical method of the stress analysis based on *Mohr's circle*. Given a slab of granular matter with fixed values of stress components $\sigma_{xx}, \sigma_{xy}, \sigma_{yx}, \sigma_{yy}$ (note that moment balance requires $\sigma_{xy} = \sigma_{yx}$), we can compute the "traction stresses" $\sigma_\theta, \tau_\theta$ for an arbitrary angle θ. Indeed, a force balance on the wedge (Fig. 3.13) yields

$$\sigma_\theta = \frac{1}{2}(\sigma_{xx} + \sigma_{yy}) + \frac{1}{2}(\sigma_{xx} - \sigma_{yy})\cos 2\theta - \tau_{xy}\sin 2\theta \qquad (3.39)$$

$$\tau_\theta = \frac{1}{2}(\sigma_{xx} - \sigma_{yy})\cos 2\theta + \tau_{xy}\sin 2\theta. \qquad (3.40)$$

Now we can define $p = (\sigma_{xx} + \sigma_{yy})/2$, $R = [(\sigma_{xx} - \sigma_{yy})^2/4 + \sigma_{xy}^2]^{1/2}$, and $\tan\psi = -2\tau_{xy}/(\sigma_{xx} - \sigma_{yy})$ and write

$$\sigma_\theta = p + R\cos(2\theta - 2\psi) \qquad (3.41)$$
$$\tau_\theta = R\sin(2\theta - 2\psi). \qquad (3.42)$$

These equations define a circle (*Mohr's circle*) with the center at $(p, 0)$ and the radius R in the plane (σ, τ), see Fig. 3.14.

Thus, given the components of the stress in any Cartesian frame, one can compute p, R, and determine the mutual position of the Mohr's circle and the yield locus (see Fig. 3.14). Consider for definiteness the normal σ and shear τ stress applied to a surface with the angle φ with respect to the x-axis. The stress components can be expressed as $\sigma = p + R\cos(2(\varphi - \theta)), \tau = R\sin(2(\varphi - \theta))$ defining a circle with the radius R and the center $(p, 0)$ on a (σ, τ) plane. Yield condition 3.38 defines a straight line on Fig. 3.14. Clearly, these two curves cannot intersect for an ideal Coulomb material. However, they can touch each other at a single point that corresponds to the state of incipient failure. The position of that point on the circle determines the orientation of the plane along which the slip will first occur. Note that in three-dimensional situations instead of a circle one has a Mohr cone in the principal stress space.

After yielding, the granular material begins to flow. The strain rate of a stationary flow can be found as a gradient of the plastic potential Z,

$$\dot{\gamma}_{ij} = \lambda \frac{\partial Z}{\partial \sigma_{ij}}, \qquad (3.43)$$

where λ is a scalar constant. In soil mechanics it is often postulated that the plastic potential coincides with the yield function,

$$Y = Z. \qquad (3.44)$$

This relationship is known as the *associated flow rule*. It is easy to see that this rule conforms to the so-called *principle of normality* which states that the strain rate is normal to the yield surface. However, the actual relationship between the yield function and the plastic potential is more complicated. In the non-associated flow

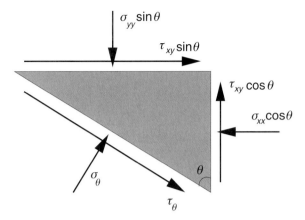

Fig. 3.13 Force diagram for a wedge under combined normal and shear stress

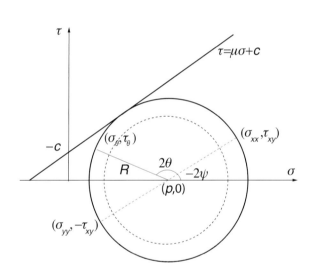

Fig. 3.14 The Mohr's circles and the Coulomb yield locus: dashed circle corresponds to a set of stresses $(\sigma_{xx}, \sigma_{xy}, \sigma_{xy})$ at which the granular material is stable, solid circle corresponds to the state of incipient failure

rule the plastic potential has the same functional form as the yield function, but the internal friction angle $\varphi = \tan^{-1}\mu$ is replaced by the angle of dilatancy $\varphi_d < \varphi$. This angle characterizes the rate of dilation of material under shear stress, $\tan\varphi_d = (\dot\gamma_{nn} + \dot\gamma_{ss})/\dot\gamma_{ns}$ (*principle of dilatancy*). Jenike (1964) proposed a more relaxed *principle of coaxiality* which postulates that the principal axes of stress and strain are co-incident, i.e. shear does not occur on planes where there is no shear stress.[3] In two-dimensional

---

[3]In subsequent work Jenike (1987) realized that the coaxiality principle predicts incorrect channel flow angles in the funnel flow and suggested several remedies for a radial flow geometry.

Cartesian coordinates, this principle yields the following relation for the strain-rate components

$$\frac{\dot{\gamma}_{xy}}{\dot{\gamma}_{xx} - \dot{\gamma}_{yy}} = \frac{\sigma_{xy}}{\sigma_{xx} - \sigma_{yy}} \quad (3.45)$$

which together with incompressibility condition $\dot{\gamma}_{xx} + \dot{\gamma}_{yy} = 0$ and boundary conditions determine the velocity field.

These classical continuum plasticity models based on the Mohr–Coulomb picture have serious problems, such as for example formation of discontinuous velocity fields. Furthermore, it is difficult to extend this model to three dimensions since a three-dimensional stress tensor has six independent components, and the force balance and incipient failure conditions only supply four equations. In the recent work by Kamrin and Bazant (2007) a new "stochastic flow rule" was suggested to replace the principle of coaxiality in classical plasticity. The stochastic flow rule takes into account two crucial features of granular materials, namely discreteness and randomness, by introducing diffusing "spots" of local plasticity. It allowed the authors to resolve many difficulties of the classical models and deduce profiles for a number of fundamental granular flows, like flow in the hopper, annular Couette cell and some others.

3.5.3 Bagnold phenomenological constitutive relation

Another approach to the constitutive relation is based on the dimensional analysis. For constant-density flows in the absence of gravity, Bagnold (1954) proposed a simple local relation between the strain rate $\dot{\gamma}$ and the shear stress τ

$$\tau = \nu d^2 \dot{\gamma}^2. \quad (3.46)$$

This relation follows from the assumption that the only relevant time scale in dense quasistatic granular flow is the deformation time scale $T_\gamma \sim 1/\dot{\gamma}$. This hypothesis spurred an avalanche of works intended to test or improve this simple relation, see Chapter 6. While the Bagnold relation (eqn 3.46) is approximately satisfied in the bulk of the flow, the boundary effects as well as dynamic inhomogeneities result in substantial deviation from this simple scaling law.

It is interesting to point out the analogy between the Bagnold scaling approach and the Prandtl model in the semiempirical theory of turbulence (Monin and Yaglom, 1971) where the turbulent Reynolds stress $\langle \mathbf{v}'_i \mathbf{v}'_j \rangle$ is written as $l_m^2 |\dot{\gamma}_{ij}| \dot{\gamma}_{ij}$, where l_m is the hypothetical "mixing length" (an analog of the particle size d here).

3.6 Phenomenological models

3.6.1 Ginzburg–Landau equation and its variants

Many pattern-forming systems are often described by models such as the Ginzburg–Landau or Swift–Hohenberg equations (Cross and Hohenberg, 1993; Aranson and Kramer, 2002). This approach allows one to study generic features of pattern formation across many different systems. For example, the complex Ginzburg–Landau equation

$$\partial_t \Psi = \epsilon \Psi - (1 + ic)|\Psi|^2 \Psi + (1 + ib)\nabla^2 \Psi \quad (3.47)$$

provides a generic description of spatiotemporal dynamics in the vicinity of Hopf bifurcation in systems with complex diffusive–dispersive coupling (for a derivation of the Ginzburg–Landau equation see, for example, Aranson and Kramer (2002)). Here, $\Psi(\mathbf{r},t)$ is the complex amplitude of an oscillating field $\psi(\mathbf{r},t) = \Psi(\mathbf{r},t)\exp(i\Omega t) + c.c.$. Similarly, the Swift–Hohenberg equation

$$\partial_t \psi = \epsilon\psi + \alpha\psi^2 - \beta\psi^3 + (1+\nabla^2)^2\psi \qquad (3.48)$$

describes systems near saddle-node or pitch-fork bifurcations occurring for perturbations with a characteristic length scale. Unlike the Ginzburg–Landau equation, the Swift–Hohenberg equation is written for the real field variable $\psi(\mathbf{r},t)$ itself. However, in many specific systems there are peculiarities that need to be taken into account. This often requires modifications to be introduced into the generic models. For example, this approach was taken in a number of works (Tsimring and Aranson, 1997; Aranson and Tsimring, 1998; Aranson et al., 1999a; Venkataramani and Ott, 1998; Crawford and Riecke, 1999) in order to describe patterns in vibrated granular layers, see Chapter 5.

3.6.2 Phase-field description of granular phase transitions

Without forcing, granular material remains in one of many metastable static configurations. But under external driving it may undergo a variety of non-equilibrium phase transitions. One example of such transitions is the fluidization of granular layer under sufficiently strong shear stress. As was explained above, the transition to fluidization occurs when the ratio of shear stress to normal stress exceeds a certain threshold (Mohr–Coulomb yield criterion). However, the continuum description that would describe this transition from static to flowing regime in the bulk of granular material is difficult (if not impossible) to obtain from first principles. A phenomenological order-parameter approach to the description of dense granular flows was suggested by Aranson and Tsimring (2001), Aranson and Tsimring (2002) who proposed to treat the shear-stress-mediated fluidization of granular matter as a non-equilibrium phase transition. As is customary in the models of phase transitions, an order parameter characterizing the local state of granular matter and the corresponding phase-field model were introduced. According to the model, the order parameter has its own relaxation dynamics and defines the static and dynamic contributions to the shear stress tensor. A similar-in-spirit approach to formulating the constitutive relation for dense granular flows was proposed by Lemaître (2002) who introduced the density of shear transformation zones (Falk and Langer, 1998) as a measure of fluidization in the granular packing. These approaches are discussed in detail in Chapter 6.

Another example of non-equilibrium phase transition is the formation of dense and cold granular clusters in externally heated granular gas (Olafsen and Urbach, 1998; Sapozhnikov et al., 2003b). The theoretical description of this phenomenon (Aranson et al., 2002) is based on the classical Lifshitz–Slyozov–Wagner theory of the first-order phase transition in equilibrium systems (Lifshitz and Slyozov, 1958), see Chapter 8 for details.

3.6.3 Saint-Venant model and its generalizations

Another popular phenomenological approach is based on the integral description of a flowing granular layer by a local mean velocity and thickness. This approach can be applied to shallow granular flows over solid substrate, for example in the description of rock avalanches or dry debris flows. Such modelling is similar in spirit to the Saint-Venant description of shallow-water dynamics with a modified "constitutive relation" relating the spreading force with the local layer thickness. This relation takes into account internal friction of the granular mass, (see, e.g. (Mangeney-Castelnau *et al.*, 2005; Larrieu *et al.*, 2006)). A similar approach can also be applied for description of near-surface flows in deep granular layers and sandpiles. It uses a two-phase description of granular flows, one phase corresponding to rolling grains and the other phase to static ones. This model was originally suggested by Bouchaud, Cates, Ravi Prakash and Edwards (BCRE) (Bouchaud, 1994; Bouchaud *et al.*, 1995) for the description of surface gravity-driven flows. Note that the BCRE and the Saint-Venant models can be derived in a certain limit from the more general order-parameter models mentioned above; for details see Chapter 6.

3.6.4 Cellular-automata-type models

In the cellular-automata approach physical variables such as height, concentration, etc., are discretized both in time and space. The variables evolve through discrete time steps according to a certain set of rules, either stochastic or deterministic, and typically, on a square (in two dimensions) or cubic (in three dimensions) grid. The best-known example of such cellular automata models is the "sandpile" model proposed by Bak *et al.* (1987) as a paradigm for the *self-organized criticality*. Conceptually similar models were used to describe granular stratification by Makse *et al.* (1997*b*), wind-blown sand ripple formation by Nishimori and Ouchi (1993) and others. In a certain limit (viz. for large-scale perturbations) cellular automata can be reduced to more traditional continuum hydrodynamic-type equations. While cellular automata are very computationally efficient, their connection to experimental conditions is somewhat remote. Consequently, with the revolutionary increase of computing power, more and more researchers rely on direct molecular dynamics simulations of granular assemblies rather than on phenomenological cellular automata.

3.7 Molecular dynamics simulations

The detailed knowledge of collisional properties of particles is essential for realistic discrete-element (or "molecular dynamics") simulations of granular matter. Due to a relatively small number of particles in granular flows as compared to atomic and molecular systems, such simulations have been very successful in capturing many phenomena occurring in granular systems.

There exist three fundamentally different classes of algorithms, so-called soft-particle simulation method; event-driven algorithm, and the contact dynamics method for rigid particles. Here we give a brief overview of these three classes of methods. For a detailed review of various molecular dynamics simulation algorithms we recommend (Rapaport, 2004; Luding, 2004; Pöschel and Schwager, 2005).

3.7.1 Soft-particle molecular dynamics

In the soft-particle algorithm, all forces acting on a particle either from walls or other particles or external forces are calculated based on the positions and velocities of particles. Once the forces are found, the particles are displaced and rotated by the explicit integration of the corresponding Newton's equations of motion over a small time step (so the displacements are small compared with the particle size),

$$m_i \frac{d^2 \mathbf{r}_i}{dt^2} = \mathbf{f}_i + m_i \mathbf{g}, \quad I_i \frac{d^2 \varphi_i}{dt^2} = \mathbf{t}_i, \tag{3.49}$$

and then the routine repeats. In eqn 3.49, $m_i, \mathbf{r}_i, \varphi_i$ are the mass, position and orientation of the ith particle (for simplicity, we consider two-dimensional case), $\mathbf{f}_i$ is the vector sum of all contact forces acting on it, and $\mathbf{t}_i$ is the total torque. Various models are used for calculating normal and tangential contact forces. In the majority of implementations, the normal contact forces are determined from the particle overlap Δ which is the difference between the some of two particles' radii and the distance between their centers. In cohesionless case, the normal force $\mathbf{f}_n$ is either proportional to Δ (linear Hookian contact) or proportional to $\Delta^{3/2}$ (Hertzian contact), provided that Δ is positive, and zero otherwise. In the spring–dashpot model, an additional dissipative force proportional to the normal component of the relative velocity is added to model inelasticity of grains. The main difficulty of this algorithm lies in the calculation of tangential forces because of the static friction indeterminacy described above in Section 3.1.2. A variety of approximations are used to model tangential forces, the most widely accepted of them being the Cundall–Strack algorithm (Cundall and Strack, 1979; Schäfer et al., 1996), in which the tangential contact is modeled by a dissipative linear spring whose force $\mathbf{F}_t = -k_t \Delta_t - m\gamma_t \mathbf{v}_t/2$ (here Δ_t is the relative tangential displacement and $\mathbf{v}_t$ is the relative tangential velocity, k_t, γ_t are model constants). When F_t reaches μF_n, the spring is replaced by the Coulomb friction force $\mathbf{F}_t = -\mu F_n \mathbf{v}_t/|\mathbf{v}_t|$. Soft-particles methods are relatively slow and used mostly for the analysis of dense flows when generally faster event-driven algorithms are not applicable; see, e.g. (Silbert et al., 2002a; Silbert et al., 2002b; Silbert et al., 2002c; Landry et al., 2003; Volfson et al., 2003a; Volfson et al., 2003b).

3.7.2 Event-driven algorithms

In the event-driven algorithms it is assumed that the particles are infinitely rigid and move freely (or are driven by macroscopic external fields) in the intervals between instantaneous collisions. Simple event-driven algorithms update positions and velocities of all particles after every collision according to Newton's laws and the two colliding particles according to collision rules (in the simplest frictionless case, according to eqn 3.9). Then the time of the next collision is calculated, after which the process repeats. Thus, the time is advanced directly from one collision to the next, and so the variable time step is dictated by the interval between the collisions. Recomputing positions and velocities of all particles after each collision is rather ineffective and in fact unnecessary. A significant improvement can be achieved if only velocities and positions of the two particles involved in a binary collision are updated (Lubachevsky, 1991). While event-driven methods are typically faster for dilute rapid granular flows, they become impractical for dense flows where collisions are very frequent or, worse, particles develop persistent contacts. Furthermore, these algorithms are essentially serial in nature, and their parallelization is difficult (but not impossible, see Miller and Luding (2004)). As a related numerical problem, event-driven methods are known to suffer from the so-called *inelastic collapse* when the number of collisions between particles in a unit time diverges (McNamara and Young, 1996). Introducing a velocity dependent restitution coefficient that approaches one as the collision velocity approaches zero, makes it possible to avoid the inelastic collapse conundrum (see, e.g. Bizon *et al.* (1998*b*)). Since the computational cost becomes prohibitively high for dense systems, event-driven methods are mostly applied to rapid granular flows; see, e.g. Ferguson *et al.* (2004), McNamara and Young (1996), Khain and Meerson (2004), Nie *et al.* (2002).

3.7.3 Contact mechanics

Contact dynamics is a discrete-element method like the soft-particles and event-driven ones, with the equations of motion integrated for each particle. Similar to the event-driven algorithm and unlike the soft-particles method, particle deformations are suppressed by treating particles as infinitely rigid. The contact dynamics method considers all contacts occurring within a certain short time interval as simultaneous, and computes all contact forces by satisfying simultaneously all kinematic constraints imposed by impenetrability of the particles and the Coulomb friction law. Imposing kinematic constraints requires contact forces (constraint forces) that cannot be calculated from the positions and velocities of particles alone. The constraint forces are determined in such a way that constraint-violating accelerations are compensated (Moreau, 1994). For a comprehensive review of the contact dynamics see Brendel *et al.* (2004).

Sometimes different molecular dynamics methods are applied to the same problem. Lois *et al.* (2005), Staron *et al.* (2002), and Radjai and Wolf (1998) applied contact dynamics methods and Silbert *et al.* (2002*a*), Silbert *et al.* (2002*b*), Volfson *et al.* (2003*a*), and Volfson *et al.* (2003*b*), used a soft-particles technique for the analysis of instabilities and constitutive relations in dense granular systems. Patterns in vibrated layers were studied by event-driven simulations by Bizon *et al.* (1998*b*), Moon *et al.* (2004) and via soft-particles simulations by Nie *et al.* (2000) and Prevost *et al.* (2002).

4
Phase transitions, clustering, and coarsening in granular gases

4.1 Clustering in freely cooling granular gases

Freely cooling or non-driven granular gas is an isolated dilute system of thermalized inelastically colliding grains in the absence of gravity. Due to the inelasticity of collisions the granular temperature T of isolated granular gas always eventually becomes zero (rest or cold state). A naïve approach to the description of this cooling regime assumes that initially uniform gas remains uniform forever. In this regime, a simple scaling argument yields the rate of the energy loss due to inelastic collisions. Indeed the energy loss dT/dt due to inelastic collisions is proportional to the mean particle energy T and the collision frequency, which in turn is proportional to the mean particle velocity, or $T^{1/2}$. Thus, $dT/dt \sim -T^{3/2}$, and the granular temperature decays as t^{-2}. More precisely, the energy decay of a uniform granular gas with zero macroscopic velocity is given by the temperature equation (eqn 3.13)

$$dT/dt = -\zeta T^{3/2}, \tag{4.1}$$

where the energy sink term $\zeta T^{3/2}$ is given by eqn 3.20. For small density ν the cooling rate $\zeta \sim (1-e^2)\nu^2$. This equation gives rise to Haff's cooling law (Haff, 1983)

$$T = \frac{T_0}{(t/t_0 + 1)^2} \quad \text{for } t \to \infty. \tag{4.2}$$

Here, T_0 is the temperature of gas at $t = 0$ and $t_0 \sim (1-e^2)^{-1} T_0^{-1/2}$ is the characteristic cooling time.

However, inelastic collisions not only extract kinetic energy from the system, but they also lead to the emergence of correlations among particle velocities and the violation of the molecular chaos approximation. This in turn leads to a non-trivial clustering instability (Goldhirsch and Zanetti, 1993) leading to spontaneous emergence of dense clusters, see Fig. 4.1. The early stages of the instability of the freely cooling state often are associated with the formation of a large-scale vortex structure (Fig. 4.1(a)) which is later taken over by the cluster state characterized by large density variations (Fig. 4.1(b)). At least at the linear stage, this clustering instability, which can be traced in many other particle systems, has a very simple physical interpretation: a local increase of the granular gas density ν results in the increase in the number of collisions, and, therefore, dissipation of energy and decrease in the granular temperature T. Due to

60 *Phase transitions, clustering, and coarsening in granular gases*

the proportionality of the pressure p to the temperature according to the equation of state (eqn 3.16), the decrease of granular temperature will reduce local pressure, which in turn will create an influx of particles towards this pressure depression, and a further increase of the density. This clustering instability has an interesting counterpart in astrophysics: self-gravitating gas of stars exhibits clustering due to irreversible dissipative processes in the course of star interactions (Shandarin and Zeldovich, 1989). Another example of clustering is the "radiative instability" in optically thin plasmas resulting in interstellar gas or dust condensation (Meerson, 1996). In this situation the gas is cooled by radiation. Condensation of the interstellar gas results in an increase of radiative losses, and, consequently, further cooling.

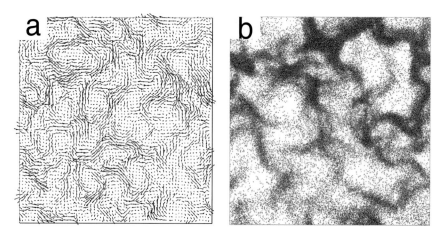

Fig. 4.1 Numerical simulation of freely cooling inelastic granular gas prepared in initially uniform state (restitution coefficient $e = 0.9$, filling fraction $\phi = 0.26$, number of particles 50,000). Panel (a) shows coarse-grained velocity flow field **v** at the moment of time $t = 80$, panel (b) shows the density field ν at $t = 160$, from Orza et al. (1997).

4.1.1 Clustering instability of the homogeneous cooling state

The initial stage of clustering can be understood as an instability of a homogeneous cooling state (uniform density ν and spatially homogeneous but time-dependent granular temperature T) within the framework of scaled granular hydrodynamics equations (3.11)–(3.13) in the dilute limit:

$$D\nu/Dt + \nu \nabla \cdot \mathbf{v} = 0, \tag{4.3}$$

$$\nu D\mathbf{v}/Dt = \nabla \cdot \boldsymbol{\sigma} \tag{4.4}$$

$$\nu DT/Dt + \nu T \nabla \cdot \mathbf{v} = \mathrm{Kn}\, \nabla \cdot (T^{1/2} \nabla T) - \mathrm{Kn} R_\mathrm{H}\, \nu^2\, T^{3/2}, \tag{4.5}$$

where $D/Dt \equiv \partial_t + (\mathbf{v}\cdot)$ is the material derivative, $\boldsymbol{\sigma} = -\nu T \mathbf{I} + (1/2)\mathrm{Kn} T^{1/2} \hat{\dot\gamma}$ is the stress tensor, $\hat{\dot\gamma}$ is the deviatoric part of the strain rate tensor $\dot\gamma = [\nabla\mathbf{v} + \nabla\mathbf{v}^T]/2$, $\mathrm{Kn} = 2/\sqrt{\pi}(dL\langle\nu\rangle)$ is the Knudsen number characterizing the ratio of mean free pass of the particle to the system size L, $\langle\nu\rangle$ is the mean density (conserved), and

$R_\mathrm{H} = 4(1-e^2)/\mathrm{Kn}^2$ is the heat-loss parameter. In eqns 4.3–4.5 the distance is measured in the units of L, the time in units of $L/\sqrt{T_0}$, and the temperature in units of T_0.

In order to examine the stability of the uniformly cooling state one has to linearize eqns 4.3–4.5 in the vicinity of the time-dependent state eqn 4.2 (Goldhirsch and Zanetti, 1993). After rescaling the time variable t by the (time-dependent) mean collision time $t_c(t) \sim T^{1/2}(t)$, $d\tau = dt/t_c(t)$ and normalizing perturbations of temperature and velocity by the time-dependent mean temperature $T(t)$ and thermal velocity $T^{1/2}(t)$, respectively, the linearized equations become time-independent and admit solutions in the form of Fourier modes $\Psi_k \sim \exp[i\mathbf{k} \cdot \mathbf{r} + \lambda(k)\tau]$, where $\mathbf{k}$ is the perturbation wave number and $\lambda(\mathbf{k})$ is the corresponding growth rate of linear perturbations. A detailed discussion of the linear stability analysis can be found, e.g., in (Orza et al., 1997; van Noije and Ernst, 2000; Brilliantov and Pöschel, 2004). Two different modes are unstable ($\mathrm{Re}\lambda > 0$). One is the shear mode with the growth rate λ_S, in which perturbations of density and temperature are absent, and the perturbations of hydrodynamic velocity $\mathbf{v}$ are perpendicular to the wavevector, $\mathbf{k} \cdot \mathbf{v} = 0$. Another is the heat mode (growth rate λ_H) in which all three perturbations (velocity, temperature, and density) grow simultaneously. Both modes grow only for sufficiently long-wave perturbations,

$$\lambda_{\mathrm{S,H}}(k) \sim (1-e^2)\left(1-(k/k_{\mathrm{S,H}})^2\right), \tag{4.6}$$

with the critical wave numbers for both modes

$$k_{\mathrm{S,H}} \sim l_0^{-1}\sqrt{1-e^2}, \tag{4.7}$$

where l_0 is the (time-independent) mean free path of particles in the cooling granular gas.[1] The eigenvalues of both modes are real, which indicates non-oscillatory behavior of growing perturbations. The length scale of the linear instability diverges in the limit of elastic particles, $e \to 1$. However, the unstable shear mode resulting in the formation of the large-scale vortex structure has a larger growth rate and remains unstable in a wider range of wave numbers (Fig. 4.2). Consequently, at the first stage of the instability the spontaneous vortex formation with little change in density precedes the heat mode that directly leads to clustering (Fig. 4.1). Eventually, the density variations grow resulting in formation of dense and cold clusters. The instability only signals the onset of clustering. Further treatment of clustering beyond the linear instability requires a solution of full non-linear granular hydrodynamics equations or molecular dynamics simulations.

4.1.2 Non-linear regime of the clustering instability

Further understanding of the dynamics of the freely cooling gas can be gained in so-called quasi-one-dimensional approximation, when particles move microscopically in two or three dimensions, but the macroscopic fields depend only on one spatial coordinate x. Such a situation can be realized, for example, in a long narrow channel with elastic sidewalls. A dilute quasi-one-dimensional gas of weakly inelastic disks can

---

[1] For the heat mode expansion 4.6 is valid only for $kl_0 = O\left((1-e^2)^{1/2}\right)$

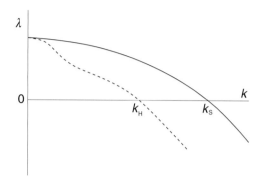

Fig. 4.2 Growth rates $\lambda_{S,H}$ for the shear (solid) and heat (dashed) modes as a function of the wave numbers k.

be described by the one-dimensional version of the dilute granular hydrodynamics equations 4.3–4.5 (Efrati *et al.*, 2005),

$$\partial_t \nu + \partial_x(\nu v) = 0 \tag{4.8}$$
$$\nu(\partial_t v + v\partial_x v) - \partial_x \sigma = 0 \tag{4.9}$$
$$\nu(\partial_t T + v\partial_x T) = \sigma \partial_x v + \mathrm{Kn}\partial_x(T^{1/2}\partial_x T) - \varrho T^{3/2}. \tag{4.10}$$

Here $\varrho = 4(1-e^2)\bar{\nu}^2 \mathrm{Kn}^{-1}$ is the rescaled cooling rate, $\sigma = -\nu T + \mathrm{Kn} T^{1/2}\partial_x v/4$ is the stress field, coordinate x is scaled by the system size L_x, time $t \to tT_0^{1/2}/L_x$, temperature $T \to T/T_0$ and density $\nu \to \nu/\langle\nu\rangle$, T_0 is the initial temperature. The numerical solutions of eqs. 4.8–4.10 indeed show strong clustering: development of multiple high and narrow density peaks accompanied by the steepening of the velocity gradients, see Fig. 4.3. The gas density grows in these peaks without limit, indicating a finite-time *hydrodynamic singularity*. Near the singularity the stress term σ becomes small, and the dynamics are described by the simple equation for inertial flow $\partial_t v + v\partial_x v = 0$. Eventually, the hydrodynamics breaks down, and some regularization needs to be introduced.

It is instructive to consider the energy balance between microscopic and macroscopic degrees of freedom in the course of cooling. In the quasi-one-dimensional gas, the total kinetic energy $E = (2N)^{-1}\sum_{i=1}^{N} v_i^2$ per particle has the following relation with the hydrodynamic variables

$$E = \frac{1}{2N}\sum_{i=1}^{N} v_i^2 = \frac{1}{L_x}\int_0^{L_x}\left(\nu T + \nu\frac{v^2}{2}\right) dx = E_T + E_{\mathrm{mac}} \tag{4.11}$$

The first term is the energy of the thermal motion of grains, and the second term is the kinetic energy of the macroscopic motion. Molecular dynamics simulations can be used to extract hydrodynamic fields ν, v, T from the particle positions x_i and velocities v_i through an appropriate coarse-graining procedure (Meerson and Puglisi, 2005). The results of simulations (Fig. 4.4) show that the total energy E decays as t^{-2}, in accordance with Haff's law, however at later stages the decay crosses over to the scaling

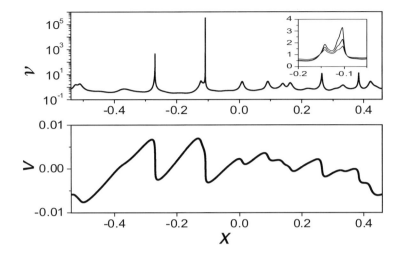

Fig. 4.3 The profiles of density ν and the hydrodynamic velocity v at scaled time $t = 7.043$, just before the major peak develops a singularity. The parameters Kn $= 4 \times 10^{-4}$, $(1 - e^2) = 0.02$. The inset shows early density evolution of a region around the main peak, from Efrati et al. (2005).

law $t^{-2/3}$, see eqn 4.12 below. Both thermal energy and macroscopic kinetic energy decay, but the thermal energy decays faster, and the long-time behavior is entirely determined by the macroscopic (or hydrodynamic) component of the mean kinetic energy. Interestingly, even in that regime, the thermal energy $E_T \sim \int \nu T \mathrm{d}x$ continues to decay as $E_T \sim t^{-2}$, i.e. it still complies with Haff's law.

4.1.3 Phenomenological modelling of the late stages of clustering

While the first stages (linear as well as non-linear) of clustering, when the relative density fluctuations are not too large, can be described within granular hydrodynamics, the late-time dynamics of the clustered state goes beyond the limits of the hydrodynamic description. As was discussed in Chapter 3, a system of grains with a constant restitution coefficient e can experience an infinite number of collisions in a finite time before coming to rest. This phenomenon is called *inelastic collapse* (McNamara and Young, 1992). In terms of granular hydrodynamics inelastic collapse translates into a finite-time singularity of the density ν. Thus, the hydrodynamic description breaks in dense cold clusters. The inelastic collapse also creates significant difficulties in some molecular dynamics methods (such as the highly efficient event-driven method) because of the divergence of the number of integration steps and the corresponding integration time. All these circumstances indicate that straightforward approaches to the description of clustering are bound to fail eventually; more delicate methods are necessary. Recent theoretical studies of the late stages of clustering instability attempted to regularize the hydrodynamic description near singularities by taking into account the finite size of particles (Nie et al., 2002; Efrati et al., 2005).

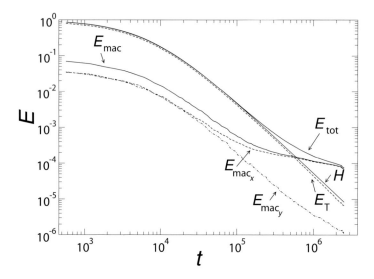

Fig. 4.4 The energy balance of the quasi-one-dimensional granular gas over time. E_{tot} is the average kinetic energy of the particles, E_T is the average thermal energy and $E_{\text{mac}_{x,y}}$ is the macroscopic energy in x and y directions, respectively, length $L_x \gg L_y$. Haff's law is indicated by the solid line marked by H, from Meerson and Puglisi (2005).

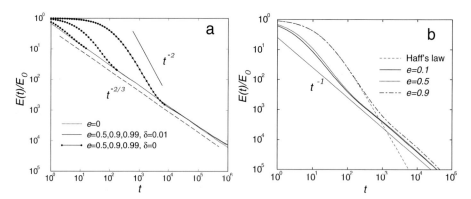

Fig. 4.5 Average kinetic energy of particles $E(t)$ normalized by the initial energy E_0 vs. time t for different values of restitution coefficient e for one-dimension (a) and two-dimensional (b) granular gas, from Ben-Naim et al. (1999), Nie et al. (2002).

The development of the heterogeneous density distribution in the course of cooling can alter the scaling for the energy dissipation at the later stages of clustering from Haff's law $T \sim t^{-2}$. Such a deviation was first observed in simulations of the freely cooling gas of point particles on a line (Ben-Naim et al., 1999; Nie et al., 2002). The problem of inelastic collapse in these simulations was circumvented by assuming that collisions become elastic ($e = 1$) below a prescribed threshold for the relative velocity of collisions $\Delta v < \delta$. The threshold δ appears to have a very small effect on the scaling behavior. The simulations (see Fig. 4.5(a)) indicate that the long-term stage of the cooling is described by a different scaling law for the average kinetic energy E,[2]

$$E \sim t^{-2/3}. \tag{4.12}$$

Analogous two-dimensional simulations yield scaling $E \sim t^{-1}$ (Fig. 4.5(b)). This asymptotic behavior is insensitive to the value of the restitution coefficient. Thus, Haff's law appears to describe only the initial (transient) stage of the cooling process. What is the underlying reason for the deviations from Haff's law at the late stages of cooling? The hint to the answer is provided by the simulations (Ben-Naim et al., 1999; Nie et al., 2002) that showed that at the late cooling stages the hydrodynamic granular velocity profile acquires a characteristic sawtooth shape reminiscent of shock waves in compressible fluids, see Fig. 4.6. Accordingly, the density ν has very sharp peaks at the locations of shocks.

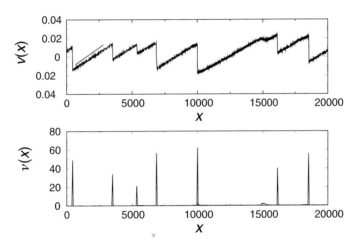

Fig. 4.6 Shock profiles formed in an inelastic one-dimensional granular gas. Velocity v (top) and density ν (bottom) are plotted at time $t = 10^5$ in a system of $N = 2 \times 10^5$ particles with restitution coefficient $e = 0.99$. A line with slope $t^{-1} = 10^{-5}$ is plotted for reference. The number of shocks N_s is consistent with the expected number $Nt^{-2/3} \approx 9$, from Ben-Naim et al. (1999).

The development of shocks in freely cooling granular gases can be understood from the analogy with so-called "sticky" gases, i.e. gases where particles stick together after the collision and move with the average velocity (Shandarin and Zeldovich, 1989; Ben-Naim et al., 1999). Formally the "sticky" limit corresponds to the zero restitution coefficient. In this limit the pressure becomes zero, and the momentum equation decouples from the density conservation equation:

---

[2] Granular temperature T coincides with the average kinetic energy E only for the case of negligible hydrodynamic velocities $\mathbf{v}$, i.e. in the beginning of cooling, see below.

$$\partial_t \nu + \partial_x(\nu v) = 0 \qquad (4.13)$$
$$\partial_t v + v\partial_x v = \eta_0 \partial_x^2 v, \qquad (4.14)$$

where η_0 is the effective viscosity (which is different from the shear viscosity in hydrodynamic equations η). Equation 4.14 is the well-known Burgers equation. In the absence of the effective viscosity η_0 the one-dimensional hydrodynamic equations (eqns 4.13–4.14) represent the simplest model of non-linear kinematic waves exhibiting discontinuous solutions and finite-time singularities of ν and gradient of v (Whitham, 1974). The non-zero effective viscosity results in the regularization of the singularities and formation of sharp shocks as in Fig. 4.6. Perhaps not surprisingly, this model is very similar to the description of the large-scale matter formation in the Universe (Gurbatov et al., 1985; Shandarin and Zeldovich, 1989).

The connection to the Burgers equation is interesting in many respects. Since the Burgers equation is reduced to a linear diffusion equation $\partial_t w = \eta_0 \partial_x^2 w$ by the Hopf–Cole transformation $v = -2\eta_0(\log w)_x$, many statistical properties of solutions of the Burgers equation with random initial conditions can be obtained analytically (Frachebourg and Martin, 2000). In particular, in the one-dimensional case the average kinetic energy $E_1 = \int v^2 dx/2$ decays as $E_1 \sim t^{-2/3}$, and the number of shocks $N_s \sim Nt^{-2/3}$, where N is the total mass (or total number of particles) in the system. These results are in excellent agreement with the simulations shown in Figs. 4.5 and 4.6.

While a qualitative similarity exists between Burgers shocks and clusters in granular materials, at least in one dimension, the applicability of the Burgers equation to the description of cooling of granular gases in general is still an open question. First, the validity of granular hydrodynamics to a strictly one-dimensional gas of particles moving on a line is questionable due to the inevitable inelastic collapse (Kadanoff, 1999). In two or three dimensions, an additional difficulty arises due to the spontaneous generation of the vorticity field $\nabla \times \mathbf{v}$, which violates the Burgers picture. Indeed, molecular dynamics simulations show the development of large-scale vortex flows in the course of clustering instability, as illustrated by Fig. 4.1 (Cattuto and Marconi, 2004; van Noije and Ernst, 2000).

4.1.4 Thermal collapse of granular gas under gravity

The main difficulty in observing freely cooling granular gas on Earth is the unavoidable gravity field. Even in microgravity conditions (e.g. on board of a space station or a parabolic flight aircraft), low gravity would break the symmetry and push grains downward. The natural question is then how does the gravity field affect the cooling process? On the one hand, gravity supplies energy to the grains and can play the role of the heat source. On the other hand, gravity forces grains to sink to the bottom of the container, where increased density enhances the collision rate and causes even faster cooling of the granulate. The quantitative description of this process is far from straightforward (Volfson et al., 2006). It turns out that, in striking contrast to Haff's law, the initially dilute granular gas eventually undergoes thermal collapse: it cools down to zero temperature and condenses on the bottom plate in a finite time $t = t_c$ exhibiting, close to collapse, a $E \sim (t_c - t)^2$ scaling of the total energy of the granular gas, see Fig. 4.7.

Equations of one-dimensional hydrodynamics (4.10) can be conveniently transformed from the Eulerian vertical z coordinate to the Lagrangian mass coordinate $m(x,t) = \int_0^z \nu(z',t) dz'$ that varies between 0 at the bottom $z = 0$ and 1 (the total rescaled mass of the gas) as $z \to \infty$. We can rescale the variables using the gravity length scale $\lambda = T_0/g$ (T_0 is the initially uniform temperature of the granular column) and the heat diffusion time $t_\mathrm{d} = \bar{\varepsilon}^{-1}(\lambda/g)^{1/2}$. The scaled parameter $\bar{\varepsilon} = \pi^{-1/2}(L_z/Nd) \ll 1$ is of the order of the inverse number of layers of grains that form after all particles settle on the bottom. The smallness of $\bar{\varepsilon}$ guarantees that t_d is much longer than the fast hydrodynamic time $t_\mathrm{f} = (\lambda/g)^{1/2}$. We measure v in units of $v_0 = \lambda/t_\mathrm{d}$, ν in units of $\nu_0 = N/\lambda L_z$, and T in units of T_0. The resulting rescaled hydrodynamic equations are (Bromberg et al., 2003; Volfson et al., 2006):

$$\partial_t(1/\nu) = \partial_m v, \qquad (4.15)$$

$$\varepsilon^2 \partial_t v = -\partial_m(\nu T) - 1 + \frac{\bar{\varepsilon}^2}{2} \partial_m(\nu T^{1/2} \partial_m v), \qquad (4.16)$$

$$\partial_t T + \nu T \partial_m v = \frac{\bar{\varepsilon}^2 \nu T^{1/2}}{2}(\partial_m v)^2 + \frac{4}{3}\partial_m(\nu \partial_m T^{3/2}) - 4\Lambda^2 \nu T^{3/2}. \qquad (4.17)$$

In addition to ε, eqns 4.15–4.17 include the parameter $\Lambda^2 = \frac{1-e^2}{4\bar{\varepsilon}^2}$, which shows the relative role of the inelastic energy loss and heat diffusion. At the boundaries $z = 0$ and $z \to \infty$ we assume zero fluxes of mass, momentum and energy (Bromberg et al., 2003; Volfson et al., 2006), which yields $v = \partial_m T = 0$ at $m = 0$ and $\nu \partial_m v = \nu \partial_m T = 0$ at $m = 1$.

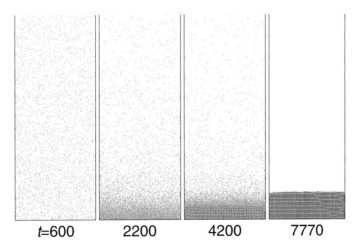

Fig. 4.7 Snapshots of an event-driven molecular dynamics simulation at different moments of time for number of particles $N = 5642$, container width $L_x = 102$, restitution coefficient $e = 0.995$, initial temperature $T_0 = 10$ and gravity acceleration $g = 0.01$. Only a part of the box in the lateral direction is shown, from Volfson et al. (2006)

Typically, the parameter $\bar{\varepsilon}$ is small. Then, after a brief transient, a *quasistatic* flow sets in. In this regime, the $\bar{\varepsilon}^2$ terms in eqns 4.16 and 4.17 can be neglected, and eqn 4.16 reduces to the hydrostatic condition $\partial_m(\nu T) + 1 = 0$, which yields $\nu T = 1 - m$. Substituting $\nu = (1-m)/T$ into eqn 4.17 and using eqn 4.15, we obtain a closed non-linear equation for a new variable $w(m,t) = T^{1/2}(m,t)$:

$$w\partial_t w = \partial_m\left[(1-m)\partial_m w\right] - \Lambda^2(1-m)w. \tag{4.18}$$

It is easy to see that the eqn 4.18 permits a separable solution

$$w(m,t) = (t_* - t)Q(m), \tag{4.19}$$

where $Q(m)$ is determined by the non-linear ordinary differential equation

$$[(1-m)Q']' - \Lambda^2(1-m)Q + Q^2 = 0 \tag{4.20}$$

(the primes denote m derivatives) and the boundary conditions $(1-m)Q' = 0$ at $m = 0$ and 1. This equation has only one parameter, Λ. What is less obvious but nevertheless true, is that a solution of eqn 4.18 starting from almost any initial condition eventually settles on the separable solution (eqn 4.19). This solution explicitly demonstrates $(t_* - t)^2$ scaling for the granular temperature $T = w^2$. The solution of eqn 4.18 is in excellent agreement with both the solution of the full hydrodynamics equations and event-driven molecular dynamics simulations (Fig. 4.8). The time of collapse, however, is not universal, as it depends on the initial conditions. For a uniform initial temperature distribution and parameter $\Lambda \gg 1$, the collapse time is $t_* \sim (1-e^2)^{-1/2}T_0^{1/2}/g$, so the higher the elasticity of grains, the longer it takes for the thermal collapse to occur. Note that in this regime, granular collapse to the bottom occurs so fast that the clustering instability does not have time to develop. The characteristic time of clustering instability is of the order of the free cooling time $t_0 \sim (1-e^2)^{-1}T_0^{-1/2}$. Comparing this expression with t_c above, one can estimate the upper bound for the gravity acceleration for which the clustering instability still has time to develop before the grains collapse on the bottom, $g < g_m \sim T_0(1-e^2)^{1/2}$.

4.2 Patterns in granular gases heated by the side walls

4.2.1 Hydrodynamic description of the clustered state

The discovery of the clustering instability stimulated a large number of experimental and theoretical studies, even experiments in low-gravity conditions (Falcon et al., 1999b). Since "freely cooling granular gas" is difficult to implement in the laboratory conditions, most experiments were performed in a rather different situation when the energy is constantly injected in the granular system in the form of vertical or horizontal vibration of a substrate, or from an oscillating sidewall. In experiments by Kudrolli et al. (1997) with a quasi-two-dimensional granular gas energized by one rapidly vibrating side wall, large clusters of almost immobile particles were observed opposite to the vibrated wall, see Fig. 4.9. This result can be understood in terms of granular hydrodynamics. Since granular particles move in a horizontal plane, the gravity is irrelevant, and so in the steady state the pressure across the system becomes constant, $p = p_0$, and the macroscopic velocity is absent, $v = 0$. The effect of a rapidly oscillating

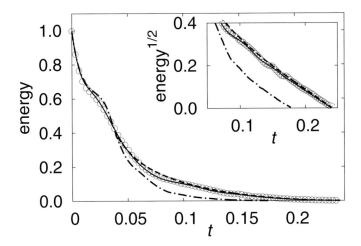

Fig. 4.8 Energy dissipation during thermal collapse of a granular gas in a gravity field. Circles: total kinetic energy of the grains, normalized by its value at $t = 0$, vs. time from an event-driven molecular dynamics simulation of 5642 grains with restitution coefficient $e = 0.995$ in an open box of width 10 particle diameters. Initial temperature $T_0 = 10$, gravity $g = 0.01$. Also shown are the results from the full hydrodynamic model (solid line), from eqn 4.18 (dashed line), and from a molecular dynamics simulation with a *different*, non-isothermal initial condition (dash-dotted line). The respective hydrodynamic parameters are $\tilde{\varepsilon} = 10^{-2}$ and $\Lambda = 5$. The inset shows the square root of the energy close to thermal collapse, from Volfson et al. (2006)

wall can be modelled by the fixed granular temperature boundary condition $T = T_0$ at the wall. A good approximation for a non-vibrating and almost elastic wall opposite to the "hot" wall is the no-flux condition $\partial_z T = 0$. Then from the equation of state (eqn 3.16) for pressure $p = \tilde{f}(\phi = \nu/\nu_0)T$ one obtains the density as a function of temperature $\nu = \nu_0 \tilde{f}^{-1}(p_0/T)$. In granular hydrodynamics function $\tilde{f}$ and therefore its inverse $\tilde{f}^{-1}$ are monotonously increasing functions. Since, due to inelasticity of collisions, temperature decays away from the hot wall, density should increase away from the hot wall. In the framework of the quasi-one-dimensional granular hydrodynamics the stationary temperature distribution is described by the one-dimensional energy balance equation

$$\partial_z(\kappa \partial_z T) - \varrho T^{3/2} = 0, \qquad (4.21)$$

with the boundary conditions $T = T_0$ at $z = 0$ (hot wall) and $\partial_z T = 0$ at $z = L_z$ (elastic wall), see Fig. 4.10. Equation 4.21 can be further simplified in the dilute limit $\nu \to 0$. Then, after normalization of length by $z \to z/L_z$, temperature $T \to T/T_0$, and density $\nu \to \nu/\nu_0$ (after this normalization the density coincides with the volume fraction ϕ), one obtains the cooling rate $\varrho = 4(1-e^2)\nu^2 \mathrm{Kn}^{-1}$ and the thermal conductivity $\kappa = \mathrm{Kn} T^{1/2}$. Excluding ϕ yields a simple steady-state equation

$$\partial_z^2 T^{3/2} - \frac{6(1-e^2)p_0^2}{\mathrm{Kn}^2 T^{1/2}} = 0. \qquad (4.22)$$

Fig. 4.9 Sample image showing dense cold cluster formed opposite the driving wall (at the bottom of the image), total number of particles 1860, from (Kudrolli et al., 1997).

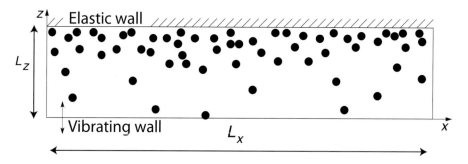

Fig. 4.10 Schematic representation of the driven granular gas system. Grains are placed in a two-dimensional box. The bottom plate is vibrated (or heated) while the other walls are assumed to be elastic.

Equation 4.22 can be integrated in quadratures giving rise to the following implicit expression for the temperature (Grossman et al., 1997; Khain et al., 2004),

$$\sqrt{T(T-T_1)} + T_1 \log\left(\sqrt{T/T_1} + \sqrt{T/T_1 - 1}\right) = \frac{4\sqrt{1-e^2}p_0}{\mathrm{Kn}}(1-z), \qquad (4.23)$$

where $T_1 < T_0$ is the temperature at the elastic wall $z = 1$. Then, this temperature can be expressed through the hot-wall temperature T_0. The dependencies of the temperature T and density $\nu = p_0/T$ on the distance from the hot wall are shown in Fig. 4.11(a). The temperature falls off from the hot wall almost linearly. Correspondingly, the density has a sharp peak opposite to the hot wall. Surprisingly, this calculation can be extended even for densities in the cold phase approaching the close-packed density ν_c (Grossman et al., 1997). In this case, the equation of state and transport coefficients can be interpolated between small density, where the dilute limit expressions are valid, almost to the close-packed density ν_c where these quantities diverge as $(\nu_c - \nu)^{-1}$. Specifically, Grossman et al. (1997) used the following equation of state (compare with eqn 3.16):

$$p = \nu T \frac{\nu + \nu_c}{\nu_c - \nu}. \tag{4.24}$$

This equation agrees with the dilute limit $p = \nu T$ expression as $\nu \to 0$ and is also valid near the close-packed density since $p = 2\nu^2 T/(\nu_c - \nu) \to \infty$ for $\nu \to \nu_c$. Figure 4.11(b) illustrates that molecular dynamics simulations for the gas of inelastic particles in the quasi-one-dimensional channel geometry are in very good agreement with the hydrodynamic solution with the "corrected" equation of state (eqn 4.24).

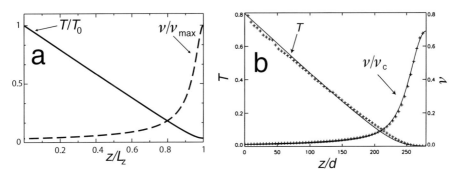

Fig. 4.11 (a) Normalized temperature T/T_0 and density $\nu/\nu_{\max}$ vs. dimensionless length z/L_z, solution to eqn 4.23. The maximum of density is located opposite to the vibrated wall; (b) Comparison of an event-driven simulation of 1000 particles at restitution coefficient $e = 0.99$ (symbols) with the numerical solution of eqn 4.21 (solid lines). The horizontal axis is the distance from the heated wall in units of the particle diameters d, from Grossman et al. (1997).

Similar expressions valid both in dilute and dense phases can be obtained for the energy sink term and the granular heat conductance from rather simple qualitative arguments. According to Grossman et al. (1997), the energy sink ε and heat conductance κ can be taken in the form

$$\varepsilon(T) \sim \frac{1-e^2}{\gamma l} \nu T^{3/2}$$

$$\kappa(T) \sim \frac{al+d}{l} \nu T^{1/2} \tag{4.25}$$

where l is the mean free path, d is the particle size, and $\alpha \approx 1.15, \gamma \approx 2.26$ are the parameters dependent on the specifics of the particle velocity distribution. The mean free path l can be expressed through the density ν. For example, for the two-dimensional close-packed value $\nu_c = 2/\sqrt{3}d^2$, one finds

$$l \approx \frac{1}{\sqrt{8}\nu d} \cdot \frac{\nu_c - \nu}{\nu_c - a\nu}, \qquad (4.26)$$

with $a = 1 - \sqrt{3/8}$. Equations 4.24–4.26 provide efficient and simple approximations for hydrodynamic equations valid in a wide range of densities. We will apply these expressions in the following sections.

4.2.2 Transversal van der Waals instability of the one-dimensional cluster

In experiments by Kudrolli et al. (1997) only one granular cluster was observed (Fig. 4.9). However, granular hydrodynamics and molecular dynamics simulations predict that in systems with large aspect ratio (ratio of the lateral dimension L_x to the distance between the hot and elastic walls L_z) the cluster should break up in a sequence of droplets through the instability reminiscent phase separation or van der Waals instability between gas and liquid (Livne et al., 2002b; Argentina et al., 2002). Event-driven molecular dynamics simulations by Argentina et al. (2002) were performed with a gas of inelastic hard spheres in a two-dimensional setting with a large aspect ratio L_x/L_z, see sketch in Fig. 4.10. The effect of the hot wall with temperature T_0 is modelled as follows: each time a grain collides with the wall, it is reflected, conserving the tangential component of the velocity, whereas the normal component is selected from a Maxwell distribution with the temperature T_0.

At low dissipation (the restitution coefficient e is very close to 1) the system develops gradients of temperature and density as described in the previous section; eventually a cold high-density cluster is formed opposite to the hot wall. At higher dissipation (smaller values of the restitution coefficient) a transversal pattern-forming instability occurs: the system exhibits coexistence of two phases: low-density gas and higher-density "liquid". As seen in Fig. 4.12, in the initially homogeneous liquid phase (appearing as a dark strip) a low-density "gas bubble" nucleates and expands. Eventually, the bubble approaches its final size (which is less than the system size), and the stable coexistence of the two phases occurs.

To characterize the phase separation quantitatively it is convenient to introduce the density ν averaged across the transversal (here vertical) coordinate z:

$$\bar{\nu}(x) = \frac{1}{L_z} \int_0^{L_z} \nu(x, z, t) dz \qquad (4.27)$$

The time evolution of the average density $\bar{\nu}$ exhibits temporal oscillations, see Fig. 4.13. The nature of these decaying oscillations is related to the propagation of density shock waves that are generated when the bubble is nucleated and propagates through the dense region faster than the bubble itself expands. Due to periodic boundary conditions, these shock waves eventually reach the bubble and excite its oscillations. In

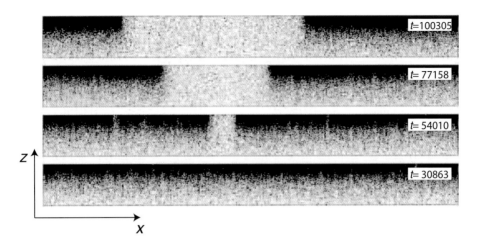

Fig. 4.12 Snapshots of a system with $N = 153,600$ particles, aspect ratio $L_x/L_z = 102.4$ (the snapshots have been laterally compressed to fit on the page), average number density $\nu_0 = 0.15$, length $L_x = 10,240$ particle diameters, restitution coefficient $e = 0.96$. The bubble appears at time $t \approx 40,000$, from Argentina et al. (2002).

sufficiently large systems simultaneous nucleation of more than one bubble is possible. Then, a slow coarsening process occurs: small bubbles shrink and large bubbles grow in their lateral size, resulting in the formation of a single bubble. The transient time, however, can be very large. This behavior is strikingly similar to the van der Waals liquid–gas transition below the critical temperature in an isolated system, where the homogeneous state is unstable and phase separation between the gas and liquid occurs.[3]

Since the phase separation occurs over a large time scale, on the phenomenological level it is possible to capture the essential features by introducing macroscopic slow variables. The large-scale evolution of the pattern can be described by the conservation laws for the z-averaged density $\bar{\nu}$ and the horizontal momentum density j

$$\partial_t \bar{\nu} = -\partial_x j$$
$$\partial_t j = -\partial_x \Phi, \tag{4.28}$$

where Φ is the momentum flux generally dependent on the density $\bar{\nu}$, momentum j and their derivatives. In order to close the description, we need to specify the expression for Φ. Following the Ginzburg–Landau ideas on phase transition, near the threshold of the phase-separation instability the momentum flux can be expanded near the uniform solution $\bar{\nu} = \bar{\nu}_0 = \text{const}$ and $j = 0$ in terms of small density variations $\tilde{\nu} = \bar{\nu} - \bar{\nu}_0$ and momentum density j. Keeping only the lowest-order terms we can write

$$\Phi = \Phi_0 + \frac{\partial \Phi}{\partial \bar{\nu}} \tilde{\nu} + \frac{\partial^2 \Phi}{\partial^2 \bar{\nu}} \frac{\tilde{\nu}^2}{2} + \frac{\partial^3 \Phi}{\partial^3 \bar{\nu}} \frac{\tilde{\nu}^3}{6} + \frac{\partial \Phi}{\partial \bar{\nu}_{xx}} \partial_x^2 \tilde{\nu} + \frac{\partial \Phi}{\partial j_x} \partial_x j + ... \tag{4.29}$$

---

[3] Another example of granular coarsening occurs in axial segregation of heterogeneous granular materials in long rotating drums; we discuss the mechanism and scaling laws of one-dimensional coarsening in Chapter 7.

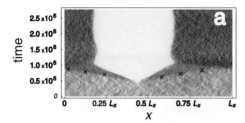
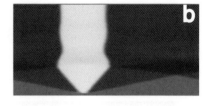

Fig. 4.13 Spatiotemporal evolution of the vertically averaged density $\bar\nu$. The density $\bar\nu$ is gray coded, with darker regions representing denser regions in the system. (a) Molecular dynamics simulation for the same parameters as in Fig. 4.12. In the final state, the density of the bubble is $\bar\nu = 0.025$, while in the dense region $\bar\nu = 0.257$. Crosses denote the location of the density shock waves. (b) Numerical solution of eqn 4.30 with $\bar\epsilon = 6.6 \times 10^{-4}$ and $\bar\zeta = 2$. The lateral system size $L_x = 5400$ and the total simulation time $t = 3.5 \times 10^5$. An initial condition (with $u = 1.4 \times 10^{-2}$) exceeding the nucleation barrier is imposed. The minimum (light) and maximum (dark) densities are $u = 2.6 \times 10^{-2}$ and $u = 2.9 \times 10^{-2}$, respectively, from Argentina et al. (2002).

Here, odd derivatives of $\tilde\nu$ and even derivatives of j are absent due to reflection symmetry: $x \to -x, j \to -j$ and translation $x \to x_0 + x, t \to t_0 + t$. Substituting expansion eqn 4.29 into eqns 4.28, after appropriate renormalization of the length $x \to x/l_m$, $t \to t/l_m$, where $l_m^2 = |\partial\Phi/\partial\bar\nu_{xx}|$,[4] at the dominant order for small deviations $u = \sqrt{\frac{\partial^3\Phi}{\partial\bar\nu^3}}\tilde\nu$, we arrive at the following equation (Argentina et al., 2002):

$$\partial_t^2 u = \partial_x^2 \left(\bar\epsilon u + bu^2 + u^3 - \partial_x^2 u + \bar\eta \partial_t u\right). \tag{4.30}$$

Here $\bar\epsilon = l_m^{-2}\partial\Phi/\partial\bar\nu|_{\bar\nu_0}$ is the control parameter, $\bar\eta = l_m^{-1}\partial\Phi/\partial j_x$ is the effective viscosity, and the parameter b is of the form $b \sim \partial^2\Phi/\partial\bar\nu^2$. Further simplification is possible if the density of homogeneous state $\bar\nu_0$ is near the Maxwell point ν_M defined by the condition

$$\left.\frac{\partial^2\Phi}{\partial^2\bar\nu}\right|_{\bar\nu=\nu_M} = 0. \tag{4.31}$$

Then, without loss of generality we can expand the density $\bar\nu$ near ν_M and introduce new variable $u = \sqrt{\frac{\partial^3\Phi}{\partial\bar\nu^3}}(\bar\nu - \nu_M)$. In this case, eqn 4.30 assumes the van der Waals normal form in which parameter $b = 0$.

Equation 4.30 is a generalization of the so-called *Cahn–Hilliard equation*

$$\partial_t u = \partial_x^2 \left(\bar\epsilon u + u^3 - \partial_x^2 u\right) \tag{4.32}$$

that serves as a paradigm model of phase-separation phenomena and spinodal decomposition. It is not surprising that both eqn 4.30 and eqn 4.32 exhibit somewhat similar

---

[4]The sign of $\partial\Phi/\partial\bar\nu_{xx}$ has to be positive in order to saturate the linear instability at the second order

behavior. However, in contrast to the Cahn–Hilliard model, eqn 4.30 also describes propagation of density shock waves.

For $\bar{\epsilon} < 0$, eqn 4.30 exhibits linear instability of the homogeneous state $u = 0$. Substituting solution $u \sim \exp[\lambda t + ikx]$ into eqn (4.30) linearized near $u = 0$, we obtain the growth rate λ as function of the modulation wave number k

$$\lambda^2 + \bar{\eta}k^2\lambda = -k^2\bar{\epsilon} - k^4. \tag{4.33}$$

It is easy to check that for any negative $\bar{\epsilon}$ the growth rate $\lambda \approx \sqrt{|\bar{\epsilon}|}|k| + O(k^2) > 0$ for $k \to 0$, i.e. the homogeneous state is always unstable. The maximum growth rate occurs at the wave number $k_m = \sqrt{|\bar{\epsilon}|/(2 + \bar{\eta})}$. The linear instability results in the formation of an almost periodic pattern at the optimal wave number k_m. Then, on the non-linear stage of evolution of eqn 4.30, coarsening begins, and the initial periodic structure evolves into a single bubble. A comparison of the results of molecular dynamics simulations and the corresponding solution of eqn 4.30 shows a good qualitative agreement, see Fig. 4.13.

A quantitative description of the van der Waals instability can be developed on the basis of quasi-two-dimensional hydrodynamic equations directly generalizing quasi-one-dimensional equations (eqns 4.10) in the dilute limit (Livne et al., 2002a; Livne et al., 2002b; Khain and Meerson, 2002; Khain and Meerson, 2004). Consider first the stationary distributions in the rectangular domain $L_x \times L_z$. The energy is injected by a hot wall located at $z = 0$, collisions with the other three walls is assumed elastic. From the momentum conservation it immediately follows that $p = p_0 = \text{const}$. From the energy-balance equation we obtain

$$\nabla \cdot (\kappa \nabla T) = \varrho T^{3/2}. \tag{4.34}$$

Equation 4.34 can be written in terms of the non-dimensional variable (scaled inverse density) $\xi(x, z) = \nu_c/\nu(x, z)$ where ν_c is the (two-dimensional hexagonal) close-packing density $\nu_c = 2/(\sqrt{3}d^2)$ and rescaled pressure $P = p/\nu_c T_0$, where T_0 is the temperature of hot the wall. In the scaled coordinates $\mathbf{r}/L_z \to \mathbf{r}$ the box dimensions become $\Delta \times 1$, where $\Delta = L_x/L_z$ is the aspect ratio of the original system. It is convenient to express temperature T, heat conductivity κ and heat loss ϱ via the inverse density ξ using the constitutive relations (eqns 4.24–4.26) under the assumption $p = \text{const}$. Then eqn 4.34 assumes the following form

$$\nabla \cdot (F(\xi)\nabla \xi)) = WQ(\xi), \tag{4.35}$$

where $F(\xi)$ and $Q(\xi)$ are certain functions of ξ. The dimensionless parameter

$$W = \frac{32}{3\gamma}(1 - e^2)\frac{L_z^2}{d^2}, \tag{4.36}$$

(where $\gamma = 2.26$) characterizes the ratio of inelastic energy dissipation to heat conductivity, see eqns 4.25 (Grossman et al., 1997). Introducing the new variable $\psi =$

$\int_0^\xi F(\xi)\mathrm{d}\xi$ and its inverse $\xi(\psi)$, we can transform eqn 4.35 into a non-linear Poisson equation

$$\nabla^2 \psi = W\tilde{Q}(\psi), \qquad (4.37)$$

where $\tilde{Q}(\psi) = Q[\xi(\psi)]$. This equation needs to be supplemented by the boundary conditions at the elastic walls and at the vibrating wall (Livne et al., 2002a). In addition, since the total number of particles is conserved, the following relation holds for the mean filling fraction ϕ:

$$\phi = \frac{1}{\Delta}\int_0^\Delta \int_0^1 \frac{\mathrm{d}x\mathrm{d}z}{\xi(\psi)} = \frac{N}{L_x L_z \nu_c} = \mathrm{const.} \qquad (4.38)$$

In order to examine linear stability of the stationary x-independent solution to eqns 4.35 or 4.37 in the framework of the full hydrodynamic equations we may seek the solution in the form $\mathbf{U} = \mathbf{U}_0(z) + \mathbf{u}_k(z)\exp(\lambda t + ikx)$, where $\mathbf{U} = (\nu, \mathbf{v}, T)$ is the vector of the hydrodynamic variables, $\mathbf{U}_0 = (\nu(z), 0, 0)$ is the unperturbed solution, $\mathbf{u}_k(z)$ is the perturbation with the growth rate λ and the modulation wave number k. The analysis simplifies considerably for the case of a non-oscillatory long-wave instability, i.e. when at the threshold $\lambda = 0$ at its maximum at $k = 0$. Then, the non-trivial critical solution corresponding to the mode with $\lambda = 0$ and $k = 0$ must satisfy the linearized equation 4.37 since all time derivatives and the terms associated with velocity in the original hydrodynamic equations (eqns 4.3–4.5) vanish. This solution can be written in the form of an infinite transversal mode expansion

$$\psi = \Psi(z) + \sum_{n=1}^{\infty} Y_n s_n(z) \exp(ik_n x) + \mathrm{c.c.}, \qquad (4.39)$$

where $k_n = \pi n/L_x$ are the wave numbers, s_n are the corresponding eigenfunctions of the linearized problem and Y_n are their amplitudes. Moreover, it is easy to see that the critical mode $s_1(z)$, satisfying the linearized one-dimensional equation (eqn 4.21) and corresponding boundary conditions, itself is a derivative of the uniform stripe solution $\psi = \Psi(z)$ with respect to the mean filling fraction ϕ, i.e. $s_1(z) = \partial_\phi \Psi(z)$. The analysis by Livne et al. (2002a) revealed that the stationary stripe solution $\Psi(z)$ becomes unstable if the aspect ratio Δ exceeds a certain critical value Δ_c.

Above the threshold of the instability, the linearized modes grow until nonlinearity comes into play and saturates the exponential growth. At this stage, the mode amplitudes Y_n reach stationary values that, near threshold, can be obtained from the nonlinear equation 4.37 sequentially via standard solvability conditions with respect to the critical mode (Fredholm alternative). This weakly-non-linear analysis yields the supercritical bifurcation for the amplitude of the lowest (most unstable) mode $Y_1 \sim (\Delta - \Delta_c)^{1/2}$ for $\Delta > \Delta_c$, see Fig. 4.14.

Thus, eqn 4.34 predicts a phase-separation instability of a one-dimensional stripe of granular material localized opposite the hot wall. This instability persists in a wide range of parameters ϕ, W above critical aspect ratio Δ_c. For an infinite in x-direction system ($\Delta \to \infty$), the threshold of the instability, following from the solvability condition, coincides with the onset of negative compressibility, characterized by the condition $\mathrm{d}P/\mathrm{d}\phi = 0$, see details of the derivation (Khain et al., 2004). In the unstable

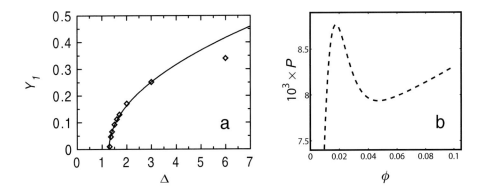

Fig. 4.14 (a) Bifurcation curve $Y_1(\Delta)$ obtained by the weakly-non-linear analysis of eqn 4.37 (solid curve) and the results of the solution of full granular hydrodynamics equations (diamonds) for $W = 1.25 \times 10^4$ and mean filling fraction $\phi = \langle \nu \rangle / \nu_c = 0.0378$, from Livne et al. (2002a). (b) Representative dependence of scaled pressure P vs. filling fraction ϕ for $W = 5 \times 10^3$ with the region of negative compressibility $dP/d\phi < 0$ (stripe phase), from Khain et al. (2004).

region, the pressure dependence has a characteristic falling part, as is illustrated in Fig. 4.14(b). This non-monotonous pressure dependence with $dP/d\phi < 0$ immediately signals the possibility of two-phase coexistence: multiple values of density ϕ correspond to the same value of pressure P. The physical interpretation of this instability is very simple. In the presence of the falling part of the $P(\phi)$ dependence, a small fluctuation, increasing the density in some region of the stripe phase, will result in a decrease of the pressure P, and, consequently, a flux of granular materials towards this region, increasing the density even further.

While the bifurcation analysis predicts the supercritical instability of the stripe state, the full solution of the granular hydrodynamics equations shows the following dynamics. For large aspect ratios Δ, the phase-separation instability proceeds in two stages, see Fig. 4.15. During the first stage, several small clusters nucleate opposite to the driving (hot) wall. Their number roughly is of the order Δ/Δ_c. At the later stage, clusters become more dense and then, as they compete for the material, coarsen. In the end, only one cluster survives. The overall phenomenology is consistent with event-driven simulations of particles with fixed restitution coefficient depicted in Fig. 4.12.

This phenomenology is strongly reminiscent of the classical spinodal decomposition in equilibrium systems. Furthermore, the phase diagram, shown for the filling fraction variable ϕ and scaled pressure P in Fig. 4.16, has the typical structure of a system near the phase-separation transition for the system at a thermodynamic (e.g. vapor–liquid) equilibrium (see, e.g. Bray (1994)). In particular, the onset of the phase-separation instability occurs at the *spinodal* curve given by the linear instability criterion of the homogeneous state. There is also a second *binodal* curve limiting the region of phase coexistence. This line can be obtained from the condition that the interface between two phases is stationary, i.e. the energies of the phases are equal. Thus, the homogeneous state is linearly stable above the spinodal, however, between spinodal and binodal, finite-amplitude perturbations may grow and lead to phase separation.

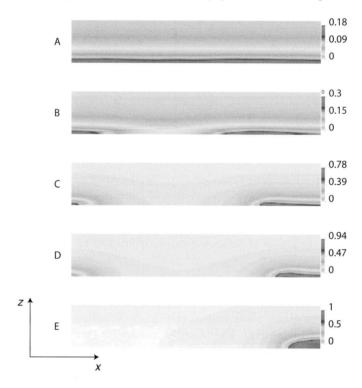

Fig. 4.15 Time evolution of the density field ν for the parameter $W = 1.25 \times 10^4$, mean filling fraction $\phi = \langle \nu \rangle / \nu_c = 0.0378$, and the aspect ratio $\Delta = 5$ at moments of scaled time $t = 650$ (A), 2100 (B), 2850 (C), 3250 (D), and 7000 (E), from Livne et al. (2002a).

The dynamics of driven granular gases are not limited to the van der Waals instability. Remarkably, hydrodynamic theory predicted a novel oscillatory instability for the position of the dense cluster for the granular gas confined between two thermal walls (Khain and Meerson, 2004); this instability indeed was found in event-driven simulations. These predictions, however, have not yet been confirmed experimentally, most likely due to the relatively small aspect ratios of the existing experimental cells.

4.3 Granular thermoconvection

One of the most familiar and well-studied pattern-forming phenomena in ordinary fluid mechanics is the Rayleigh–Bénard convection that occurs in a layer of fluid heated from below (Cross and Hohenberg, 1993). As the temperature difference ΔT between the bottom and the top plates exceeds a certain threshold value, depending on the thickness of the layer L_z, viscosity η, heat conductivity κ, and the heat expansion coefficient β of the fluid, the density inversion becomes unstable, and a fluid flow in

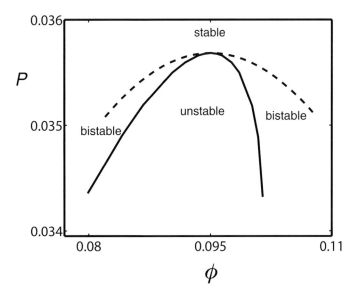

Fig. 4.16 Phase diagram for a driven granular gas system in a large aspect ratio system ($\Delta \gg 1$) in the vicinity of a critical point. The spinodal $dP/d\phi = 0$ is shown by the solid curve, and the binodal line (dashed) limits the region of bistability. From Khain et al. (2004).

the form of spatially periodic rolls or hexagons emerges. The convenient dimensionless control parameter for the Rayleigh–Bénard convection is the Rayleigh number

$$\mathrm{Re} = \frac{g\beta}{\eta\kappa}\Delta T L_z^3.$$

Based on the similarity of the equations of granular hydrodynamics to the standard hydrodynamics, one can expect that a granular layer excited by a rapidly vibrated bottom can develop a density inversion that under certain circumstances may exhibit a similar convective instability. There are, however, important differences between granular thermoconvection and the Rayleigh–Bénard convection. First, in contrast to ordinary fluids, granular materials are "self-cooled" due to the inelastic character of particle collisions. Second, in the framework of granular hydrodynamics the Prandtl number $\mathrm{Pr} = \eta/\kappa$ characterizing the ratio of viscosity to heat conductivity, is always fixed at the value of the order one, whereas for fluids it may vary within a very broad range (for example, $\mathrm{Pr} = 0.7$ for air and $\mathrm{Pr} \sim 10^4$ for engine oil).

Density inversions and multiple convection rolls were reported in molecular dynamics simulations with hard spheres by several groups (Ramírez et al., 2000; Sunthar and Kumaran, 2001; Paolotti et al., 2004; Risso et al., 2005), see Fig. 4.17. Moreover, theoretical analysis based on the solution of the granular hydrodynamics equations (eqns 3.11–3.13) supports the existence of a convective instability in a certain range of parameters (He et al., 2002; Khain and Meerson, 2003).

Early experiments on convection in vibrofluidized granular matter were not conclusive enough (Wildman et al., 2001; Wildman et al., 2005). As we mentioned above,

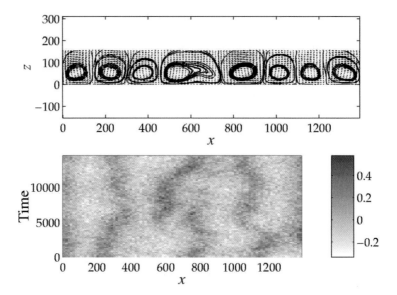

Fig. 4.17 Multiple convection rolls in an open cell heated from below. The upper panel shows velocity vectors and streamlines. The height of the roll is close to the vertical scale of the density variations. The bottom panel shows the time evolution of density deviations from a homogeneous initial state, from Sunthar and Kumaran (2001).

there are multiple mechanisms that may lead to convective flows in granular systems. In particular, it is difficult to discriminate experimentally between convection induced by vibration and convective flows induced by friction with sidewalls (Wildman *et al.*, 2001; Pak and Behringer, 1993; Garcimartin *et al.*, 2002; Wildman *et al.*, 2005). However, in later experiments, the inverse density profiles of vibrated two-dimensional granular material, the pre-requisite of convection instability, were indeed observed experimentally (the so-called granular Leidenfrost effect, i.e. levitation of dense clusters above vibrating (hot) plane, (Eshuis *et al.*, 2005)), and, finally, multiple convection rolls were indeed reported for vigorously vibrated (about $\Gamma = 60$) layers or granular matter by Eshuis *et al.* (2007), see Fig. 2.5.

4.3.1 Density inversion

The density inversion in vibrated layers can be described by the equations of granular hydrodynamics in the dilute limit (eqns 4.3–4.5) with an additional driving term due to gravity (Khain and Meerson, 2003):

$$D\nu/Dt + \nu \nabla \cdot \mathbf{v} = 0, \tag{4.40}$$

$$\nu D\mathbf{v}/Dt = \nabla \cdot \boldsymbol{\sigma} - \mathrm{Fr}\nu \mathbf{e}_z, \tag{4.41}$$

$$\nu DT/Dt + \nu T \nabla \cdot \mathbf{v} = \mathrm{Kn}\, \nabla \cdot (T^{1/2} \nabla T) - \mathrm{Kn} R_\mathrm{H}\, \nu^2\, T^{3/2}, \tag{4.42}$$

where $\mathbf{e}_z$ is the unit vector in the vertical (z) direction, $\mathrm{Fr} = mgL_z/T_0$ is the Froude number, characterizing the ratio of the typical flow velocity to thermal velocity of grains (m is the mass of a single particle, T_0 is the temperature of the hot wall). In eqns 4.40–4.42 the distance is measured in units of vertical plate separation L_z, the time in units of $L_z/\sqrt{T_0}$, the temperature in units of T_0, and the Knudsen number Kn is based on L_z and $\langle \nu \rangle$, the number of particles per unit length in the horizontal direction.[5]

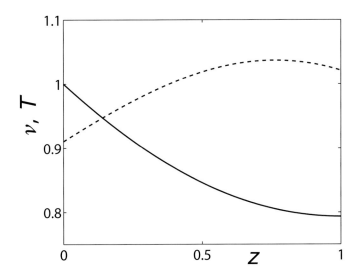

Fig. 4.18 Density inversion in granular thermoconvection. Static temperature (solid line) and density (dashed line) profiles for Froude number Fr = 0.1 and heat-loss parameter R_H = 0.5. Density is minimal at the base ($z = 0$) and reaches its maximum in the bulk of the layer, from Khain and Meerson (2003).

The boundary conditions for the temperature $T(x, z)$ are $T = 1$ at the base $z = 0$ and $\partial T/\partial z = 0$ at the upper plate $z = 1$. As for the velocity $\mathbf{v}$, zero normal component and slip at the boundaries are assumed. In the simplest steady-state (no flow) case eqns 4.40–4.42 yield stationary density $\nu_s(z)$ and temperature $T_s(z)$ distributions

$$(\nu_s T_s)' + \text{Fr}\, \nu_s = 0 \quad \text{and} \quad (T_s^{1/2} T_s')' - R_H \nu_s^2 T_s^{3/2} = 0, \tag{4.43}$$

(primes denote differentiation with respect to z). The boundary conditions become $T_s(0) = 1$ and $T_s'(1) = 0$. The representative static profiles shown in Fig. 4.18 have a well-pronounced density inversion for large enough heat-loss parameter R_H, indicating the possibility of convection instability.

4.3.2 Convection instability

In order to obtain the conditions for convection instability, we have to examine the evolution of small perturbations with respect to stationary solutions of eqns 4.43 along the x-direction with the wave number k_x. The corresponding linearized equations are

---

[5] Here we neglect the small viscous heating term in the heat balance equation (eqn 4.42).

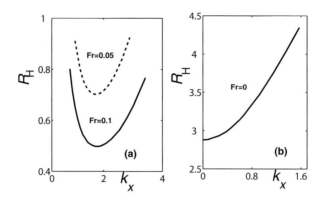

Fig. 4.19 The critical values of the heat-loss parameter R_H for the convection instability (non-zero gravity, a) and phase-separation instability (zero gravity, b) versus the horizontal wave number k_x for the Knudsen number Kn = 0.02, from Khain and Meerson (2003).

$$\frac{\partial \tilde{\nu}}{\partial t} + \nu_\mathrm{s}\frac{\partial v_x}{\partial x} + \frac{\partial}{\partial z}(\nu_\mathrm{s} v_z) = 0, \tag{4.44}$$

$$\nu_\mathrm{s}\frac{\partial \mathbf{v}}{\partial t} = -\nabla(\nu_\mathrm{s}\tilde{T} + T_\mathrm{s}\tilde{\nu}) + \frac{1}{2}\mathrm{Kn}\nabla\cdot(T_\mathrm{s}^{1/2}\hat{\dot{\gamma}}) - \mathrm{Fr}\tilde{\nu}\,\mathbf{e}_z, \tag{4.45}$$

$$\frac{\partial \tilde{T}}{\partial t} + T'_\mathrm{s} v_z + T_\mathrm{s}\nabla\cdot\mathbf{v} = \frac{\mathrm{Kn}}{\nu_\mathrm{s}}\nabla^2(T_\mathrm{s}^{1/2}\tilde{T}) - \mathrm{Kn}R_\mathrm{H}\nu_\mathrm{s}T_\mathrm{s}^{3/2}\left(\frac{2\tilde{\nu}}{\nu_\mathrm{s}} + \frac{3\tilde{T}}{2T_\mathrm{s}}\right), \tag{4.46}$$

where $\tilde{\nu},\tilde{T}$ are perturbations to the density and temperature, and $\hat{\dot{\gamma}}$ is the deviatoric part of the strain rate tensor $\dot{\gamma} = [\nabla v + \nabla v^T]/2$. The solutions to the linearized problem are sought in the form $\tilde{\nu},\tilde{T},\mathbf{v} \sim \exp(\lambda t + ik_x x)f_i(z) + \text{c.c.}$, where λ is the growth rate, and $f_i(z)$ are corresponding vertical profiles of each perturbation mode.

The linear stability analysis of eqns 4.44–4.46 yields the positive growth rate λ of periodic perturbations for some wave numbers k_x for large enough values of the parameter R_H (analogous to the Rayleigh number in the Rayleigh–Bénard convection). The location of the marginal curve $\lambda = 0$ is shown in Fig. 4.19(a). The curve has a minimum at certain (k_x^*, R_H^*) similar to the convection in classical fluids. Therefore, in infinite systems granular convection begins at $R_\mathrm{H} = R_\mathrm{H}^*$. Figure 4.20 depicts convection rolls found by numerically solving eqns 4.44–4.46. For small Froude number Fr convection rolls occupy the whole layer of granular gas and are elongated in the horizontal direction. In contrast, for large Fr values, corresponding to an open container, the rolls are located near the base and their aspect ratio is close to one.

As was discussed in Section 4.2, in the absence of gravity driven granular gases also exhibit symmetry breaking or a van der Waals instability. This instability, corresponding to the Fr = 0 limit, results in the formation of a static symmetry-broken state without flow, with only density variations. For Fr $\neq$ 0 both instabilities may coexist.

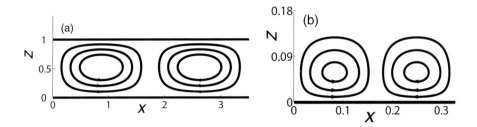

Fig. 4.20 Convection cells at the instability onset $k_x = k_x^*$ and $R_H = R_H^*$. The Knudsen number is Kn = 0.02. Panel (a) corresponds to the solid curve of Fig. 4.19(a). Here Fr = 0.1, $k_x^* = 1.8$ and $R_H^* = 0.49$. Panel (b) corresponds to the large-Fr limit when the grains are localized at the base (open container). Here, Fr = 5, $k_x^* = 19$, and $R_H^* = 2.86$, from Khain and Meerson (2003).

4.3.3 Convection instability of a rapid granular flow along a rough incline

Experiments by Forterre and Pouliquen (2001), Forterre and Pouliquen (2002) show the development of longitudinal vortices in rapid chute flows; see Fig. 2.13. The vortices develop for large inclination angles and large flow rates in the regime of accelerating flow when the flow thickness decreases and the mean flow velocity increases along the chute. Forterre and Pouliquen (2001) proposed a qualitative explanation of this phenomenon based on granular thermoconvection. Indeed, a rapid granular chute flow has a high shear near the rough bottom that may lead to the local increase of granular temperature and consequently may create a density inversion. In turn, the density inversion may trigger a convection instability similar to that in ordinary fluids. The instability wavelength L_c would then be determined by the depth of the layer h (in experiments $L_c \approx 3h$).

In a subsequent work Forterre and Pouliquen (2002) studied the formation of longitudinal vortices and the stability of granular chute flows quantitatively in the framework of granular hydrodynamics eqns 3.11–3.13. When a heuristic boundary condition is introduced at the bottom relating slip velocity and the heat flux, the steady-state solution of eqns 3.11–3.13 indeed yields a density inversion (Fig. 4.21) which turns out to be unstable with respect to short-wavelength perturbations for sufficiently large flow velocities; see Fig. 4.22. While the linear stability analysis captures many important features of the phenomenon, there are still open questions. In particular, Forterre and Pouliquen (2002) performed the stability analysis for a steady flow whereas the instability in experiments occurs in the accelerating flow regime. Possibly due to this discrepancy the linear stability analysis yielded an oscillatory instability near the onset of vortices when for the most part vortices in the experiment appear to be steady. Another factor that is neglected in the theory is the air drag. The high flow velocity in the experiment (about 1–2 m/s) is of the order of the terminal velocity of individual grains in air, and therefore air drag may, in principle, affect the granular flow. However, recent experiments conducted in vacuum by Borzsonyi and Ecke (2006) indicate that the air drag only has a quantitative effect on the pattern formation: vortices were observed with and without air.

84 *Phase transitions, clustering, and coarsening in granular gases*

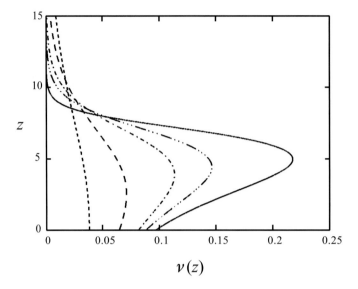

Fig. 4.21 Density profiles $\nu(z)$ as a function of distance form the chute bottom, z, for different values of mean flow velocity, from Forterre and Pouliquen (2002).

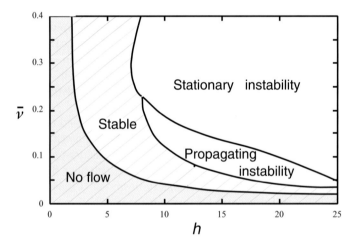

Fig. 4.22 Phase diagram in mean density ($\bar{\nu}$) and flow thickness (h) plane delineating various flow instabilities. Smaller $\bar{\nu}$ corresponds to faster flow, from Forterre and Pouliquen (2002).

4.4 Clustering of grains on a vibrated substrate

In this section we discuss a different class of experiments with dilute vibrofluidized granular matter. Unlike the previous case where energy was injected into the system at the boundary, here we focus on systems in which the energy is injected directly in the bulk. Experimentally, this excitation can be realized in thin layers of granular materials on a vibrated rigid plate, see Fig. 2.1. When the frequency of vibration ω is so high that the characteristic time of particle bounce is much larger than the period of vibrations, the effect of the vibrating plate can be approximated as an effective "thermal bath". Gravity limits the amplitude of vertical displacement of grains, and for sufficiently small amplitude of vibration $\Gamma\omega^{-2}$, the motion of particles is mostly confined to the surface. Dilute systems can be modelled by working with very thin layers, so thin that they do not cover the surface completely. We call such layers *submonolayers* as their mean effective thickness is less than one particle diameter. The amount of granular materials in such systems can be characterized by their area fraction ϕ, i.e. area density normalized by the density of a close-packed monolayer; the dilute limit corresponds to $\phi \ll 1$. Experiments with quasi-two-dimensional submonolayers of vertically vibrated particles revealed a surprising phenomenon: formation of a dense close-packed cluster coexisting with dilute granular gas; see Fig. 2.1. This phenomenon bears a strong resemblance to the first-order solid–liquid phase transition in equilibrium systems: hysteresis and phase coexistence, see Fig. 4.23. Immobile solid-like clusters coexist with rapidly moving particles (gas) in a rather narrow range of accelerations: $0.85 < \Gamma < 1.2$. A similar experiment by Losert *et al.* (1999*b*) revealed propagating fronts between gas-like and solid-like phases in vertically vibrated submonolayers. Such fronts are expected in extended systems in the vicinity of the first-order phase transition, e.g. solidification fronts in supercooled liquids. Qualitatively similar phase coexistence occurs in vibrated granular submonolayers confined between two plates (Prevost *et al.*, 2004). In this situation particles are energized by collisions with both upper and lower plates of the container. One of the unexpected features of this experiment is the formation of a *square* lattice in the dense cluster, see Figure 4.24. In contrast to the early experiments by Olafsen and Urbach (1998) where the cluster had a hexagonal structure, the crystalline cluster in the experiment by Prevost *et al.* (2004) was composed of two layers of balls, each with a square symmetry. The balls in the second layer are above the centers of the squares formed by the balls in the bottom layer, as in ionic crystals. The crystalline clusters are not densely packed; the particles jitter around in the cage formed by their neighbors.

4.4.1 Energy injection and bistability

The cluster formation in vertically vibrated submonolayers shares many common features with processes in freely cooling granular gases and in gases driven from the boundary, because it is also caused by the energy dissipation due to the inelasticity of collisions. However, there is a significant difference: the instability described in Section 4.1 is insufficient to explain the phase separation. A very important additional factor is *bistability* and persistent coexistence of states with different temperatures and densities: a cold dense cluster state and a more dilute and hot granular gas. The instability of the homogeneous state and the formation of clusters in a vertically vibrated granular submonolayer shares a certain degree of similarity with the van der Waals instability of the one-dimensional dense cluster described in Section 4.2.2. However, the specific physical mechanisms leading to clustering are very different. As we showed in Section 4.2.2, the van der Waals instability of clusters occurs due to the

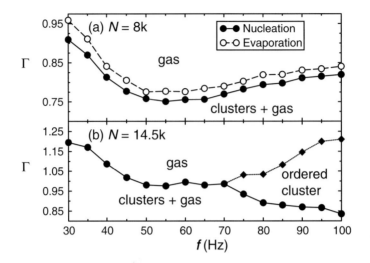

Fig. 4.23 The phase diagram of vibrated submonolayers for $N = 8000$ particles (a) and $N = 14,500$ particles (b). The filled circles indicate the values of acceleration Γ where the clustering starts, open circles indicate the points where clustering disappears. The diamonds in panel (b) show the transition to the ordered cluster state as the acceleration is reduced, from Olafsen and Urbach (1998).

non-monotonous dependence of the pressure on density, see Fig. 4.14(b), and it is not very sensitive to the way the energy is supplied from the vibrating sidewall.

For the case of vertically vibrated submonolayers there is an evidence that the bistability leading to the cluster formation and the phase separation are caused mostly by the mechanism of the energy injection from the vibrated substrate into the granulate. In fact, the bistable nature of the energy injection can already be seen in the dynamics of a *single* inelastic particle on a vibrated plate (Losert et al., 1999b; Geminard and Laroche, 2003).

Geminard and Laroche (2003) conducted systematic experimental and numerical studies of a single ball interacting with the periodically vibrated plate. The experiments showed that the energy of vertical motion of the ball E exhibits bistability and hysteresis in the range of vibration acceleration Γ, see Fig. 4.25.

The dynamics of an inelastic ball bouncing on the vibrated plate can be described by the simple Newton's equations of motion. The bead, after it collides with the plate, moves under gravity according to

$$h(t) = h_n + v_n(t - t_n) - \frac{1}{2\Gamma}(t - t_n)^2, \qquad (4.47)$$

where h_n, v_n are, respectively, the initial height and velocity of the particle after the nth collision with the plate. Here height h is normalized by the amplitude of plate displacement A, time t is normalized by the vibration period $T = 2\pi/f$, and velocity is normalized by $A\omega$. In rescaled variables the plate position is given by $r(t) = \sin t$. The velocity of the bead after the collision with the plate v_n is given by the following

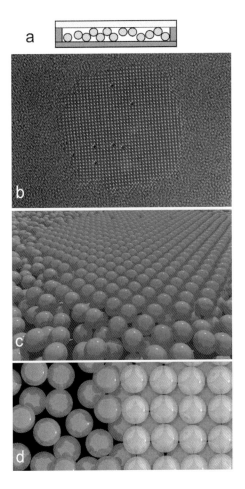

Fig. 4.24 Two-phase coexistence in the steady state. (a) Schematics of the experimental setup; (b) Experiment: time-averaged image of particles positions, only the top layer is visible; (c,d) Molecular dynamics simulations, three-dimensional rendering of particle positions, (d) a closeup of the crystal edge, from Prevost *et al.* (2004).

relation

$$v_n = (1+e)v_p - ev_n^-, \qquad (4.48)$$

where v_n^- is the velocity prior to collision, e is the bead/plate restitution coefficient, and $v_p = \cos t_n$ is the plate velocity at the instant of collision. The condition of collision is given by $h(t) = r(t)$. Assuming instant collisions, one can solve eqns 4.47 and 4.48 either analytically for simple periodic orbits or numerically for non-periodic orbits.

Equations 4.47 and 4.48 possess a variety of periodic as well as chaotic solutions. The solutions can be characterized by the associated vertical average kinetic energy E (or the vertical temperature T_z). According to Geminard and Laroche (2003), for the periodic orbits the average kinetic energy is quantized and is of the form

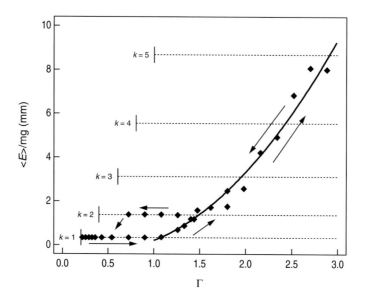

Fig. 4.25 Energy of the ball E vs. acceleration Γ for the frequency of vibration 60 Hz. Diamonds show experimental data, dashed lines indicate energy of periodic orbits E_k. Solid line shows solution to eqn 4.50 for non-periodic (chaotic) orbits for $e = 0.88$, from Geminard and Laroche (2003).

$$\langle E \rangle = m \frac{g^2}{\omega^2} \left(\frac{\pi^2 k^2}{2} + \sqrt{\Gamma^2 - \Gamma_k^2} \right), \qquad (4.49)$$

where k is the period of the orbit in terms of periods of vibration $T_0 = 2\pi/\omega$, and the minimum acceleration $\Gamma_k = k\pi(1-e)/(1+e)$. The average energy of chaotic (non-periodic) orbits is approximated by the expression

$$\langle E \rangle = am \frac{g^2}{\omega^2} \frac{(\Gamma - \Gamma_s)\Gamma}{(1-e) + b(1-e)^2}, \qquad (4.50)$$

with constants $a \approx 3.8$, $b \approx 4.45$ and critical acceleration $\Gamma_s \approx 0.85$. Accordingly, the energy of chaotic orbits is not quantized and increases continuously with acceleration Γ, as shown in Fig. 4.25. The hysteresis is observed due to synchronization of the bead motion to the plate oscillation. When acceleration Γ is decreased, the bead first synchronizes for $\Gamma = 1.5$ to a period-doubled trajectory with $k = 2$, then it synchronizes to a single-period orbit below $\Gamma < 0.7$, and for $\Gamma < 0.2$ it settles on the bottom. By contrast, when the acceleration is increased, the bead experiences periodic oscillations with $k = 1$ up to $\Gamma \approx 1.07$, after which it switches to chaotic oscillations with continuous energy increase. These results provide an insight into the nature of hysteresis and phase separation in vertically vibrated granular materials.

4.4.2 Horizontal granular temperature in many-particle systems

Direct experimental measurements of the horizontal granular temperature T_h in the system with many particles show a very similar dependence of T_h on acceleration Γ with that obtained for an isolated particle, compare Figs. 4.25 and 4.26(a). The horizontal temperature exhibits a very weak dependence on the filling fraction.[6] A similar behavior is obtained in soft-particle simulations of vertically vibrated submonolayers by Nie et al. (2000), see Fig. 4.26(b). The simulation results show that while there is no exact energy equipartition, the horizontal temperature T_h follows the vertical temperature T_z: $T_h \propto T_z$. There is a small but noticeable jump in the vertical temperature T_z at the critical acceleration Γ_c corresponding to the onset of horizontal motion (experimental value $\Gamma_c \approx 0.8-0.9$ depending on the frequency of vibration). This jump is an indication of a hysteresis and possible bistable behavior and is consistent with the behavior of a single bouncing ball, see Fig. 4.25. This bistability leads to the nucleation of small clusters. However, their subsequent growth and coarsening are caused by the interparticle interactions. The basic mechanism of horizontal clustering is similar to the mechanism of clustering instability in freely cooling and driven granular gases – the cold regions have smaller pressure, and therefore "attract" hot particles. However, clustering here is even stronger than in freely cooling gases because of the bistability: as soon as the initially hot particle cools down below a certain threshold, it loses energy much more rapidly and joins the cold cluster.

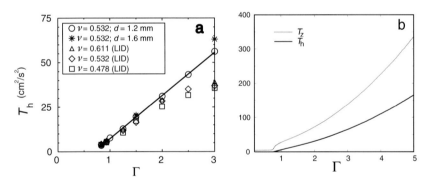

Fig. 4.26 (a) Horizontal granular temperature T_h vs. acceleration Γ for different values of number density ν, with and without a confining lid, experiment by Urbach and Olafsen (2001). (b) Vertical T_z and horizontal T_h granular temperatures obtained in soft-particle simulations, number of particles $N = 2000$, from Nie et al. (2000).

4.4.3 Phase separation and multi-cluster states

One of the most important questions regarding the dynamics of phase coexistence is the scaling for the number of macroscopic clusters as a function of time. It is of special interest because it tests the similarity between equilibrium thermodynamic systems

---

[6]Direct experimental measurements of the vertical temperature are more technically difficult

and non-equilibrium granular systems. This question has not been resolved decisively for vibrofluidized granular media. Since mechanical shaking experiments were performed with relatively large grains (about 1 mm) and therefore had a relatively small aspect ratio, only one or a few clusters were observed by Urbach and Olafsen (2001). Experiments with much smaller particles (about 120 micrometers) showed coexistence of multiple (up to 10) clusters (Sapozhnikov et al., 2003a). In a certain range of parameters these clusters exhibit dynamics similar to the *Ostwald ripening* in equilibrium systems (Meerson, 1996): small clusters evaporate and large clusters grow, resulting in the formation of a single cluster. While Ostwald ripening of two-dimensional clusters is qualitatively similar to the coarsening of one-dimensional clusters considered in the context of the van der Waals instability in Section 4.2.2, there are also significant differences. In particular, in two dimensions coarsening kinetics are driven by surface tension: the system tends to minimize surface energy by decreasing the number of clusters, typically leading to a power-law decay of the number of clusters with time. In contrast, in one-dimensional systems, the surface tension does not play a role, and the number of clusters decays much more slowly.

However, as was shown by Aranson et al. (2000), more reliable statistical information on out-of-equilibrium Ostwald ripening can be obtained in electrically driven granular media. This system operates with very small particles and allows one to observe a large number of macroscopic clusters. In this system the number of macroscopic clusters N was shown to decay with time as $1/t$. This law is consistent with the interface-controlled Ostwald ripening in two dimensions (Wagner, 1961; Meerson, 1996). While the mechanisms of energy injection in vibrofluidized and electrically driven systems are different, both show qualitatively similar behavior: macroscopic phase separation, coarsening, transition from two- to three-dimensional cluster growth, etc (Sapozhnikov et al., 2003a). This system will be discussed in detail in Chapter 8.

5
Surface waves and patterns in periodically vibrated granular layers

In this chapter we focus on pattern-forming phenomena in thin layers of granular materials subjected to periodic vertical vibrations with an acceleration amplitude greater than the acceleration of gravity, $\Gamma > 1$. Unlike the submonolayers discussed in the previous chapter, we will explore multi-layers of at least a few particle diameters thick. Compared to submonolayers that resemble driven granular gases, dense multilayers of granular materials under sufficiently strong excitation exhibit fluid-like motion. The most spectacular manifestation of the fluid-like behavior of granular layers is the occurrence of surface gravity waves. Surprisingly, fluid-like surface gravity waves can be observed in layers as thin as ten particle diameters. Sufficiently strong periodic vibration of a thin granular layer can excite standing surface gravity waves that form patterns that are quite similar (however, with some important differences discussed below) to the corresponding patterns in ordinary fluids. To understand the nature of these collective phenomena, their similarities to and differences from the regular fluids, many theoretical and computational approaches have been developed. The most straightforward is to use molecular dynamics simulations, which is feasible for relatively small systems. On the other hand, the fact that the scale of the observed pattern typically is much greater than the size of individual grains calls for a continuum description of the observed patterns, especially in large systems where molecular dynamics simulations are prohibitively expensive. A variety of theoretical models, ranging from phenomenological Ginzburg–Landau-type theories to granular hydrodynamics have been proposed for such continuum description of the pattern-forming behavior of vibrated granular layers.

5.1 Heaping in vertically vibrated layers

As mentioned in Chapter 2, experimental studies of vibrated layers of sand have a long and illustrious history, beginning from the seminal works of Chladni (1787) and Faraday (1831) in which they used a violin bow and a membrane to excite vertical vibrations in a thin layer of grains. The main effect observed in those early papers was "heaping" of granular matter, or formation of mounds near the nodal lines of membrane oscillations. In fact, this phenomenon was used to visualize and study the vibrational modes of membranes of different geometrical shapes at different frequencies, see Fig. 2.2. The behavior of not too fine powders on a vibrated plate was immediately (and correctly) attributed to the non-linear "demodulation" of the non-uniform excitation of grains by membrane modes. Another non-trivial effect discovered in early studies was "acoustic streaming": a vibrating membrane generates convective flows in the air above it, and these flows carry very thin powders towards antinodal lines of the vibrated plate.

In subsequent years the focus of attention shifted from dynamical properties of thin layers of vibrated sand, and only in the last third of the twentieth century physicists returned to this old system equipped with new experimental capabilities. The dawn of the new era was marked by the studies of heaping by Jenny (1974). The main difference from the early studies of Chladni (1787) and Faraday (1831) was that in the new experiments the vibration was provided by a powerful electromagnetic shaker attached to a thick metal bottom plate, rendering the membrane-type modes of the bottom plate irrelevant. This also allowed researchers to use relatively thick layers of grains. A number of studies of heaping have been performed with and without interstitial gas, with somewhat controversial conclusions on the role of the ambient gas (Walker, 1982; Dinkelacker et al., 1987; Evesque and Rajchenbach, 1989; Evesque et al., 1990; Douady et al., 1989; Laroche et al., 1989; Clément et al., 1992).

Careful measurements of heaping in a sealed annulus cell by Pak et al. (1995) showed that heaping indeed disappeared upon reducing the pressure of the ambient gas or increasing the particle size. Furthermore, experiments helped to identify the specific mechanism of the gas-induced heaping among two competing candidates: viscous drag of the gas permeating through the granular layer, and the pressure buildup under the compressing layer. As the viscous drag force was shown to be small compared with the weight of the layer and having little pressure dependence, one is led to the conclusion that the convection due to the gas trapping is the primary mechanism of heap generation. For an independent confirmation, in some experiments the outer Plexiglas sidewall of the annulus was replaced by a thin porous steel sheet with an open area fraction ~ 0.4, which allowed ambient gas to escape freely, and thus avoid trapping. No heap formation was detected in this modified system for various granular materials. This conclusion agreed with molecular dynamics simulations (Taguchi, 1992; Gallas et al., 1992a; Gallas et al., 1992c; Gallas et al., 1992b; Gallas and Sokołowski, 1993; Luding et al., 1994) which showed no heaping without interstitial gas effects. More recent studies of deep layers ($50d < h < 200d$) of small particles ($10\mu m < d < 200\mu m$) by Falcon et al. (1999a), Duran (2000), Duran (2001) found a number of interesting patterns and novel instabilities caused by the interstitial air. In particular, Duran (2001) observed formation of isolated droplets of grains after periodic taping similar to the Rayleigh–Taylor instability in ordinary fluids, see Fig. 5.1.

Besides molecular dynamics simulations of heaps, a number of simple cellular automata models was put forth to explain heap formation. Jia et al. (1999) proposed a simple model for the heap formation. In a discrete lattice version of the model, the decrease in local density due to vibrations is modelled by random creation of empty sites in the bulk with the rate α. Discrete exchange rules are applied between the void and its neighbors, see Fig. 5.2. When an empty site (void) reaches the top of the pile, it disappears. The simulation cycle consists of one empty-site generation step followed by γ steps of slope relaxation. In each relaxation step, each void attempts to exchange with its neighbor. Thus, γ can be interpreted as the toppling rate of the grains.

While the bulk flow is simulated by the dynamics of empty sites, the surface flow

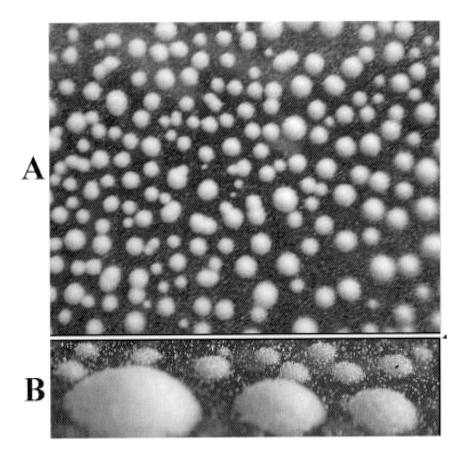

Fig. 5.1 (a) A "bird's eye" view of the pattern obtained after 40 taps of the layer of 30-μm powder. The mean separation between the heaplets is about 5 mm. (b) A closeup of several heaps, from Duran (2001).

is modelled by rules similar to the sandpile model in the theory of the self-organized criticality (SOC), see Section 6.6. This model reproduced both the convection inside the powder and the heap formation for a sufficiently large rate of empty-site formation, see Fig. 5.3.

A continuum model mimicking the behavior of this cellular automata model was proposed by Jia *et al.* (1999) in the form of a non-linear reaction–diffusion equation for the local height of the sandpile h,

$$\partial_t h = D\nabla^2 h + \alpha h - \beta h^2, \tag{5.1}$$

where D is the diffusion constant accounting for the rate of surface relaxation of height h, and the term αh models the effect of the volume increase due to vibration. The last term βh^2 accounts for the non-linear dissipation of energy. Note that while eqn 5.1

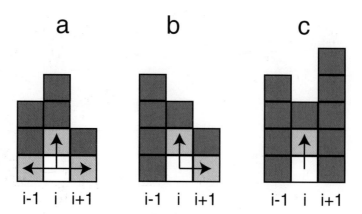

Fig. 5.2 Sketch illustrating the dynamic rules for displacements of an empty site in the lattice model (Jia et al., 1999). "Grains" are shown in gray and "voids" are shown in white. (a) The void can exchange positions with grains in three directions, but since the right stack is lower, the probability of the right hop is higher than the left hop; (b) The void cannot exchange the position with the grain on the left because the left stack is higher than the middle stack; (c) Both left and right stacks are higher than the middle on, so the void only can move upward.

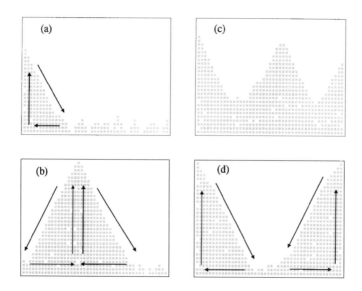

Fig. 5.3 Steady-state configurations of an initially flat layer, for the system width $L = 45$ and the toppling rate parameter $\gamma = 10$, arrows indicate the direction of convective rolls. (a) number of grains $N = 225$ and the voids creation rate $\alpha = 0.1$; (b–d) $N = 675$ and $\alpha = 0.05, 0.1$ and $\alpha = 0.3$, respectively, from Jia et al. (1999).

indeed shows dynamics similar to the discrete system, it is not a rigorous continuous limit of the discrete cellular automata model described above.

Overall, both the discrete and the continuous versions of the model reproduce, at least on a qualitative level, many features of experimentally observed heap formation, however, they are rather schematic to capture more subtle physical processes, such as the effects of the ambient air friction with sidewalls, convection within heaps, etc.

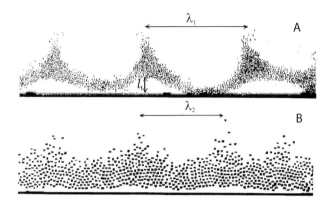

Fig. 5.4 The side view of a 9-particle deep granular layer at acceleration $\Gamma = 3.4$ and frequencies $f = 7.8$ Hz (a) and $f = 12$ Hz (b). The standing wavelength λ and maximum layer elevation l decrease with the increase of frequency f (not shown in the same scale), from Clément et al. (1996).

5.2 Standing wave patterns

A variety of standing wave patterns emerge in vertically vibrated granular multilayers at a high enough vertical acceleration (typically $\Gamma > 2-3$). They were first reported in a quasi-two-dimensional geometry (Fauve et al., 1989; Douady et al., 1989; Clément et al., 1996), see Fig. 5.4. These waves oscillate at half of the driving frequency, i.e. two cycles of driving are necessary to restore the original waveform. This indicates the subharmonic resonance pointing to a parametric instability as the mechanism of the pattern formation. The pattern wavelength λ appears to decrease with the increase of the vibration frequency f, similar to the dispersion relation of the surface gravity waves in fluids $\lambda \sim f^{-2}$, while there are deviations from this law, see below.

These first observations spurred a number of similar experimental studies of standing waves in thin granular layers that were also extended to three-dimensional geometry (Melo et al., 1994; Melo et al., 1995; Umbanhowar et al., 1996; Mujica and Melo, 1998; Aranson et al., 1999a). Importantly, these studies were performed in evacuated containers, which allowed reproducible results not contaminated by heaping to be obtained. Figure 2.3 shows a selection of regular patterns observed in vibrated granular layers under vibration (Melo et al., 1994).

5.2.1 Phase diagram of vibrated multilayers

A particular pattern is determined by the interplay between the driving frequency f and the dimensionless acceleration of the container Γ. For $\Gamma < 1$ the layer rests on the plate, and particles do not lose contact with the bottom plate. Furthermore, up to acceleration values $\Gamma \sim 2.4$ the layer remains flat more or less independently of the driving frequency, even though it leaves the plate for a part of every vibration cycle. At higher acceleration Γ various patterns of standing waves emerge that correspond to spatiotemporal variations of the layer thickness and height (see also Fig. 5.4 where these waves are shown from the side). These variations are subharmonic: crests are replaced by troughs and vice versa at successive periods of plate vibrations. At small frequencies $f < \tilde{f}$ (for the typical experimental conditions of Melo et al. (1995), $\tilde{f} \approx 45$ Hz), the transition is subcritical, leading to the formation of square wave patterns; see Fig. 2.3(b). For higher frequencies $f > \tilde{f}$, the selected pattern is quasi-one-dimensional stripes (Fig. 2.3(a)), and the transition becomes supercritical. In the intermediate region $f \sim \tilde{f}$, stable localized excitations were observed within the hysteretic region of the parameter plane. The simplest radially symmetrical localized structure, an *oscillon*, is shown in Fig. 2.4. This particle-like structure also oscillates at one half of the driving frequency, so at one oscillation cycle it has a mound shape, and at the next cycle it forms a crater. These oscillons weakly interact with each other and form various bound states; see Fig. 2.3(f). The frequency $\tilde{f}$ corresponding to the strip-square transition depend on the particle diameter d as $d^{-1/2}$. This scaling suggests that the transition is controlled by the relative magnitude of the energy influx from the vibrating plate $\propto f^2$ and the gravitational dilation energy $\propto gd$. At higher acceleration ($\Gamma > 4$), stripes and squares become unstable, and hexagons appear instead; see Fig. 2.3(c). Further increase of acceleration at $\Gamma \approx 4.5$ converts hexagons into a domain-like structure of flat layers oscillating with frequency $f/2$ with opposite phases. Depending on parameters, interfaces separating flat domains, are either smooth or "decorated" by periodic undulations; see Fig. 2.3(e). For $\Gamma > 5.7$ various quarter-harmonic patterns emerge. The complete phase diagram of different regimes observed in a three-dimensional container is shown in Fig. 5.5. For even higher acceleration ($\Gamma > 7$) the experiments reveal chaotic patterns oscillating at approximately one-fourth of the driving frequency and so-called phase bubbles, see Figure 5.6. In the phase bubbles the granular layer oscillates with the opposite phase (different by π) with respect to the surrounding pattern. Subsequent investigations revealed that periodic patterns share many common features with convective rolls in Rayleigh–Bénard convection, for example skew-varicose and cross-roll instabilities (de Bruyn et al., 1998).

5.2.2 Dispersion relation for standing waves

The wavelength of periodic patterns has a non-trivial dependence on the frequency of vibration f and and the layer thickness h. Detailed studies by Umbanhowar and Swinney (2000) revealed that below some critical frequency the dispersion relation is similar to that of surface gravity waves in inviscid fluids in the absence of surface tension in the shallow-water limit ($\lambda \gg h$): $\lambda_* = f_*^{-1}$, where we introduce the non-dimensional wavelength $\lambda_* = \lambda/h$ and frequency $f_* = f/\sqrt{g/h}$. As one may expect, for higher frequencies this scaling crosses over to the dispersion relation similar to

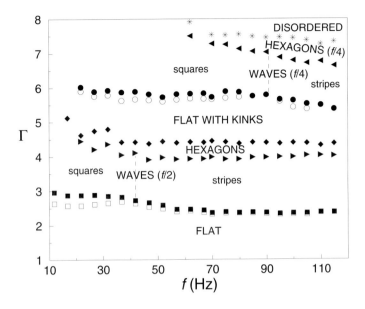

Fig. 5.5 Phase diagram of various regimes in a vibrated granular layer, from Melo *et al.* (1995).

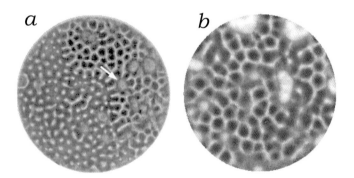

Fig. 5.6 Phase "bubbles" in a quarter-harmonic ($f/4$) pattern. A top view of the experimental pattern observed in vibrated layers at the high acceleration, $\Gamma = 7.3$ (a) and $\Gamma = 7.2$ (b). The bubble in the panel (a) is indicated by a white arrow, from Moon *et al.* (2001).

that for the surface gravity waves in deep water, $\lambda \sim f^{-2}$ independent of the layer thickness. The dependence of the wave number λ on the frequency f for different layers thicknesses is shown in Fig. 5.7.

Umbanhowar and Swinney (2000) proposed a simple fit that describes well the experimental data in the range of performed measurements (see Fig. 5.7(b)):

$$\lambda_* = 1.0 + 1.1 f_*^{-1.32 \pm 0.03}. \tag{5.2}$$

While eqn 5.2 is not a "true" or an asymptotic scaling relation between the frequency,

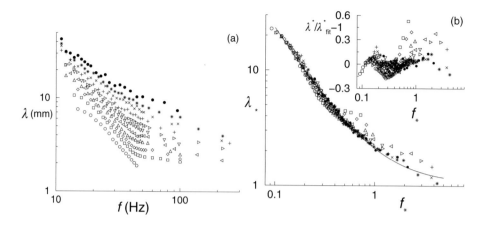

Fig. 5.7 Dispersion relations for standing waves in vibrated granular layers in the range of layer thickness, h, for acceleration $\Gamma = 3$ and particle diameter $d = 0.17$ mm: (a) wavelength as a function of frequency: $h = 2.2d$ (○), $h = 3.1d$ (□), $h = 4.0d$ (◇), $h = 5.4d$ (△), $h = 6.8d$ (◁), $h = 8.7d$ (▽), $h = 10.7d$ (▷), $h = 13.8d$ (+), $h = 18.1d$ (×), $h = 22.8d$ (∗), and $h = 30.7d$ (●); (b) the same data plotted with dimensionless variables. The dimensionless data in (b) are fitted to eqn 5.2; the inset shows the residuals. As h/d increases, the f^* range over which the data collapse occurs increases, from Umbanhowar and Swinney (2000).

depth and wavelength, it does provide a useful approximation of experimental data in the range of frequencies up to $f \approx 100$ Hz.

5.3 Simulations of vibrated granular layers

While the general understanding of the standing wave patterns in thin granular layers can be gained by the analogy with ordinary fluids, there are also important differences. The Faraday instability in fluids and corresponding pattern-selection problems have been studied theoretically and numerically in great detail (see, e.g. Zhang and Viñals (1997)). The primary mechanism of the instability leading to patterns is the parametric resonance between the spatially uniform periodic driving at frequency f and two counterpropagating gravity waves at frequency $f/2$. However, this instability in ordinary fluids leads to a supercritical bifurcation (i.e. the onset of pattern bears the features of a continuous, second-order phase transition), and square wave patterns near the onset, and as a whole the phase diagram lacks the richness of the granular system. Of course, this can be explained by the fact that there are many qualitative differences separating granular matter from ordinary fluids, such as strong and non-linear dissipation, dry friction, and the absence of surface tension. Interestingly, however, localized oscillon-type objects were subsequently observed in vertically vibrated layers of non-Newtonian fluids (Lioubashevski et al., 1999), and stripe patterns were also observed in a highly viscous fluids (Kiyashko et al., 1996). The theoretical understanding of the pattern formation in a vibrated granular system presents a challenge, since unlike fluid dynamics there is no universal theoretical description of dense granular flows analogous to the Navier–Stokes equations. In the absence of this common base, theoretical and computational efforts in describing these patterns followed several different directions. Aoki et al. (1996) were the first to perform molecular dynamics simulations of patterns in the vibrated granular layer. They concluded that grain–grain friction is necessary for pattern formation in this system. However, as noted by Bizon et al. (1997), this conclusion was a direct consequence of the fact that the algorithm of Aoki et al. (1996), which was based on the Lennard–Jones interaction potential and velocity-dependent dissipation, led to the restitution coefficient of particles approaching unity for large collision speeds rather than decreasing according to experiments.

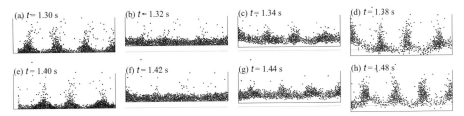

Fig. 5.8 A sequence of snapshots of a typical two-dimensional simulation over two periods from time $t = 1.30$ s to $t = 1.48$ s. The parameters are: number of particles $N = 600$, box size $L = 100d$, frequency of vibration $f = 10$ Hz, acceleration $\Gamma = 3.6$, restitution coefficients between particles $e = 0.4$, and between wall and particles $e_w = 0.2$, friction coefficient between particles and with the walls $\mu = 0.2$. The dashed line indicates the initial plate position, from Luding et al. (1996).

A number of detailed simulations were performed using the event-driven molecular dynamics algorithm (Luding et al., 1996; Bizon et al., 1998a; Bizon et al., 1998b). This method assumes that most of the grains are involved in free flights with only short collisions with the bottom and each other, and none of the grains form dense immobile clusters with prolonged contacts among them. Early two-dimensional simulations reproduced key experimental observations of standing subharmonic waves; compare Figs. 5.4 and 5.8. More recent full three-dimensional simulations demonstrated that even without friction, patterns do form in the system, however, only the supercritical bifurcation to stripes was found in this way. It turned out that friction is still necessary to produce other patterns observed in experiments, such as subcritical squares and $f/4$ hexagons. Simulations with frictional particles reproduced the majority of patterns observed in experiments and many features of the experimental phase diagram. Due to the intrinsic technical limitations of the event-driven method, which were mentioned above, the localized oscillons, coexisting with flat and dense layers where particles form prolonged contacts with each other, were not found in these simulations. Bizon et al. (1998b) set out to match an experimental cell and a numerical system, maintaining exactly the same geometry and the number of particles in simulations and physical experiment. After fitting only two parameters of the numerical model, a very close quantitative agreement between various patterns in the experimental cell and in simulations throughout the parameter space (frequency of driving, amplitude of acceleration, thickness of the layer) was found, see Fig. 5.9.

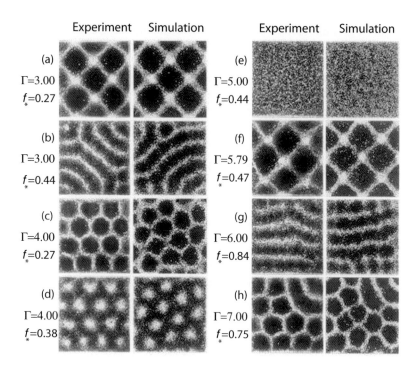

Fig. 5.9 A comparison between subharmonic patterns in experiment (left) and three-dimensional molecular dynamics simulations (right) of 30,000 particles in a square vibrated container for different values of the non-dimensional frequency $f_* = f/\sqrt{g/h}$ and different amplitudes of vibration acceleration Γ, from Bizon et al. (1998b).

Shinbrot (1997) proposed a model that combined ideas from molecular dynamics and continuum modelling. Specifically, this model ignores the vertical component of the particle motion and assumes that the impact with the plate adds a certain randomizing horizontal velocity to individual particles. The magnitude of the random component being added at each impact can serve as a measure of the impact strength. After the impact particles are allowed to travel freely in the horizontal plane for a certain fraction of a period after which they inelastically collide with each other (a particle acquires momentum averaged over all particles in its neighborhood). This model does reproduce a variety of patterns seen in experiments (stripes, squares, and hexagons) for various values of control parameters (frequency of driving and impact strength), however, it does not describe some of the experimental phenomenology (localized objects as well as interfaces), besides, in certain ranges of parameters it produces a number of intricate patterns not seen in experiments.

5.4 Phenomenological continuum theories

5.4.1 Amplitude equations description

Amplitude equations (or Ginzburg–Landau-type equations) are widely used to describe patterns in extended out-of-equilibrium systems whose amplitudes evolve slowly in space and time compared with the characteristic "microscopic" time and space scales of the underlying pattern. In certain cases these equations can be rigorously derived from first principles (such as the Navier–Stokes equations in fluid mechanics) near the threshold of a pattern-forming instability. However, when first-principle continuum equations either do not exist or are not well established (as often is the case in granular physics), the amplitude equations can be formulated on phenomenological grounds incorporating key experimental observations and fundamental symmetries of the problem, see Section 3.6.1 and the review by Aranson and Kramer (2002).

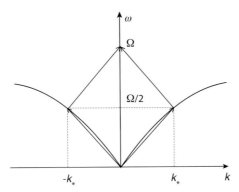

Fig. 5.10 Schematics of a parametric excitation of a pair of counterpropagating waves with wave numbers $\pm k$ and frequency $\Omega/2$, where $\Omega = 2\pi f$ is the vibration frequency, solid curve shows the dispersion relation $\omega(k)$ for surface waves.

Localized and periodic patterns. The first continuum models of pattern formation in vibrating sand were purely phenomenological. In the spirit of weakly non-linear perturbation theories one can introduce the complex amplitude $\psi(x, y, t)$ of subharmonic oscillations of the layer surface, $h = \psi \exp(i\pi f t) + c.c.$ for a pair of counterpropagating parametrically excited waves (Tsimring and Aranson, 1997). In this situation the external driving at the frequency $\Omega = 2\pi f$ generates a pair of waves with wave numbers $k_{1,2}$ and frequencies $\Omega_{1,2}$ when the resonance conditions for the wave vectors (momentum) and frequencies (energies) are satisfied: $k_1 + k_2 = 0$, $\Omega_1 + \Omega_2 = \Omega$. In an isotropic medium one obtains immediately that a spatially uniform driving at the frequency Ω generates a pair of counterpropagating waves with wave numbers $k_1 = -k_2 = k_*$, and at one half of the driving frequency $\Omega_1 = \Omega_2 = \omega_* = \Omega/2$. The value of the wave number k_* is determined by the driving frequency according to the dispersion relation $\tilde{\Omega}(k_*) = \Omega/2$, see Fig. 5.10.

The equation for the slowly varying complex amplitude of the wavefunction ψ in the lowest order can be written as

$$\partial_t \psi = \gamma \psi^* - (1 - i\omega_*)\psi + (1 + ib)\nabla^2\psi - |\psi|^2\psi - \nu\psi. \tag{5.3}$$

Here, γ is the normalized amplitude of the forcing at the driving frequency f, and parameter b characterizes the ratio of dispersion to diffusion. The linear terms in eqn 5.3 correspond to the two leading terms in the expansion of the complex growth rate $\Lambda(k) = -\Lambda_0 - \Lambda_1 k^2 + O(k^4)$ for infinitesimal periodic perturbations of the layer height $h \sim \exp[\Lambda(k)t + ikx]$. It is easy to see that the coefficients on the linear terms in eqn 5.3 are related to the expansion coefficients via $b = \text{Im}\Lambda_1/\text{Re}\Lambda_1$ and $\omega_* = -(\text{Im}\Lambda_0 + \pi f)/\text{Re}\Lambda_0$, where b characterizes the ratio of dispersion to diffusion and parameter $\omega = \Omega/2$ characterizes the frequency of the driving.

The only difference between this equation and the Ginzburg–Landau equation for the parametric instability (Coullet et al., 1990) is the coupling of the complex amplitude ψ to the "slow mode" ν, where ν can be interpreted as the coarse-grained layer number density. The additional term $-\nu\psi$ in eqn 5.3 is responsible for the increase of local dissipation rate in the granular layer due to the increase of the layer thickness. Apparently there are no analogs of this kind of coupling in conventional Newtonian fluids. The slow mode obeys its own dynamical equation

$$\partial_t \nu = \alpha \nabla \cdot (\nu \nabla |\psi|^2) + \beta \nabla^2 \nu. \tag{5.4}$$

Equation 5.4 describes redistribution of the averaged thickness due to the standard diffusive flux $\propto -\nabla \nu$ and the additional flux $\propto -\nu \nabla |\psi|^2$ that is caused by the spatially non-uniform vibrations of the granular material. This coupled model was used by Tsimring and Aranson (1997) to describe the pattern selection near the threshold of the primary bifurcation.

Slightly above the threshold of the instability the pattern-selection problem can be solved in the framework of a weakly non-linear analysis. For a large enough amplitude of driving, the trivial (flat state) $\psi = 0$, $\nu = \bar{\nu} = \text{const}$ loses its stability with respect to periodic perturbations in the form $\{\text{Re}\psi, \text{Im}\psi\} \sim \exp[i\mathbf{k}\mathbf{r}] + \text{c.c.}$, where $\mathbf{k}$ is the modulation wave number.[1] Simple linear analysis yields that the instability first occurs at the most unstable wave number $k = k_c$ above critical driving $\gamma > \gamma_c$ given by the conditions

$$\gamma_c^2 = \frac{[\omega_* + b(1 + \bar{\nu})]^2}{1 + b^2}$$
$$k_c^2 = \frac{\omega_* b - 1 - \bar{\nu}}{1 + b^2}. \tag{5.5}$$

The instability first occurs at non-zero k_c if $\omega_* b > 1 + \bar{\nu}$. The critical driving γ_c depends on the mean number density $\bar{\nu}$ (or mean layer thickness). The instability leads to the exponential growth of a periodic pattern with the wave number close to the critical wave number k_c and arbitrary orientation of the modulation wavevector $\mathbf{k}$. The resulting pattern selection, however, is determined by the interplay between non-linearity, driving, and dissipation.

---

[1] Since eqn 5.3 includes complex-conjugated function ψ^*, we cannot use the standard linear anzatz $\psi \sim \exp[i\mathbf{k}\mathbf{r}]$

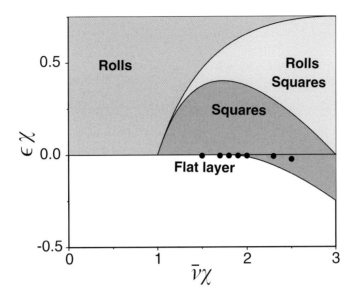

Fig. 5.11 Phase diagram showing stable patterns derived from eqns 5.3 and 5.4. Points indicate stable oscillons obtained by numerical solution of eqns 5.3 and 5.4, $\chi = \alpha/\beta$, $\bar{\nu}$ is the average density, and $\epsilon \sim \gamma - \gamma_c$ is the supercriticality parameter, from Tsimring and Aranson (1997).

For small supercriticality parameter $\epsilon \sim \gamma - \gamma_c$ various periodic patterns can be sought in the form of a superposition of linear solutions. For example, both stripes and square patterns can be represented in the form

$$\psi = [A_x(t)\sin(k_c x) + A_y(t)\sin(k_c y)] \exp[i\bar{\Phi}] + w, \tag{5.6}$$

where $A_x(t), A_y(t)$ are the real amplitudes of two orthogonal waves, the constant phase $\bar{\Phi}$ is given by the solution of the linearized problem, and w is a small correction to the solution that we demand remain small as $\gamma - \gamma_c \to 0$. Stripe patterns correspond to either $A_x = 0$ or $A_y = 0$, and square patterns to $A_x = A_y$. Near the threshold, the variable ν becomes enslaved to amplitude $|\psi|^2$, and we can neglect the time derivative in eqn 5.4 and obtain the solution in the form

$$\nu = \tilde{\nu}\exp[-\chi|\psi|^2], \tag{5.7}$$

where $\chi = \alpha/\beta$ and function $\tilde{\nu}$ is obtained from the mass conservation condition

$$\frac{1}{S}\int \nu \mathrm{d}x\mathrm{d}y = \bar{\nu} = \mathrm{const}, \tag{5.8}$$

with average density $\bar{\nu}$ and total area S. Substituting eqns 5.6 and 5.7 into eqn 5.3 and applying standard orthogonality conditions with respect to periodic functions $\sin(k_c x), \sin(k_c y)$ in order to keep correction w small (Aranson and Kramer, 2002), one derives close to the threshold of parametric instability a closed set of equations for $A_{x,y}$:

$$\dot{A}_x = A_x \left[\epsilon + \frac{\bar{\nu}\chi - 3}{4}(A_x^2 + 2A_y^2) - \frac{\bar{\nu}\chi^2}{2}(A_y^2 A_x^2 + 2A_y^2) \right]$$
$$\dot{A}_y = A_y \left[\epsilon + \frac{\bar{\nu}\chi - 3}{4}(A_y^2 + 2A_x^2) - \frac{\bar{\nu}\chi^2}{2}(A_x^2 A_y^2 + 2A_x^2) \right]. \quad (5.9)$$

Here, $\epsilon = \gamma_c(\gamma - \gamma_c)/(1 + \bar{\nu} + k_c^2)$ is the normalized supercriticality parameter, by assumption $\epsilon \ll 1$.[2] As it follows from eqns 5.9, the hysteretic transition to squares ($A_x = A_y$) occurs for $\bar{\nu}\chi > 9/5$. In contrast, stripe solutions ($A_x \neq 0, A_y = 0$), or ($A_x = 0, A_y \neq 0$) exhibit supercritical bifurcation at $\bar{\nu}\chi > 3$.[3]

The phase diagram of various patterns is summarized in Fig. 5.11. At small $\bar{\nu}\chi$, the primary bifurcation is subcritical and leads to the emergence of square patterns: there is a noticeable hysteresis at the onset of the square pattern. For large $\bar{\nu}\chi$ the transition is supercritical and leads to stripe patterns. While parameter χ does not have direct connection to the driving frequency, one can expect that parameter α is responsible for vibration-induced transport of grains, and therefore $\chi = \alpha/\beta$ decreases with Ω since at large frequencies vertical displacements of particles become small and so the horizontal transport should become small as well. So, a transition from supercritical rolls to subcritical squares with χ agrees with the experimentally observable bifurcation with decreasing driving frequency.

At intermediate frequencies stable localized solutions of eqns 5.3 and 5.4 corresponding to isolated *oscillons* and a variety of bound states were found in agreement with experiment. The mechanism of oscillon stabilization is related to the oscillatory asymptotic behavior of the tails of the oscillon (see Fig. 5.12), since this underlying periodic structure provides pinning for the circular front forming the oscillon. Without such pinning, the oscillon solution is not robust: it could only exist at a certain unique value of the driving parameter γ, and would either collapse or expand otherwise.

Interfaces and more complex patterns. The phenomenological model (eqns 5.3 and 5.4) also provides a good qualitative description of patterns away from the primary bifurcation – hexagons and interfaces. In the high-frequency limit both α and β parameters in the ν equation become large, and ν again becomes enslaved by ψ. Then the dynamics can be described by a single parametric Ginzburg–Landau equation 5.3 (Aranson et al., 1999b).

It is convenient to shift the phase of the function ψ via $\tilde{\psi} = \psi \exp[i \arcsin(\omega_*/\gamma)/2]$. The equations for the real and imaginary parts $\tilde{\psi} = A + iB$ are:

$$\partial_t A = (s-1)A - 2\omega_* B - (A^2 + B^2)A + \nabla^2(A - bB),$$

---

[2] Additional non-linear terms $O(A_x^5, A_y^5)$ in eqn 5.9 arising from the coupling of the primary mode with its higher harmonics can be neglected in the limit $\chi \gg 1$.

[3] Note that eqns 5.9 are valid not only for orthogonal waves forming square patterns, but also for any oblique pair of waves forming, in general, rhombic patterns. The mechanism of selection of square patterns among the family of rhombic patterns is not captured by this model and presumably requires taking into consideration higher-order terms in the perturbative analysis.

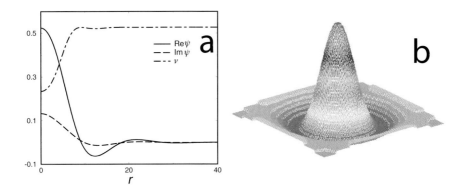

Fig. 5.12 The radially-symmetric oscillon solution of eqns 5.3 and 5.4. (a) Profiles of density ν and wavefunction ψ for $\gamma = 1.8, \bar{\nu} = 0.567, b = 2, \omega_* = \alpha = 1, \chi = 5/\gamma$; (b) Surface plot of Re$\psi$, from Tsimring and Aranson (1997).

$$\partial_t B = -(s+1)B - (A^2 + B^2)B + \nabla^2(B + bA), \quad (5.10)$$

where $s^2 = \gamma^2 - \omega_*^2$. At $s < 1$, eqns 5.10 have only one trivial uniform stationary state $A = 0$, $B = 0$. But at $s > 1$, two new uniform stationary states appear, $A = \pm A_0, B = 0, A_0 = \sqrt{s-1}$. The onset of these states corresponds to the period doubling of the layer flights sequence, observed in experiments and predicted by the simple inelastic ball model (Melo et al., 1994; Melo et al., 1995; Metha and Luck, 1990). Signs $\pm$ reflect two relative phases of layer flights with respect to the bottom vibrations.

A weakly non-linear analysis reveals that the uniform states $\pm A_0$ lose their stability with respect to finite-wave number perturbations at $s < s_c$, and the non-linear interaction of growing modes leads to hexagonal patterns. The reason for this is that the non-zero base state $A = \pm A_0$ lacks the up-down symmetry $\psi \to -\psi$ and the corresponding amplitude equations contain quadratic terms that are known to favor hexagons close to the onset (see, e.g. Cross and Hohenberg (1993)). In the regime when the uniform states $A = \pm A_0, B = 0$ are stable, there exists an interface solution connecting these two asymptotic states. Due to symmetry, the interfaces should be immobile.

However, breaking the symmetry of driving can lead to interface motion. This symmetry breaking can be achieved by additional subharmonic driving at frequency $f/2$, i.e. $\sim q_s \sin(\pi f t + \Phi)$, where q_s is the amplitude of the subharmonic driving component and Φ is the relative phase between the primary driving frequency f and its subharmonic $f/2$.

In experiments, however, even in the absence of additional driving the interface was found to slowly drift toward the middle of the cell (Aranson et al., 1999a). This phenomenon can be explained via a subtle feedback mechanism between interface motion and the response of the vibration system (Aranson et al., 1999b). Due to the finite size (and mass) of the container, collisions of a granular layer containing the interface with the bottom plate induce the subharmonic response in the vibration system: the vibrating cell can acquire subharmonic motion from the periodic impacts of the granular layer on the bottom plate at half the driving frequency. This subharmonic component of driving breaks the symmetry and leads to the interface motion. If the interface is located in the middle of the cell, the masses of material on both sides of the interface are equal, and, due to the antiphase character of the layer motion on both sides, an additional subharmonic driving is minimized. The displacement of the interface X from the center of the cell leads to a mass difference Δm on opposite sides of the interface, and so in two consecutive phases of vibrations, the total mass of sand falling on the plate is different, which in turn causes a subharmonic response of the plate proportional to Δm. In a rectangular cell, $\Delta m \sim X$. The speed of the interface is proportional to the amplitude of the subharmonic component of driving, or X,

$$\frac{dX}{dt} = -X/\tau_0. \tag{5.11}$$

i.e. the interface indeed drifts to the center. The relaxation time constant τ_0 depends on the mass ratio, and can therefore be varied by changing the mass of the vibrating cell. Note that one can use this effect to control the interface position by adding a subharmonic component to the driving acceleration: $a = \Gamma \sin(\Omega t) + q_s \sin(\Omega t/2 + \Phi)$. Then, the equation for the interface position generalizes to

$$\frac{dX}{dt} = -X/\tau + q_s \tilde{\alpha} \sin \Phi \tag{5.12}$$

Parameter $\tilde{\alpha}$ is a "susceptibility" of the system with respect to subharmonic driving. By varying the phase of the small subharmonic component Φ one can move the interface through the system.[4]

The interfaces may become transversally unstable and produce undulations (see experimental Fig. 2.3(e)). The problem of transversal instability and the transition between flat and decorated interfaces appears to be rather subtle. While eqn 5.3 predicts the transverse instability of the flat interface with the decrease of driving frequency, the resulting instability does not saturate and leads to a proliferation of the interfaces and formation of labyrinthine patterns. In order to account for the saturation of the interface instability, an additional non-local term has to be added to eqn 5.3,

$$\partial_t \psi = \left[\gamma + \frac{\bar{\epsilon}}{S} \int d\mathbf{r}' \exp\left(-\frac{|\mathbf{r}-\mathbf{r}'|^2}{r_0^2}\right) |\nabla \psi(\mathbf{r}')|^2\right] \psi^* \\ -(1-i\omega_*)\psi + (1+ib)\nabla^2 \psi - |\psi|^2 \psi. \tag{5.13}$$

The nonlocal term in eqn 5.13 models the effects of convection induced by the interface, where r_0 is the characteristic scale of the convection flow, and the parameter $\bar{\epsilon}$ is the magnitude of coupling to the convection. Since $|\nabla \psi|$ is non-zero only at the interface, the integral is proportional to the total length of the interface. As the length of the interface grows, the effective forcing increases as well, and eventually leaves the instability domain. Thus, one can obtain stable steady-state "decorated" interfaces (Fig. 5.13(c)).

[4] As was noted by Aranson et al. (1999a) and Moon et al. (2004), moving interface can be used to separate granular material of different sizes.

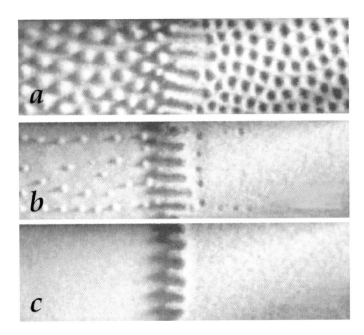

Fig. 5.13 Representative high-acceleration patterns in the rectangular 4×12 cm² cell, vibration frequency $f = 40$ Hz: (a) the interface between up- and down-hexagons at $\Gamma = 3.75$; (b) "superoscillons" in the flat layer pinned near the interface at $\Gamma = 3.94$; (c) isolated decorated interface at $\Gamma = 4$, from Blair *et al.* (2000).

In addition to decorated interfaces, a variety of new localized structures such as "superoscillons" were observed both experimentally and theoretically at higher acceleration levels, see Figs. 5.13 and 5.14. In contrast to conventional oscillons existing on the flat background oscillating with driving frequency f, i.e. in our notation $\psi = 0$, superoscillons exist on the background of the flat period-doubled solution ($\psi = \text{const} \neq 0$).

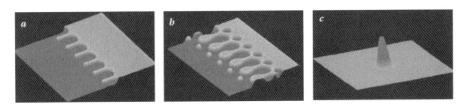

Fig. 5.14 Stationary solutions to eqn 5.13. Surface plots show Reψ for different stationary solutions: (a) decorated interface solution; $\omega_* = 1.2, \gamma = 1.85, b = 4, \bar{\epsilon} = 0.002, r_0 = 40$, and $L = 160$; (b) decorated interface with bound superoscillons; $\omega_* = 1.2, \gamma = 1.75, b = 4, \bar{\epsilon} = 0.002, r_0 = 40$, and $L = 160$; (c) single superoscillon at $\omega_* = 1.2, \gamma = 2.2, b = 4, \bar{\epsilon} = 0.002, r_0 = 40$, and $L = 100$, from Blair *et al.* (2000).

Generalized Swift–Hohenberg model.. An alternative description of the primary pattern-forming bifurcation can be achieved in the framework of another generic model, the generalized Swift–Hohenberg equation (Crawford and Riecke, 1999)

$$\partial_t \vartheta = R\vartheta - (\partial_x^2 + 1)^2 \vartheta + b\vartheta^3 - c\vartheta^5 + \epsilon \nabla \cdot [(\nabla \vartheta)^3]$$
$$- \beta_1 \vartheta (\nabla \vartheta)^2 - \beta_2 \vartheta^2 \nabla^2 \vartheta. \qquad (5.14)$$

Here, the (real) function ϑ characterizes the amplitude of the oscillating solution, so implicitly it is assumed that the whole pattern always oscillates in phase. The standard Swift–Hohenberg equation (i.e. without terms $\propto \epsilon, \beta_{1,2}, c$ and with $b < 0$) was first introduced for the description of convective rolls (see, e.g. Cross and Hohenberg (1993)). It exhibits a supercritical bifurcation to square patterns when $R > 0$. The extended model (eqn 5.14) with higher-order terms exhibits a subcritical bifurcation for $b > 0$, i.e. finite-amplitude patterns exist even for $R < 0$. This equation also describes both square and stripe patterns depending on the magnitude of ϵ, and for negative R it has a stable oscillon-type solution.

The success of the Swift–Hohenberg-type description of the pattern formation in vibrated granular layers is not accidental. Since the primary bifurcation to the patterned state occurs via a finite wave number isotropic instability ($k_c \neq 0$), the Swift–Hohenberg model is the universal amplitude equation which can be asymptotically deduced for this type of instability (see, e.g. Aranson and Kramer (2002) for the classification of amplitude equations). Moreover, eqns 5.10 can be reduced to the Swift–Hohenberg model in the limit of $s \to 1$. In the vicinity of $s = 1$ the following relations hold: $A \sim (s-1)^{1/2}, B \sim (s-1)^{3/2} \ll A$. In the leading order one finds $B \approx b\nabla^2 A/2$, and eqns 5.10 yield the Swift–Hohenberg model:

$$\partial_t A = (s-1)A - A^3 + (1 - \omega_* b)\nabla^2 A - \frac{b^2}{2}\nabla^4 A. \qquad (5.15)$$

In the one-dimensional case this equation describes a supercritical transition to stripes near the onset of an instability of the trivial state $A = 0$. Moreover, for $s > 1$ it has solutions in the from of an interface connecting two symmetric states $A = \pm\sqrt{s-1}$.

5.4.2 Distributed map approach

The amplitude equation technique is not the only phenomenological approach applied to the analysis of patterns in vibrated granular layers. Spatiotemporal dynamics of patterns generated by parametric forcing can be understood in the framework of even simpler, discrete-time, continuous space distributed map system that locally exhibits a sequence of period-doubling bifurcations and whose spatial coupling operator selects a certain spatial scale (Venkataramani and Ott, 1998; Venkataramani and Ott, 2001). Note that the distributed map approach generalizes coupled map lattices (Kaneko, 1993) where time as well as space are discrete, and cellular automata models where the physical variable also takes discrete values, see Section 3.6.4.

In this approach the discrete-time local dynamics is modelled by a one-dimensional map $\xi_{n+1} = M(\xi_n)$ at each point of space, where the discrete time index n corresponds to the number of periods of external driving. It is assumed that the mapping M

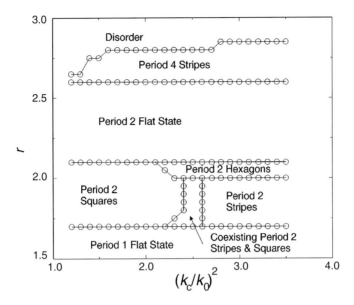

Fig. 5.15 Phase diagram illustrating various stable patterns obtained from the solution of eqn 5.16, from Venkataramani and Ott (1998).

exhibits a generic cascade of period-doubling bifurcations with the change of a control parameter r that corresponds to the amplitude of the periodic driving.

The spatial coupling is introduced via a generalized diffusion operator $\mathcal{L}$,

$$\xi_{n+1}(\mathbf{x}) = \mathcal{L}[M(\xi_n(\mathbf{x}))] \qquad (5.16)$$

Specifically, the local mapping $M(\xi)$ is modelled by a Gaussian map

$$M(\xi) = r \exp[-(\xi - 1)^2/2] \qquad (5.17)$$

and the linear spatial operator $\mathcal{L}$ has an azimuthally symmetric Fourier transform

$$f(k) = \text{sign}[k_c^2 - k^2] \exp[k^2(1 - k^2/2k_0^2))/2]. \qquad (5.18)$$

Here, k is the wave number, k_c, k_0 are two inverse length scales characterizing the spatial coupling. On the phenomenological level the parameter r can be interpreted as the acceleration magnitude Γ, and the ratio of two characteristic length scales is related to the driving frequency f. With an appropriate choice of the control parameters, this model leads to a phase diagram on the plane $(k_c/k_0, r)$ which is similar to the experimental one, see Fig. 5.15. Since the model (eqns 5.16–5.18) does not yield a subcritical bifurcation to squares, an additional modification of the local mapping M is necessary in order to account for the subcritical transition. For example, the map

$$M(\xi) = -(r\xi + \xi^3) \exp[-\xi^2/2] \qquad (5.19)$$

demonstrates a subcritical transition to squares and localized oscillon-like solutions in a certain range of parameters. A hybrid map combining features of maps (eqn 5.18)

110 *Surface waves and patterns in periodically vibrated granular layers*

and (eqn 5.19) can in principle describe on the qualitative level the topology of the entire phase diagram.

While this distributed map system is obviously a greatly simplified model of the pattern formation in vibrated granular layers, it nevertheless captures the important phenomenology of the interplay of period doubling and spatial coupling. Compared to the amplitude equations described in the previous section, this approach can be applied beyond the first subharmonic bifurcation. On the other hand, it is difficult to establish the connection between the distributed map systems and experimental conditions, or to derive the distributed map system from more general hydrodynamic equations.

5.5 Hydrodynamic description of pattern formation in vibrated layers

Description of the pattern formation in the framework of full three-dimensional granular hydrodynamic equations (eqns 3.11–3.13) is the most physical approach to the problem; however, this approach is possibly the most difficult to implement. This difficulty prompted several research groups to employ a somewhat simpler quasi-two-dimensional Saint–Venant-like continuum description of the vibrated sand patterns (Cerda *et al.*, 1997; Eggers and Riecke, 1999; Park and Moon, 2002). These models deal with depth-averaged mass and momentum conservation equations that are augmented by specific constitutive relations for the mass flux and pressure. Since for acceleration $\Gamma > 1$ the granular layer periodically lifts off and collides with the vibrating plate, Cerda *et al.* (1997) assumed that during impact particles acquire horizontal velocities proportional to the gradient of the local thickness, during the flight they move freely with these velocities and redistribute mass, and then after the layer lands back on the plate, during the remainder of the cycle the layer diffusively relaxes on the plate. It was found that in a periodic sequence of free flights, the layer becomes unstable and as a result square patterns develop, however, the transition appears to be supercritical. In order to account for the subcritical character of the primary bifurcation to square patterns, Cerda *et al.* (1997) postulated the existence of a certain critical slope (related to the angle of repose) below which the free flight initiated by the impact does not occur. They also observed localized excitations (oscillons and bound states), however, these excitations appeared only as transients. This model was later generalized by explicitly writing the momentum conservation equation and introducing the equation of state for the hydrodynamic pressure that was proportional to the square of the velocity divergence (Park and Moon, 2002). This effect provides saturation of the pattern-forming instability and leads to a squares-to-stripes transition at higher frequencies, which was missing in the original model. By introducing multiple free-flight times and contact times hexagonal patterns and superlattices were also reproduced.

5.5.1 Shock propagation

Going beyond the shallow-water approximation involves understanding processes occurring as a result of a granular layer hitting a vibrated bottom plate. Ignoring the

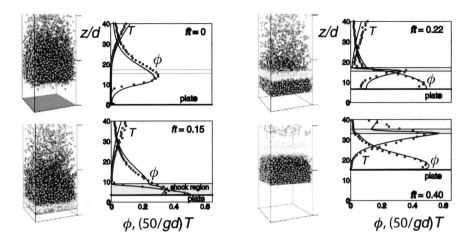

Fig. 5.16 Shock propagation in a vertically vibrated granular layer at four times ft within the oscillation cycle. Left column: snapshots from a molecular dynamics simulation; the particles are graycoded according to temperature (white corresponds to higher values of temperature). Right column: horizontally averaged volume fraction $\phi = \nu/\nu_0$ and temperature $T/(gd)$ as functions of the vertical coordinate z/d for the same four moments of times. The plate position is denoted by a horizontal solid black line; results from molecular dynamics simulation are shown as dots, continuum theory results are shown in solid lines. The width of the shock (thin horizontal lines) varies throughout the cycle, from Bougie et al. (2002).

transversal non-uniformities related to pattern formation, these dynamics can be addressed in the framework of quasi-one-dimensional granular hydrodynamics for the vertical profiles of density $\nu(z,t)$, the z-component of velocity $v_z(z,t)$ and the granular temperature $T(z,t)$ (eqns 3.11–3.13) in the frame moving with the vibration plate. Then the problem can be closed by setting boundary conditions on the plate at fixed $z = 0$ and at $z \to \infty$. Plate vibration enters eqn 3.12 as periodic modulation of gravity acceleration, $\Gamma \sin(\Omega t)$. Even this deceptively simple problem turns out to be highly non-trivial. In addition to perennial issues of granular hydrodynamics, such as the formal validity of the equations only for the restitution coefficient $e \to 1$, and for not too high a density, the application of granular hydrodynamics to the problem of a vibrated layer meets an additional difficulty: the hydrodynamic equations develop an artificial numerical instability for the density $\nu \to 0$. The dilute (essentially empty) region naturally appears when the layer takes off the plate. In order to deal with this numerical artifact at $\nu \to 0$, a certain regularization procedure had to be introduced in the hydrodynamic equations (Bougie et al., 2002).

Nevertheless, even with all these limitations, both hydrodynamic equations and molecular dynamics simulations produce consistent and robust results on shock propagation in vibrated layers, see Fig. 5.16. At each cycle of vibration, high-temperature dense shocks are created when the layer collides with the plate. These shocks rapidly propagate upward within only a quarter of the vibration cycle. Both molecular dynamics simulations and hydrodynamic modelling show that the details of the shock propagation are not very sensitive to boundary conditions, the value of the restitution coefficient e, and the specific form of the constitutive relation (eqn 3.16). Remarkably, the shocks are essentially supersonic: the corresponding Mach number is of the order of 10.

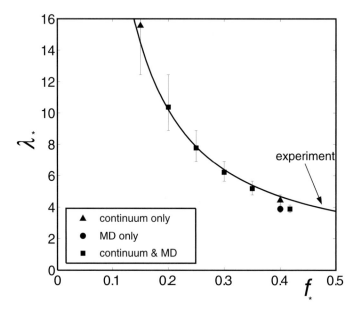

Fig. 5.17 Dispersion relation for stripes near the onset from granular hydrodynamics (▲) and molecular dynamics simulations ● compared with experimental fit (solid line). Points where continuum and MD simulation results coincide are shown by ■. Here $f_* = f/\sqrt{g/h}$ and $\lambda_* = \lambda/h$ are normalized frequency and wavelength, respectively, from Bougie et al. (2005).

5.5.2 Analysis of dispersion relation for surface waves

The dispersion relation for surface waves in vibrated granular layers can in principle be found in the framework of granular hydrodynamics. For this purpose a periodic in time one-dimensional solution to the granular hydrodynamics equations $\nu(x,t), v_z(z,t), T(z,t)$, such as discussed in Section 5.5.1, needs to be examined with respect to periodic in x, y small perturbations. The dispersion relation for the surface waves $\omega(k)$ is then obtained from the Floquet analysis (Morse and Feshbach, 1953). However, to date this problem has not been solved.

Bizon et al. (1999) considered a somewhat simpler problem: an instability of an incompressible isothermal "fluid" layer with zero surface tension that undergoes periodic collisions and separations with the vibrating plate. Using viscosity as a fitting parameter, the authors obtained a reasonable agreement with experimental data (eqn 5.2), see Fig. 5.17. However, as one can see from the analysis of Section 5.5.1, the approximation of the granular layer by an incompressible isothermal fluid ignores some important details of granular physics.

The dispersion relation can also be obtained by the direct solution of full three-dimensional granular hydrodynamics equations (eqns 3.11–3.13) in a large domain (Bougie et al., 2005). A quantitative agreement was found between this approach and event-driven molecular dynamics simulations and experiments in terms of the wavelength dependence on the vibration frequency (Fig. 5.17). Since standard granular hydrodynamics does not take into account friction among particles, the simulations only yielded stripe patterns, in agreement with earlier molecular dynamics simulations. Furthermore, a small but systematic difference ($\sim$10%) was found between the critical value of plate acceleration in fluid-dynamical and molecular dynamics simulations that could be attributed to the role of fluctuations near the onset. A proper account of the interparticle friction and fluctuations within the full hydrodynamics description still remains an open problem (a detailed discussion of dense granular flows can be found in Chapter 6).

5.5.3 Effects of fluctuations at the onset of patterns

Fluctuations are expected to play a significantly greater role in granular systems than in normal fluids, because the total number of particles involved in the dynamics per characteristic spatial scale of the problem is many orders of magnitude smaller than the Avogadro number. An example of fluctuations near the transition to square patterns in a vibrating granular layer is shown in Fig. 5.18.

Due to similarities between granular hydrodynamics and ordinary fluid mechanics as well as similarities in pattern formation, it is tempting to apply the apparatus developed for the analysis of fluctuations in hydrodynamic phase-transition phenomena, such as Rayleigh–Bénard convection (Swift and Hohenberg, 1977), to the description of granular patterns (Goldman et al., 2004; Bougie et al., 2005). The Swift–Hohenberg theory is based on the equation for the scalar order parameter ϑ,

$$\partial_t \vartheta = [\epsilon - (\nabla^2 + k_0^2)^2]\vartheta - \vartheta^3 + \xi(\mathbf{x}, t), \tag{5.20}$$

where ϵ is the bifurcation parameter, k_0 is the wave number corresponding to the most unstable perturbations, and ξ is the Gaussian δ-correlated noise term with intensity F, $\langle \xi(\mathbf{x},t)\xi(\mathbf{x}',t')\rangle = 2F\delta(\mathbf{x}-\mathbf{x}')\delta(t-t')$. This equation, without the fluctuation term, describes a second-order phase transition in a uniform two-dimensional system to one-dimensional spatially periodic (roll) patterns in Rayleigh–Bénard convection. Using perturbation theory developed earlier by Brazovskii (1975) for three-dimensional systems exhibiting a second-order phase transition, Swift and Hohenberg (1977) showed that a weakly *first-order* phase transition occurs as a result of fluctuations. While the "jump" of the magnitude of the order parameter ϑ at the transition may be too small to be detected in experiments, the Swift–Hohenberg theory also predicts that noise offsets the bifurcation value of the control parameter from the mean-field value $\epsilon_{\mathrm{MF}} = 0$ to the critical value $\epsilon_c \propto F^{2/3}$. Furthermore, far below the bifurcation point, a linear regime is expected to apply, in which the magnitude of noise-excited modes scales as $|\epsilon - \epsilon_c|^{-1/2}$, while the time coherence of fluctuations (width of the spectral peak) decays as $|\epsilon - \epsilon_c|^{-1}$. Fitting the Swift–Hohenberg model (eqn 5.20) to match the transition in

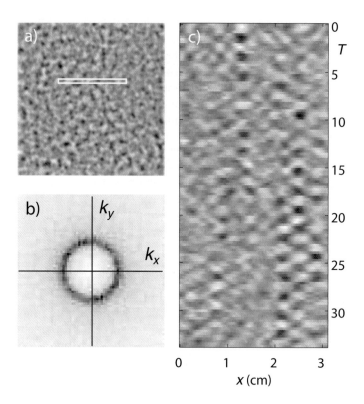

Fig. 5.18 Fluctuations in vibrated granular layer near the onset of square patterns: (a) Snapshot of an area 6.25×6.25 cm^2 in a container oscillating with $\Gamma = 2.6$. (b) The spatial power spectrum of (a). The dark ring corresponding to randomly oriented spatial structures with a length scale 0.52 cm. (c) Space-time diagram for the row of pixels within the box in (a); the period of the localized transient oscillations is roughly twice the period of plate vibrations, from Goldman et al. (2004).

a vibrated granular layer, a good agreement with molecular dynamics simulations and experiments was found for the dependence of the noise power and the magnitude of the structure factor $C(\theta)$ [5] on the acceleration Γ (Fig. 5.19), as well as for the scaling of the noise power and the correlation time with the distance to the threshold, Fig. 5.20 (Goldman et al., 2004; Bougie et al., 2005). Interestingly, the magnitude of the fitted noise term in eqn 5.20 $F \approx 3.5 \times 10^{-3}$ turned out to be an order of magnitude greater than for the convective instability in a fluid near a critical point (Oh and Ahlers, 2003). This difference from fluid dynamics probably stems from the fact that granular systems are not thermal, and fluctuations have a much higher effective temperature than predicted by thermodynamics. An additional difference from fluids is that the classical Swift–Hohenberg theory developed for ordinary fluids is formally valid for

---

[5]Structure factor $C(\theta)$ is defined as the radially averaged power spectrum of the spatial patterns $C(\theta) = \langle S(\mathbf{k}) \rangle_k$, where $k = |\mathbf{k}|, \theta = \arg(\mathbf{k})$.

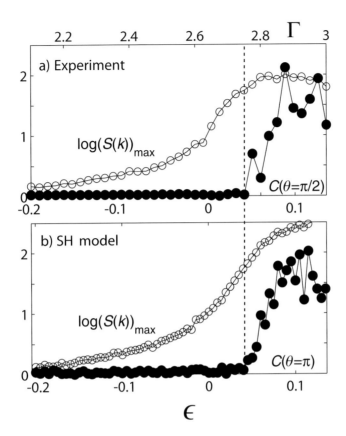

Fig. 5.19 Comparison of fluctuation parameters near the onset of the long-range order in (a) experiment and (b) Swift–Hohenberg model (eqn 5.20). The maximum of $S(k)$ (○) increases through the mean field onset ($\epsilon_{MF} = 0$), while the onset of long-range order, indicated by appearance of angular correlations of the radially averaged structure factor [$C(\theta = \pi/2)$ for the experiment and $C(\theta = \pi)$ for the Swift–Hohenberg equation (●)], is delayed to $\epsilon_c \approx 0.04$. The solution to eqn 5.20 is obtained on a 128×128 grid with $k_0 = 1$ and the integration time step 0.5, from Goldman et al. (2004).

second-order phase transitions, at least in the mean-field approximation, whereas in vibrated granular layers the transition to square patterns is known to be first order (Goldman et al., 2004; Bougie et al., 2005). As we mentioned earlier, in the presence of noise the transition to patterns in the framework of the Swift–Hohenberg equation becomes only weakly subcritical (Swift and Hohenberg, 1977). Consequently, the nonlinear terms can be important close to the transition point and may distort the scaling of the fluctuation power. Thus, it would be interesting to examine the fluctuations scaling in the framework of the subcritical Swift–Hohenberg equation

$$\partial_t \vartheta = [-\epsilon - (\nabla^2 + k_0^2)^2]\vartheta + \beta\vartheta^2 - \vartheta^3 + \xi(\mathbf{x}, t), \tag{5.21}$$

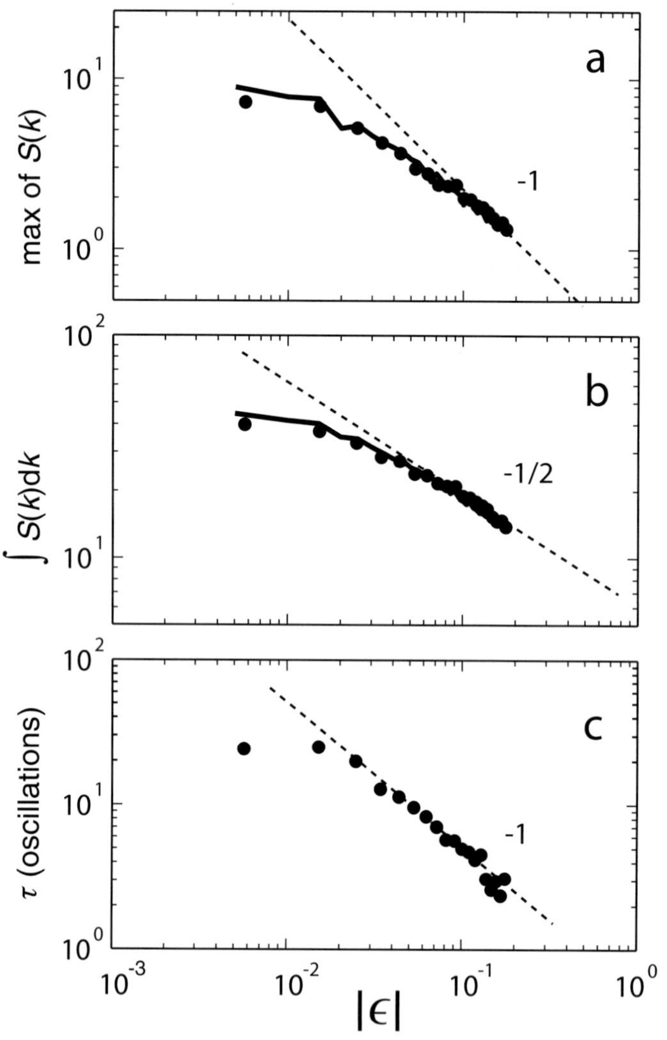

Fig. 5.20 Comparison between the Swift–Hohenberg theory and the experiment for the noise peak intensity (a), total noise power (b), and the correlation time (c).● - experiment, solid lines - Swift–Hohenberg model, dashed lines - linear theory for small noise magnitude, from Goldman *et al.* (2004).

with the parameter $\beta \neq 0$. While far below the threshold we expect the same scaling with the noise strength F as for the supercritical eqn 5.20, the non-linearity should introduce qualitative differences near the transitions, such as nucleation and growth of seeds of patterned states, the hallmark of first-order phase transitions (Landau *et al.*, 1981).

6
Patterns in gravity-driven granular flows

The focus of this chapter is the overview of mechanisms and theoretical models of pattern formation in dense gravity-driven granular flows. We will consider a wide range of phenomena, from stable and unstable avalanches flowing upon steep slopes, flows in rotating drums, to self-organized criticality and statistics of granular avalanches. In the majority of gravity-driven granular flows, the motion is confined at the surface of the granular system. This is easy to understand since below the surface grains are compressed by gravity and are harder to dilate. There are a number of different approaches to the description of surface flows. Most of them are based on granular hydrodynamic equations (mass and momentum conservation) for the flowing material (Savage and Hutter, 1989). However, this type of modelling is restricted to the description of flows over fixed bottom topography (e.g. the side of a mountain). After integration over the thickness of the flowing layer, depth-averaged or Saint-Venant equations are obtained that allow the calculation of the flow thickness and the mean velocity.

Another common experimental situation occurs when the bottom of the flowing layer is erodible, such as in avalanches on snow, sand dunes, or deep sandpiles. In this case, the position of the border between flowing and resting grains is not fixed during the avalanche, since flowing grains colliding with initially immobile grains can either bring them into motion (erosion of the bottom by the flow), or, on the contrary, get trapped. Thus, the position of the static/rolling interface (the "effective bottom" position) can be thought of as a new dynamical variable of the problem. Its dynamics is controlled by certain conversion rules for the erosion/deposition processes that in turn depend on the local slope and the rolling speed of grains. This approach was adopted in the models introduced by Mehta (1994) and Bouchaud, Cates, Ravi Prakash and Edwards (Bouchaud et al., 1994) (the latter is known as the BCRE model), and was later developed in a number of other papers (e.g. Aradian et al. (2002)).

In real granular flows there is no sharp distinction between a "flowing phase" and a "static phase": the transition from static to flowing regime is continuous. To account for this, one can of course rely on direct molecular dynamics simulations of individual grains. However, this approach is computationally expensive and limited to relatively small systems. In continuum modelling, one can also attempt to introduce "hybrid" constitutive relations into the hydrodynamic description that would work in the "partially fluidized" regime. One way to do this is to assume that the "granular fluid" viscosity depends strongly on the local density and diverges as the density approaches the close-packed limit (Bocquet et al., 2002). Alternatively, in analogy with the theory of phase transitions one can introduce a new "phase field" or an order parameter that controls the local state of the granular material (Aranson and Tsimring, 2001).

We begin this chapter with a brief description of experimental data concerning the rheological properties of dense granular flows, and then present an overview of theoretical models and their applications to patterns formed by gravity-driven granular flows. In most of this chapter (unless otherwise stated) we assume two-dimensional flow geometry.

6.1 Rheological properties of dense granular flows

The beginning of serious scientific studies of the rheology of dense granular flows is usually associated with the famous researcher and explorer Ralph Bagnold. For constant-density flows in the absence of gravity, from heuristic dimensional arguments Bagnold (1954) proposed a simple local relation between the strain rate $\dot{\gamma}$ and the shear stress, $\sigma_{\text{shear}} \sim d^2 \nu \dot{\gamma}^2$, see eqn 3.46 in Section 3.5.3. The Bagnold relation is approximately satisfied in the bulk of the flow, while the boundary effects as well as dynamic inhomogeneities often result in substantial deviation from this simple scaling law.

The dynamics of dense slowly sheared granular media is very different from the behavior of dilute rapid granular gases. One of the distinct features of flows in dense granular assemblies is the occurrence of yield and fluidization transition as the shear stress exceeds a certain finite critical value.

The incompleteness of the knowledge about the dense-liquid regime and the transition between the different regimes has motivated many experimental, numerical, and theoretical works. Different flow configurations have been investigated from confined flows in channels to free surface flows on piles, both experimentally and numerically. Although a wealth of important information is now available about the flow characteristics, it is often difficult to extract common features and general trends for various granular flows because the experimental or numerical conditions vary from one study to another. A comprehensive survey of experimental and numerical results by various groups on dense granular flows, such as chutes with rough or sticky bottom conditions (which precludes the plug-flow regime when the layer slides down as a whole), rotating drums, heap flows, and other geometries, was conducted by a French research network, the Groupement De Recherche Milieux Divisés (MiDi, 2004). The surveyed experiments and simulations for chute flows are summarized in Table 6.1, and some of the experimental results are presented in Figs. 6.1 and 6.2.

6.1.1 Granular flow in a chute

The most studied example of dense granular flows is a relatively thin layer of grains flowing down on a surface of a plane with a fixed inclination angle $\bar{\varphi}$, see Fig. 6.1(a). As the inclination of the plane is slowly increased, an initially static granular layer of uniform thickness h on a rough incline plane starts flowing when the slope angle φ exceeds a critical value φ_{start} (which depends on h). When the inclination is slowly reduced from high values corresponding to the flowing regime, the flow is sustained down to a second critical angle φ_{stop} (also h dependent) that is *smaller* than φ_{start}. Sometimes these two critical angles φ_{start} and φ_{stop} are called static and dynamic *repose angles*, respectively. The existence of these two repose angles provides compelling evidence of the hysteretic nature of granular flows. Alternatively, one can introduce critical thicknesses $h_{\text{start}}(\varphi)$ and $h_{\text{stop}}(\varphi)$ for a fixed inclination φ. In fact, the measurement of h_{stop} is easier as it corresponds to the thickness of the deposit remaining on the plane once the flowing sand has left the chute. However, an experimental measurement of h_{start} is more complicated because the initiation of the flow is very sensitive to the preparation conditions, density fluctuations, apparatus vibration, etc. An example of the phase diagram measured for a layer of glass beads on an inclined plane covered by velvet (which makes grains to stick to the bottom) is shown in Fig. 6.1(b). The two curves $h_{\text{stop}}(\varphi)$ and $h_{\text{start}}(\varphi)$ divide the phase diagram (h, φ) into three regions: a region where no flow is possible ($h < h_{\text{stop}}$), a bistable region where both static and flowing regimes can coexist ($h_{\text{stop}} < h < h_{\text{start}}$), and a region where only a flowing regime can be observed $h > h_{\text{start}}$.

Critical curves measured for different materials and bottom conditions are plotted in Figs. 6.1(b) and (c). As demonstrated by Fig. 6.1(c), changing the bottom surface from glued particles to velvet cloth dramatically shifts the critical curves h_{start} and h_{stop} toward higher repose angles. This strong stabilizing effect can be possibly attributed to long velvet fibers permeating granular matter near the bottom and thus creating strong bonds between the granulate and the substrate. In the flowing regime when $h > h_{\text{stop}}(\varphi)$, the flow is steady and uniform for moderate inclination, but continuously accelerates along the plane for larger inclinations.

6.1.2 Flow rheology

In dense granular flows a very short collision time scale has no obvious effect on macroscopic flow properties. Accordingly, the grain size d is the natural length scale (besides the size of a container) of the problem. For a single-phase granulate, the grain mass can also be scaled out, implying that the characteristics of granular flows are independent on the material density ν. In homogeneous steady granular flows, the strain-rate tensor contains only one variable, the strain rate $\dot{\gamma}$, and the stress tensor contains two independent variables, the normal stress or pressure $p = -\sigma_{zz}$ and shear stress σ_{xz}. These three quantities define two dimensionless parameters, the effective frictions coefficient, μ_{eff}

$$\mu_{\text{eff}} = \frac{\sigma_{xz}}{p}, \qquad (6.1)$$

and the dimensionless shear rate, I

$$I = \frac{\dot{\gamma} d}{\sqrt{p/\nu}}. \qquad (6.2)$$

The (inertial) parameter I can be interpreted as the ratio of two different time scales, the deformation scale $T_\gamma = 1/\dot{\gamma}$, and the microscopic or confinement time scale $T_p = d\sqrt{\nu/p}$.

The volume fraction ϕ of granular flows is another important parameter affecting granular flows. However, in many situations such as open chute flows, it is itself determined by the flow regime. One reasonable assumption is that the volume fraction is

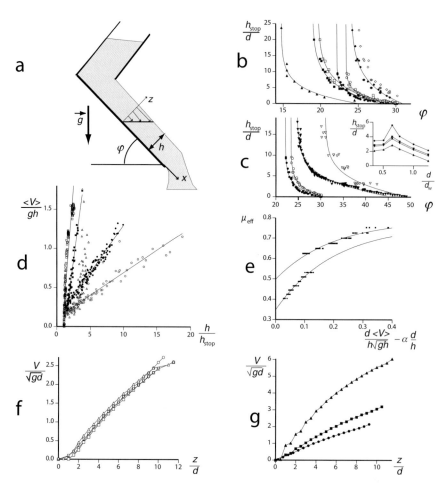

Fig. 6.1 Summary of experimental and numerical results for granular flow down rough chutes. (a) Schematics of a typical experimental setup. (b) $h_{stop}(\varphi)$ (black filled symbols) and $h_{start}(\varphi)$ (white open symbols) from simulations IP3 (▲), experiments IP5 with glass beads (○,●), IP6 with mustard seeds (◇, ◆) or with glass beads on carpet (□, ■). (c) Same as (b) for experiments IP5 with glass beads (○,●), IP7 with glass beads on velvet (▽, ▼). Inset: h_{stop} for different roughness condition from experiment IP8 $\varphi = 27°$ (●), $\phi = 28°$ (■), $\phi = 28.3°$ (▼), $\varphi = 30°$ (▲). (d) Froude number $\langle V \rangle / \sqrt{gh}$ as a function of $h/h_{stop}(\varphi)$ from simulation IP3 (○), experiments IP5 with glass beads on glass beads (●), sand on sand (□), IP6 sand on moquet (■), IP7 glass beads on velvet (△). Lines are fits by eqn 6.7. (e) Effective friction deduced from the flow rule (see text). Experiment IP5 with glass beads (●), with sand (■). Continuous lines are deduced from eqn 6.8 and fit of $h_{stop}(\varphi)$. (f) Velocity profiles from simulations IP4 ($\varphi = 14.4°$, $\mu = 0$.) for different restitution coefficients $e = 0.4$ (△), $e = 0.6$ (▽), $e = 0.7$ (□) and $e = 0.8$ (○). (g) Velocity profiles from simulations IP4 ($\varphi = 18°$, $e = 0.6$) for different friction coefficients $\mu = 0$. (▲), $\mu = 0.25$ (■) and $\mu = 0.5$ (●), reproduced from MiDi (2004).

Table 6.1 Short description of experimental and numerical setups in studies of inclined plane flow. E and N stand for experiment and numerics, D is the dimensionality, MD stands for molecular dynamics simulations and CD for contact dynamics simulations. d is the particle diameter, μ_p is the interparticle friction coefficient, e is the normal restitution coefficient, SW stands for sidewalls, see MiDi (2004) and references therein.

#	E/N	D	Material and geometry	h/d	Walls
IP1	E	2D	aluminium beads: $d = 3$ mm $e = 0.5$; plane: 2 m long	≈ 10	SW: glass; bottom: glued grains
IP2	E	2D	polystyrene disks: $d = 8$ mm $e = 0.4$; plane: 2 m long	≈ 10	SW: glass; bottom: glued grains
IP3	N/CD	2D	disks: $e = 0$, $\mu_p = 0.5$	≈ 50	SW: none; bottom: glued grains
IP4	N/MD	2D	disks: $e = 0.4 \rightarrow 0.8$, $\mu_p = 0 \rightarrow 0.5$	≈ 10	sidewalls: none; bottom: glued grains
IP5	E	3D	glass beads $d = 0.5$ mm and sand $d = 0.8$ mm; plane: 2 m long, 0.7 m wide.	< 20	sidewalls: none; bottom: glued grains
IP6	E	3D	glass beads $d = 1.5$ mm, sand $d = 1$ mm, mustard seeds $d = 2$ mm; plane: 2 m long, 0.7 m wide	≈ 10	sidewalls: none; bottom: carpet
IP7	E	3D	glass beads $d = 0.24$ mm; plane: 1.35 m long, 0.6 m wide	≈ 10	SW: none; bottom: velvet cloth
IP8	E	3D	glass beads $d = 0.14 \rightarrow 0.53$ mm; plane: 1.3 m long, 0.6 m wide	< 20	SW: none; bottom: glued grains
IP9	E	3D	glass beads $d = 1$ mm; plane 2 m long, 0.05 m wide	< 100	SW: Plexiglas; bottom: glued grains

slaved to the parameter I. In this situation one can expect a unique local relationship between the effective friction μ_{eff} and rescaled shear rate I.

Assuming a simple local relation between μ_{eff} and I, $\mu_{\text{eff}} = \mu(I)$, one can obtain the steady-state profile of the x-component of velocity $V(z)$ from the following simple argument. For a steady-state flow on an inclined plane, the stress distribution is given by the expression

$$\sigma_{zz} = -p = \nu g(h-z)\cos\varphi, \quad \sigma_{xz} = \nu g(h-z)\sin\varphi. \tag{6.3}$$

Thus, the effective friction coefficient $\mu_{\text{eff}} = \sigma_{xz}/\sigma_{zz} = \tan\varphi$ in the steady–state is independent of depth and is a function of the inclination angle φ only. Therefore, the strain rate can be inferred from the parameter I by inverting the relation $\mu_{\text{eff}} = \mu(I) = \tan\varphi$:

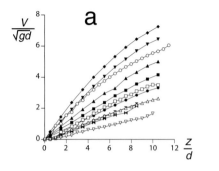 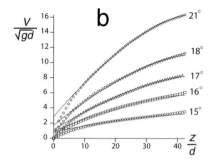

Fig. 6.2 (a) Velocity profiles for thin flows from experiment IP2 for $\varphi = 34°, 36°$ (black bow ties), IP1 for $\theta = 21°$ (●), $22°$ (■), $23°$ (▲), $25°$ (▼), $27°$ (♦), and simulations IP4 ($e = 0.6$) with $\varphi = 12.6°$ (▽), $14.4°$ (△), $16.2°$ (□), $18°$ (○); (b) Velocity profiles for thick flows from numerical simulations IP3, reproduced from MiDi (2004).

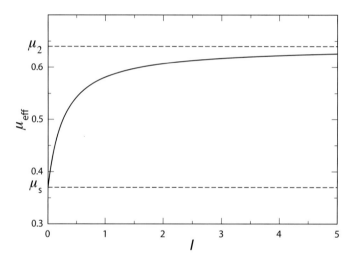

Fig. 6.3 Typical dependence of the effective friction coefficient μ_{eff} on scaled shear rate I.

$$I = \frac{\dot{\gamma}d}{\sqrt{P/\nu}} = \mu^{-1}(\tan\varphi) \equiv \text{const}. \quad (6.4)$$

Integration of this equation using the strain rate $\dot{\gamma} = dV/dz$, and adopting the no-slip boundary condition at the bottom: $V = 0$ at $z = 0$, yields the following expression for the velocity V:

$$\frac{V(z)}{\sqrt{gd}} = A(\varphi)\frac{h^{3/2} - (h-z)^{3/2}}{d^{3/2}}, \quad (6.5)$$

where $A(\varphi)$ is some dimensionless pre-factor. Note that for the chute flow geometry a similar expression (up to the pre-factor) can be obtained assuming Bagnold-type rheology (eqn 3.46). Indeed, from the conditions $\sigma_{xz} \sim \dot{\gamma}^2$ (see eqn 3.46), we readily obtain $\dot{\gamma} = dV/dz \sim \sqrt{h-z}$, which yields eqn 6.5.

Figures 6.2(a) and (b) display the velocity profiles for different inclinations and for thin and thick flows, respectively. In both cases, increasing the inclination φ increases the average shear rate and leads to more and more concave profiles. Detailed experimental and numerical studies confirm that for flow parameters (h,φ) far enough from the flow arrest curve $h_{\text{stop}}(\varphi)$ the velocity roughly obeys a Bagnold-like profile (eqn 6.5) (Pouliquen and Chevoir, 2002; Silbert et al., 2001).

The continuous lines in Figure 6.2(b) depict the velocity profiles obeying this rheology. The Bagnold profile (eqn 6.5) fits the numerical data in the core region but not at the base nor at the free surface, where the data exhibit a non-zero shear rate. These regions of discrepancies enlarge when the inclination angle φ decreases. Close to the flowing threshold, for thin layers and/or small inclinations, the velocity profile becomes more linear, see Fig. 6.2(a) (Silbert et al., 2003). The volume fraction profile $\phi(z)$ remains almost constant across the layer, except close to the free surface. This constant value appears to be independent of the flow thickness but decreases with the inclination.

Experiments and numerical simulations indeed support the existence of a local relation between the effective friction μ_{eff} and the scaled strain rate I for a variety of flows. In most experiments the effective friction μ_{eff} is a monotonously increasing function of I, however, it is limited by the "static" friction value μ_s for $I \to 0$ and the dynamic friction value μ_2 for high I, see Fig. 6.3.

Accordingly, Jop et al. (2006) proposed a useful phenomenological relation between μ_{eff} and I, which takes into account static friction,

$$\mu_{\text{eff}}(I) = \frac{\sigma_{xz}}{p} = \mu_s + \frac{\mu_2 - \mu_s}{1 + I_0/I} \qquad (6.6)$$

where $I_0 \approx 0.279$ is a constant, and parameters μ_s, μ_2 depend on the materials properties of grains. As shown in Fig. 6.1(e), this expression fits well the available experimental data.

6.1.3 Flow rule

As we discussed above, the inclined plane configuration gives information about the effective friction coefficient μ_{eff} between the flowing layer and the rough bottom. The stress distribution for steady uniform flows implies that μ_{eff} defined as the ratio between tangential and normal stress has to be equal to $\tan\varphi$. Thus, choosing an inclination for the plane is equivalent to imposing the effective friction. The flow adjusts its velocity so that the friction is equal to $\tan\varphi$. One can then deduce how the effective friction evolves with the velocity and the thickness by measuring the flow rule or the mean velocity $\langle V \rangle = h^{-1} \int_0^h V(z)dz$ of the granular layer as a function of the inclination φ and thickness h. Fig. 6.1(d) shows experimental and numerical measurements of the relation $\langle V \rangle(\varphi, h)$. The Froude number $\text{Fr} = \langle V \rangle/\sqrt{gh}$ is plotted versus the ratio $h/h_{\text{stop}}(\varphi)$. Each set of data collapses on a single curve, indicating that the influence of the inclination seems to be encoded in the function $h_{\text{stop}}(\varphi)$, while the data collapse is reasonable but not perfect. This correlation between the flow velocity and the granular layer thickness h is observed for different materials and different bottom coverage both in experiments and simulations. The following scaling relation approximately holds for all experiments

$$\mathrm{Fr} = \frac{\langle V \rangle}{\sqrt{gh}} = \alpha + \beta \frac{h}{h_{\mathrm{stop}}(\varphi)}. \tag{6.7}$$

Let us discuss this result. The Bagnold velocity profile (eqn 6.5) implies that $\mathrm{Fr} = \langle V \rangle/\sqrt{gh} = h^{-1}\int_0^h V(z)\mathrm{d}z/\sqrt{gh} \sim h$, i.e. we should anticipate a linear dependence of the Froude number Fr on the flow thickness h, at least for large enough h. However, the coefficients α and β are system dependent. For example, for experiments using glass beads on glued glass beads (IP5), α is close to zero, and $\beta \approx 0.2$, i.e. the glass beads comply with the Bagnold rheology relatively well. The same is observed in recent three-dimensional simulations using spheres (Jop et al., 2006). For rough sand particles one infers from the flow rule $\beta \approx -\alpha = O(1)$, see Fig. 6.1(d). Thus, for sand particles Bagnold rheology prediction is not accurate, at least in the vicinity of $h = h_{\mathrm{stop}}$.

The experimentally obtained flow rule can be used to extract the dependence of the effective friction μ_{eff} on the parameter I. Since for the chute flow $\mu_{\mathrm{eff}} = \tan\varphi$, the effective friction coefficient can be obtained by inverting relation 6.7 in order to express $\tan\varphi$ as a function of $\langle V \rangle$ and h. It is straightforward to show that according to eqn 6.7, μ_{eff} should be a function of a single parameter (Jop et al., 2006):

$$\mu_{\mathrm{eff}}(\langle V \rangle, h) = \mu_{\mathrm{eff}}\left(\frac{\langle V \rangle d}{h\sqrt{gh}} - \alpha\frac{d}{h}\right). \tag{6.8}$$

Figure 6.1(e) shows the effective friction coefficient obtained by this procedure for two different materials. The continuous lines are μ_{eff} deduced from eqn 6.7 using fits of $h_{\mathrm{stop}}(\varphi)$. In both cases, the effective friction coefficient increases with the shear rate $\dot{\gamma}$. Once again, it is important to note that this relation is not valid for thicknesses close to the critical thickness $h = h_{\mathrm{stop}}$.

As we mentioned above, the flow rule (eqn 6.7) provides a reasonable but not perfect scaling of the experimental data for various flow thicknesses h and inclination angles φ. On the basis of experiments with glass beads and sand on sandpaper Borzsonyi and Ecke (2007) noticed that plotting the surface Froude number $\mathrm{Fr_s} = V_\mathrm{s}/\sqrt{gh}$ against $h/\tilde{h}_{\mathrm{stop}}$ yields a much better data collapse than the original flow rule (eqn 6.7), where V_s is the surface flow velocity and the modified stop height is of the form $\tilde{h}_{\mathrm{stop}} = h_{\mathrm{stop}}\tan^2\varphi/\tan^2\varphi_{\mathrm{stop}}$. This flow rule was suggested by Jenkins (2007) on the basis of a phenomenological modification of the hydrodynamic equations for dense flows of identical, frictionless, inelastic disks. This modification involves the incorporation of a length scale other than the particle diameter in the expression for the rate of collisional dissipation and thus allows for a qualitative estimation of h_{stop}. However, the direct comparison between the flow rule (eqn 6.7) and the modified flow rule suggested by Jenkins (2007) and Borzsonyi and Ecke (2007) might be difficult because the experimental data correspond to different flow characteristics: the original flow rule (eqn 6.7) is defined for the *depth-averaged* flow velocity, whereas the experiment by Borzsonyi and Ecke (2007) operates with the *surface* flow velocity. Consequently, a difference in the vertical flow profile will contribute differently to these flow rules.

6.1.4 Effects of bottom roughness and materials properties of grains

The roughness of the bottom plane strongly influences the properties of a granular flow above it. A systematic study has been carried out by Goujon et al. (2003) who glued beads of diameter d_w and gradually changed the flowing bead diameter d. In the inset to Fig. 6.1(c), the deposit thickness h_{stop} is plotted versus the beads diameter ratio d/d_w for different inclinations. This work points out the existence of a given ratio d/d_w for which the deposit is maximal, which might correspond to a maximum of the effective bottom friction. This optimal ratio, independent of φ, is mainly determined by the surface fraction of glued beads on the bottom plane.

Parameters affecting particle–particle interactions, such as the interparticle friction, restitution coefficient, etc., also affect the flow properties. It is difficult to compare the roles of different parameters in physical experiments since changing grain material affects all of them at once. However, this information can be obtained in numerical simulations where one can independently vary the internal friction coefficient μ_p or the restitution coefficient e. Figure 6.1(f) shows the velocity profiles for the same φ and h but for different e. The interesting result is that in the range $0.4 < e < 0.8$, the profiles do not depend on e. This is in sharp contrast with what is observed in a kinetic regime dominated by binary collisions. The dependence of the velocity profile on the friction coefficient μ_p is also weak, as shown in Fig. 6.1(g).

6.2 Continuum models of gravity-driven granular flows

In this section we review various approaches to the continuum modelling of dense granular flows. We start with the simplest class of these models (so-called depth-averaged hydrodynamic or Saint-Venant models) which is usually applied to shallow granular layers in engineering practice. In deep granular layers, the flow is often confined to a thin near-surface layer, and in that case the Saint-Venant approach can be generalized into a two-layer description that deals with a coupled set of equations for the thicknesses of the static and rolling (or flowing) phases of granular flow. At the end of this section we will introduce the "partial fluidization" theory in which no clear spatial separation between flowing and static grains is assumed.

6.2.1 Depth-averaged (Saint-Venant) models

In this section we present a rather detailed derivation of the Saint-Venant equations for granular flows. In the course of the derivation we will discuss necessary assumptions and difficulties related with this approach.

The Saint-Venant-type models are formally obtained by depth–averaging of granular hydrodynamic equations 3.11 and 3.12. The energy equation (eqn 3.13) is usually neglected, and the flow is considered incompressible, i.e. $\nabla \cdot \mathbf{v} = 0$. The coordinate system is chosen so the x-axis is directed down the slope (with the "mean" angle of the bottom φ_0; we reserve the angle φ for the *local* slope, since the bottom can generally be non-flat), and the z-axis points upward (see Fig. 6.4). The granular hydrodynamic equations need to be complemented by the boundary conditions at the static base $z = z_S(x)$ (which here is assumed to be rigid and inerodible), and the free surface of the flow $z = h(x,t)$. We denote the local thickness of the granular layer

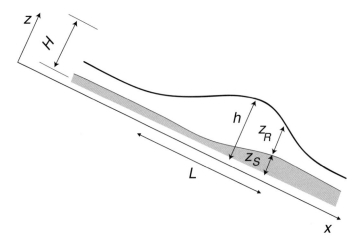

Fig. 6.4 Schematic representation of a finite–mass avalanche moving along a curved rigid surface. Parameters L and H characterize the size and the maximum height of the avalanche.

$z_R(x,t) = h(x,t) - z_S(x)$. For rigid and inerodible bottom, the normal velocity has to be zero. The tangential velocity at the bottom depends on the roughness of the substrate: for a rough bottom it is zero, while for a smooth bottom, the shear stress at the bottom is absent. The evolution equation for the free surface is obtained by the differentiation of the free surface position assuming the surface moves with the local hydrodynamic velocity $\mathbf{v}$ (the so-called kinematic boundary condition):

$$\partial_t h + v_x \partial_x h - v_z = 0. \tag{6.9}$$

In addition, one needs to impose the zero stress at the free surface

$$\sigma = 0. \tag{6.10}$$

Since the base is rigid and inerodible, the hydrodynamic velocity $\mathbf{v}$ is assumed to be tangential to the base, i.e. the normal velocity $v_n(x, z_S(x), t) = 0$. However, a number of additional simplifying and non-trivial assumptions are necessary to specify the components of the stress tensor at the base. For example, one can assume a solid friction law relating local shear traction $\mathbf{S}$ (force parallel to the basal surface) to the local friction coefficient μ and normal stress $\mathbf{N}$, $\mathbf{S} = \pm \mu \mathbf{N}$, where the sign is given by the direction of sliding velocity (Savage and Hutter, 1989).

The momentum conservation equation (eqn 3.12) needs to be integrated over $z_S(x) < z < h(x)$. In general, this procedure leads to rather complicated expressions. Further simplification is possible assuming that the typical longitudinal scale of the granular flow L is large compared to the typical height H, see Fig. 6.4. Introducing a small parameter $\epsilon = H/L$ and rescaling the length coordinates $x \to x/L$, $z \to z/H$, velocity $v_x \to v_x/\sqrt{gL}$, $v_z \to v_z/\sqrt{gL}/\epsilon$, time $t \to t\sqrt{g/L}$, and stress-tensor components: $\sigma_{ii} \to \sigma_{ii}/(\nu g \cos \varphi_0 H)$, $\sigma_{xz} \to \sigma_{xz}/(\nu g \sin \varphi_0 H)$, we recast eqn 3.12 in the following dimensionless form:

$$\partial_t v_x + v_x \partial_x v_x + v_z \partial_z v_x = \sin \varphi_0 (1 + \partial_z \sigma_{xz}) + \epsilon \cos \varphi_0 \partial_x \sigma_{xx} \tag{6.11}$$

$$\epsilon(\partial_t v_z + v_x \partial_x v_z + v_z \partial_z v_z) = -\cos\varphi_0(1 - \partial_z \sigma_{zz}) + \epsilon \cos\varphi_0 \partial_x \sigma_{xz}. \quad (6.12)$$

In the limit of small ϵ, i.e. in the "shallow-water" or long-wave approximation, eqn 6.12 in the lowest order is reduced to the simple hydrostatic equilibrium equation for the pressure $p = -\sigma_{zz}$,

$$p = -\sigma_{zz} = h(x,t) - z. \quad (6.13)$$

Now, eqn 6.11 has to be integrated over the interval $z_S(x) < z < h(x,t)$. After applying the kinematic condition 6.9 and using the incompressibility condition $\partial_x v_x + \partial_z v_z = 0$ and identity $v_z \partial_z v_x = \partial_z(v_x v_z) - v_x \partial_z v_z = \partial_z(v_x v_z) + v_x \partial_x v_x$, we obtain in the lowest order in ϵ

$$\partial_t \int_{z_S}^{h} v_x dz + \partial_x \int_{z_S}^{h} v_x^2 dz = \sin\varphi_0 [(h - z_S) + \sigma_{xz}(z = h) - \sigma_{xz}(z = z_S)]$$

$$+ \epsilon \cos\varphi_0 \left[\partial_x \int_{z_S}^{h} \sigma_{xx} dz - \sigma_{xx}(h) \partial_x h + \sigma_{xx}(z = z_S) \partial_x z_S \right]. \quad (6.14)$$

This equation can be further simplified assuming certain constitutive relations for stresses σ_{ij} and the velocity profile $v_x(z)$. Following Savage and Hutter (1989), we can assume the following phenomenological relation between stress components $\sigma_{xx} = k\sigma_{zz}$ with the coefficient $k \approx 1$ (for isotropic fluid $k = 1$). Taking into account the no-stress condition at the free surface $\sigma_{zz} = \sigma_{xx} = \sigma_{xz} = 0$ for $z = h(x,t)$, we obtain from eqn 6.14

$$\partial_t (\langle V \rangle z_R) + \partial_x (\langle V^2 \rangle z_R) = \sin\varphi_0 (z_R - \sigma_{xz}|_{z=z_S})$$
$$- \epsilon k \cos\varphi_0 [\partial_x (\langle p \rangle z_R) - \sigma_{xx}(z = z_S) \partial_x z_S], \quad (6.15)$$

where the following depth-averaged quantities are introduced:

$$\langle V \rangle = z_R^{-1} \int_{z_S}^{h} v_x dz; \quad \langle V^2 \rangle = z_R^{-1} \int_{z_S}^{h} v_x^2 dz; \quad \langle p \rangle = -z_R^{-1} \int_{z_S}^{h} \sigma_{zz} dz \quad (6.16)$$

The depth-averaged value of $\langle V^2 \rangle$ can be approximately expressed through the value of $\langle V \rangle$ assuming a certain velocity profile, $\langle V^2 \rangle \approx \alpha_1 \langle V \rangle^2$. For a parabolic velocity profile one obtains $\alpha_1 = 6/5$ (Savage and Hutter, 1989). For a Bagnold-type profile typically observed in dense granular flows (eqn 6.5), one obtains a slightly higher value, $\alpha_1 = 5/4$. For a plug flow $v_x(z) = \text{const}$, and so $\alpha_1 = 1$. Thus, the particular shape of the flow velocity profile has a relatively small effect on the relation between $\langle V^2 \rangle$ and $\langle V \rangle$.

Integration of the continuity condition $\nabla \cdot \mathbf{v} = 0$ using the kinematic condition at the free surface (eqn 6.9) after simple algebraic transformations gives rise to the equation for the evolution of the flow depth z_R:

$$\partial_t z_R + \partial_x (\langle V \rangle z_R) = 0. \quad (6.17)$$

In order to close the description one needs to supplement eqns 6.15 and 6.17 with the relation between the shear stress component σ_{xz} at the bottom and the average velocity $\langle V \rangle$, flow depth z_R, and mean inclination angle φ_0. That task is rather non-trivial as it involves the approximation of a complicated and poorly understood rheology of dense granular flows. Savage and Hutter (1989) proposed a relation between σ_{xz} and the hydrostatic pressure $p = -\sigma_{zz}$ based on the Mohr–Coulomb yield criterion (see Chapter 3). Assuming a local friction law between normal and shear components of the stress, one arrives at the following relation for the stress σ_{xz} at the base $z = z_S(x)$:

$$\sigma_{xz} = \text{sign}(\langle V \rangle) p \frac{\mu}{\tan \varphi_0} + O(\epsilon^2). \tag{6.18}$$

Here, we need to emphasize that while such a simple constitutive law is attractive, it is insufficient to describe a number of important features of the dense granular flows, such as hysteresis expressed in the existence of two different repose angles, the role of the boundary conditions at the bottom, etc. In subsequent work, there have been a number of attempts to modify and generalize this expression to accommodate the effects noted above, at least on the phenomenological level.

Combining now eqns 6.15 and 6.18 and returning to the original scaling of variables x, h, t and v, we obtain the following equation for the depth-averaged hydrodynamic velocity (for the sake of simplicity we set $\alpha_1 = 1$):

$$\partial_t \langle V \rangle + \langle V \rangle \partial_x \langle V \rangle = g \left(\sin \varphi_0 - \mu \cos \varphi_0 \text{sign}(\langle V \rangle) \right) - gk \cos \varphi_0 \partial_x z_R. \tag{6.19}$$

While the Saint-Venant equations 6.17 and 6.19 are much simpler than the full set of hydrodynamic equations, their analytical solution remains challenging. Savage and Hutter (1989) succeeded in finding only some narrow classes of symmetric self-similar localized solutions propagating along a planar bottom, a parabolic cup, $h \sim L_0(t)^2 - \xi^2$, and M-wave, $h \sim a_0 + \zeta^2 - L_0(t)^2$, $a_0 = \text{cost} > 0$. These solutions are defined in the interval $|\xi| \leq L_0(t)$, while $h = 0$ for $|\xi| > L_0(t)$, where $\xi = x - v_0 t$ is the similarity variable, $L_0(t)$ is the time-dependent half-width of the solution, and $v_0 = g(\sin \varphi_0 - \mu \cos \varphi_0)$ is the translation velocity. The numerical analysis of eqns 6.17 and 6.19 is complicated by the singular (discontinuous) behavior of the velocity and its derivatives at the front and rear edges of the localized solution, see Fig. 6.4. Various regularization and modification schemes to remedy this problem were proposed in subsequent publications, see, e.g. (Douady et al., 1999; Khakhar et al., 2001; Borzsonyi et al., 2005).

6.2.2 Two-phase models of surface granular flows

As we mentioned above, many granular flows are localized in a narrow near-surface zone. It is therefore tempting to apply the ideas of the Saint-Venant approach by applying it to a layer of "rolling" grains above the layer of static grains. However, one has to take into account that the bottom of this thin layer of rolling grains is no longer fixed by externally imposed boundary conditions. One has to take into consideration the erosion/deposition processes occurring at this interface. This "two-phase" approach was pioneered by Bouchaud et al. (1994) (and therefore often called the BCRE model) and subsequently developed in a number of publications; see, e.g. (Bouchaud et al., 1995; Boutreux et al., 1998; Douady et al., 1999; Mehta, 1994; Khakhar et al., 2001). For a recent review on these models of near-surface flows see Aradian et al. (2002). All these models distinguish rolling and static phases of granular flow and operate with the set of coupled equations for the thicknesses of both phases, z_R (rolling) and z_S (static), respectively, see Fig. 6.4, where now z_S is a function of horizontal coordinate x *and* time t.

In the most compact form these equations can be written as

$$\partial_t z_S = -\Xi(z_S, z_R) \qquad (6.20)$$
$$\partial_t z_R = V_d \partial_x z_R, +\Xi(z_S, z_R) \qquad (6.21)$$

where $\Xi(z_S, z_R)$ is the *exchange term*, or a conversion (erosion) rate between rolling and static grains, and V_d is the downhill velocity of rolling grains. The physical meaning of the BCRE equations is the mass conservation of the incompressible granular fluid: eqn 6.20 expresses the change of the local thickness of the static layer due to erosion/deposition processes between static and rolling grains, and eqn 6.21 in addition to the inverse of this process describes the downhill advection of the rolling fraction by the flow with velocity V_d.

Bouchaud et al. (1994) suggested the following phenomenological expression for the conversion rate Ξ as a function of the model variables z_S, z_R:

$$\Xi(z_S, z_R) = \gamma z_R (\tan \varphi - \tan \varphi_c). \qquad (6.22)$$

Note that here φ is the *local* slope of the static layer. This expression has a simple physical interpretation. Constant γ is the typical collision rate of rolling grains with the static phase. The total number of collisions experienced by the static phase is proportional to the thickness of the rolling fraction z_R, i.e. $\Xi \propto \gamma z_R$. On the other hand, the probability of the dislodgement of static grains or trapping of rolling grains in the static phase should depend on the local slope angle φ. Bouchaud et al. (1994) further assumed that the probability of the dislodgement equals the probability of trapping when the slope angle equals a certain *neutral slope angle* φ_c. Thus, when $\varphi > \varphi_c$ there is a net erosion of the static phase and vice versa. For the slope angle close to φ_c one can use a linear approximation in the form of eqn 6.22. Thus, eqns 6.21 and 6.21 for z_S and z_R become coupled.

Let us now briefly discuss simple solutions to the BCRE model. First, we consider spatially uniform solutions (no x dependence). From eqns 6.20–6.22 we obtain for the width of the rolling phase z_R:

$$\partial_t z_R = \gamma z_R (\tan \varphi - \tan \varphi_c). \qquad (6.23)$$

The solution to this equation for all supercritical slopes ($\tan \varphi > \tan \varphi_c$) describes unlimited exponential growth of the rolling phase depth z_R, i.e. unlimited erosion. This obviously unphysical effect is an artifact of the assumption in eqn 6.22 that the conversion rate $\Xi(z_S, z_R)$ is proportional to z_R. The issue of unlimited erosion will be addressed in the next section where we introduce a more elaborate version of the BCRE model.

Now we consider evolution of small perturbations $\tilde{z}_S, \tilde{z}_R$ with respect to the steady-state flow at the neutral slope angle, $\partial_x z_S = \tan \varphi = \tan \varphi_c, z_R = z_0 = $ const. The linearized BCRE equations are then of the form:

$$\partial_t \tilde{z}_S = -\gamma z_0 \partial_x \tilde{z}_S \qquad (6.24)$$

$$\partial_t \tilde{z}_R = V_d \partial_x \tilde{z}_R + \gamma z_0 \partial_x \tilde{z}_S. \tag{6.25}$$

These equations describe two wave modes: the wave propagating uphill with the velocity $V_{up} = \gamma z_0$ (see eqn 6.24), and a downhill wave with the velocity V_d, eqn 6.25. The existence of the uphill wave is the result of erosion/deposition kinetics in the BCRE model, while the downhill wave is caused by simple advection due to gravity.

6.2.3 Generalized two-phase models

The BCRE model has a clear physical interpretation and is convenient for analytical and numerical studies, but it does not account for a number of complexities of real near-surface granular flows. One important effect ignored by the simple form of the conversion rate (eqn 6.22) is the screening of the erosion/deposition process in thicker granular layers. In flowing layers thicker than a few grain diameters, grains from the upper layers of the flow have no opportunity to collide with the static layer because they are "screened" by the lower flowing grains. Thus, the number of collisions and the exchange rate should not be simply proportional to z_R when the thickness of the flowing layer z_R exceeds a certain "screening" length z_m. Following this argument Boutreux et al. (1998) suggested that the exchange rate should saturate for thicker flows, i.e.

$$\Xi(z_S, z_R) \to \gamma z_m (\tan \varphi - \tan \varphi_c) \text{ for } z_R > z_m. \tag{6.26}$$

With the above modification the BCRE equations assume the form

$$\begin{aligned} \partial_t z_S &= -V_{up}(z_R)(\tan \varphi - \tan \varphi_c) \\ \partial_t z_R &= V_d \partial_x z_R + V_{up}(z_R)(\tan \varphi - \tan \varphi_c), \end{aligned} \tag{6.27}$$

where the uphill velocity V_{up} has the following dependence on the flow thickness: $V_{up} \to \gamma z_R$ for $z_R \ll z_m$ and $V_{up} \approx \gamma z_m$ for $z_R > z_m$ (compare with eqn 6.23). The modified set of equations 6.27 has certain advantages compared to the original formulation of the BCRE model. In particular, the above-mentioned screening effect regularizes the unlimited growth of the erosion rate present in the original model illustrated by eqn 6.23. Moreover, for larger z_R the set of equations becomes linear in both z_R and z_S, and the general solution can be formulated in terms of uphill and downhill waves for arbitrary slope angles $\tan \varphi > \tan \varphi_c$ (as opposed to the unique neutral slope $\tan \varphi_c$ in the original BCRE model).

Another serious shortcoming of the BCRE model is the assumption of a single critical angle φ_c at which the conversion rate changes sign. This in turn translates into a single repose angle for the granular slope. Thus, the BCRE model in the form of eqn 6.27 cannot account for the bistability and the hysteresis observed near the onset of flow. Yet another limitation of the BCRE model is related to the implementation of the momentum conservation law, in particular, the assumption that the downhill velocity V_d does not depend on z_R and the slope angle φ (Aradian et al., 2002).

In order to incorporate the bistability of surface granular flows and a non-trivial dependence of the flow velocity on the flow thickness, Douady et al. (2002) suggested the following two-phase model to describe avalanches in thin granular layers:

$$\partial_t z_R + 2 \langle V \rangle \partial_x z_R = \frac{g}{\Gamma} (\tan \varphi - \mu(z_R)) \tag{6.28}$$

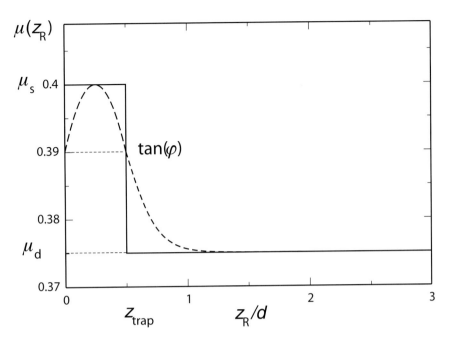

Fig. 6.5 Dependence of the effective friction coefficient μ on the depth of the flowing layer z_R. Below a "trapping" height z_{trap}, which is of the order of a grain diameter, the friction coefficient $\mu(z_R)$ changes from the dynamic value μ_d to the static value μ_s.

$$\partial_t h + 2\langle V \rangle \partial_x z_R = 0, \qquad (6.29)$$

where $h = z_S + z_R$ is the position of the free surface, $\bar{\Gamma}$ is the mean velocity gradient in the flowing layer, and $\langle V \rangle = 1/2\bar{\Gamma} z_R$ is the depth-averaged velocity of the flow. Equations 6.28 and 6.29 include two phenomenological functions, $\bar{\Gamma}$ and $\mu(z_R)$, that describe depth-dependent friction with the static layer. The schematic dependence of the friction coefficient $\mu(z_R)$ is shown in Fig. 6.5. As one sees from the figure, the friction coefficient $\mu(z_R)$ exchanges its value from the dynamic value μ_d to the static value μ_s below the "trapping" depth of the order of a grain diameter. Thus, for z_R in the range $0 < z_R < z_{trap}$ each value of the friction $\mu(z_R)$ corresponds to two values of the flow depth z_R.

For the mean velocity gradient $\bar{\Gamma}$ the following phenomenological expression was proposed:

$$\bar{\Gamma} = \frac{\gamma_0 \sqrt{g/d}}{\log\left((\mu_\infty - \mu_d)/(\tan\varphi - \mu_d)\right)}. \qquad (6.30)$$

Here, and $\gamma_0 \approx 0.4$ and $\mu_\infty \approx 0.72$. Equation 6.30 was motivated by the analogy between the flow of rolling grains over the static layer and the motion of a single grain down a rough inclined plane. This analogy can be justified at least in the dilute limit when collisions of rolling grains with the underlying static layer can be regarded as independent. In the spirit of this analogy, μ_∞ can be interpreted as the critical slope above which the grain begins unlimited accelerated motion. The values of constants γ_0, μ_∞ can be obtained by fitting to the experimental data (Douady et al., 2002). Modified in this way, the BCRE model provides a much better agreement with experiments (see below).

6.2.4 Partial fluidization theory

While the depth-averaged and two-phase description of granular flow are simple and intuitive, they can be problematic when there is no clear separation between rolling and static phases, especially near the onset of motion. Often, grains can be static and involved in rolling motion intermittently. Within the same mesoscopic volume of granular material, stress can be transmitted both through short collisions and long-lived quasistatic contacts. In this situation, a more general approach is needed. Since the transition from the static to fluidized phases possesses many features of a first-order transition, an analogy with the classical Ginzburg–Landau theory of phase transitions (see, e.g. Landau and Lifshitz (1980)) can be exploited. A continuous distribution of grains between flowing and static "extremes" can be characterized by introducing a "phase" variable characterizing the phase state of the flow or the degree of fluidization. This phase field (which we call the *order parameter*) is assumed to have its own dynamics controlled not by the temperature (as, e.g., in classical melting–freezing or magnetization transitions) but by the local distribution of stresses inside granular matter. This extended theory of dense granular flows can be applied both in static and flowing phases, as well as in the transition zone, and therefore it can address the situations not captured by two-phase models, e.g. the onset of motion and flow arrest.

The order parameter ρ is introduced in such a way that in granular solid $\rho = 1$ and in a well-developed flow (granular liquid) $\rho \to 0$. On the "microscopic level" the order parameter can be defined as a fraction of the number of static (or persistent) contacts among particles Z_s to the total number of the contacts Z, $\rho = \langle Z_s/Z \rangle$ within a mesoscopic volume that is large with respect to the particle size but small compared with the characteristic size of the flow.[1] The order parameter defined in such a way is difficult to measure experimentally but it can be extracted from soft-particle molecular dynamics simulations (Volfson et al., 2003b; Volfson et al., 2003a).

Due to a strong dissipation in dense granular flows the order parameter ρ is assumed to obey purely relaxational dynamics controlled by the Ginzburg–Landau-type equation for a generic first-order phase transition,

$$\tau \left(\partial_t \rho + \mathbf{v} \nabla \rho \right) = l^2 \nabla^2 \rho - \partial_\rho F(\rho, \delta). \tag{6.31}$$

Here, τ is some characteristic time scale of the problem, and l is the order parameter "correlation length", which is typically of the grain size. These parameters can be scaled away by renormalization of length, time and velocity,

$$t_{\text{new}} = t/\tau, \mathbf{r}_{\text{new}} = \mathbf{r}/l, \mathbf{v}_{\text{new}} = \mathbf{v}\tau/l. \tag{6.32}$$

---

[1] As usual, the definition of the order parameter is not unique. An alternative candidate for the order parameter could be, for example, the ratio of local elastic energy E_{el} of the grains to the total energy $E_{\text{tot}} = E_{\text{el}} + E_{\text{kin}}$, $\rho = \langle E_{\text{el}}/E_{\text{tot}} \rangle$, where E_{kin} is local the kinetic energy. The expectation is however that all *relevant* order parameters would exhibit similar dynamic properties and scale in the same way near a transition.

In the following, we drop the subscript *new*.

The local dynamics of the order parameter is governed by the *free energy* $\mathcal{F}(\rho, \delta)$. The free energy depends on the *control parameter* δ that in turn is determined by the local stress tensor. The simplest assumption consistent with the Mohr–Coulomb yield criterion (Section 3.5.2) is to take δ to be a function of the local ratio of shear to normal stresses, i.e. the effective friction coefficient $\mu = \max |\sigma_{mn}/\sigma_{nn}|$, where the maximum is sought over all possible orthogonal directions m and n locally in space. Furthermore, there are two angles that characterize the fluidization transition in the bulk of granular material, the "dynamic angle of repose" $\varphi_0 = \tan^{-1} \mu_0$ such that at $\varphi < \varphi_0$, the "dynamic" phase is unstable, and the internal friction angle $\varphi_1 = \tan^{-1} \mu_1$ such that if $\varphi > \varphi_1$ the static equilibrium is unstable. Values of φ_0 and φ_1 depend on microscopic properties of the granular material, and in general they do not coincide. Thus, the free-energy density should have such a functional form that for $\varphi < \varphi_0$ it has only one minimum at $\rho = 1$ (static phase), at $\varphi > \varphi_1$ it has a single minimum at $\rho = 0$ (fluidized phase), and at $\varphi_0 < \varphi < \varphi_1$ there are two coexisting minima corresponding to a bistable regime. The simplest form of the free energy $\mathcal{F}$ satisfying the above condition is the quartic polynomial of ρ,

$$\mathcal{F}(\rho) = \int^{\rho} \rho(\rho - 1)(\rho - \delta) d\rho, \tag{6.33}$$

with an algebraic form of the control parameter δ,

$$\delta = (\tan^2 \varphi - \tan^2 \varphi_0)/(\tan^2 \varphi_1 - \tan^2 \varphi_0). \tag{6.34}$$

(it is quadratic in $\tan \varphi$ since the sign of the slope is irrelevant). By this choice, the interval $\varphi_0 < \varphi < \varphi_1$ is mapped on the unit interval $0 < \delta < 1$. Within this interval, eqn 6.31 has two stable uniform solutions $\rho = 0$ and 1 corresponding to the fluid and solid states, and one unstable solution $\rho = \delta$, see Fig. 6.6. Of course, the choice of the free-energy function 6.33 is rather arbitrary and was originally (Aranson and Tsimring, 2001; Aranson and Tsimring, 2002) dictated by simplicity and convenience of analysis. In subsequent studies the specific form of the free energy was extracted from two-dimensional soft-particle molecular dynamics simulations (Volfson et al., 2003b; Volfson et al., 2003a).

In addition to the new order-parameter equation, the dynamics of the granular matter has to conform to the usual mass, momentum, and energy conservation laws that are expressed in the form of granular hydrodynamics eqns 3.11–3.13. Since we are mostly interested in the dense flow regime, the density of the granular matter ν is assumed to be everywhere close to the random close-packed limit, $\nu = \nu_c$, and the granular matter can be considered incompressible. Therefore, the mass conservation equation is reduced to $\nabla \cdot \mathbf{v} = 0$. Furthermore, for simplicity it is assumed that the "granular temperature" T of the granulate does not have a direct impact on the dynamics (effectively, it is slaved to the order parameter near the transition), and so the energy equation (eqn 3.13) is also neglected.

The coupling between the momentum conservation equation (eqn 3.12) and the order parameter equation (eqn 6.31) is achieved through the constitutive relation that

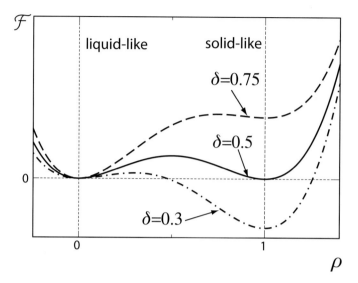

Fig. 6.6 A sketch illustrating "free energy" density $\mathcal{F}$ as a function of the order parameter ρ for three values of the control parameter δ.

relates the stress distribution and the order parameter. We separate the stress tensor σ into two parts, a static (rate-independent) part σ^s, and a fluid part σ^f.[2]

The fluid part of the stress σ^f is taken in a purely Newtonian form

$$\sigma^f_{\alpha\beta} = -p_f \delta_{\alpha\beta} - \eta_f \dot{\gamma}_{\alpha\beta}, \tag{6.35}$$

where p_f is the "partial" fluid pressure, η_f is the viscosity coefficient associated with the fluid stress tensor that is *different* from the shear viscosity η introduced for the full stress tensor in the hydrodynamic description, see eqn 3.14. As we shall see in the following, unlike η, η_f does *not* diverge as $\nu \to \nu_c$ (compare with eqn 3.17). The viscosity of the fluid phase can also be a function of the pressure. We discuss this rather non-trivial issue later in this chapter.

The main idea of the partial fluidization theory is that the fluid part of the off-diagonal components of the stress tensor is proportional to the off-diagonal components of the full stress tensor with the proportionality coefficient q being a function of the order parameter ρ,

$$\sigma^f_{xz} = q(\rho)\sigma_{xz}, \tag{6.36}$$

and the same for σ^f_{zx}.

Equation 6.36 stipulates the equation for the static part of the off-diagonal stress component,

$$\sigma^s_{xz} = (1 - q(\rho))\sigma_{xz}. \tag{6.37}$$

---

[2] For the case of a free-surface granular flow down an incline the representation of total stress as the linear sum of a rate-independent, dry friction part plus a strain-rate-dependent "viscous" part was suggested by Savage (1983), see also Savage (1998).

Both fluid and solid parts of the stress tensor are assumed symmetric, $\sigma_{xz}^{f,s} = \sigma_{zx}^{f,s}$. This assumption is confirmed by the two-dimensional soft-particles molecular dynamics simulations (Volfson et al., 2003b; Volfson et al., 2003a). Since we defined the order parameter such that it has a fixed range between $\rho = 0$ in a completely fluidized state and $\rho = 1$ in a completely static regime, the function $q(\rho)$ has the property $q(0) = 1$, $q(1) = 0$. Correspondingly, Aranson and Tsimring (2001), Aranson and Tsimring (2002) made the simplest choice $\eta_f = $ const and $q(\rho) = 1 - \rho$. However, as we will see below, the molecular dynamics simulations lead to slightly different expressions for the function $q(\rho)$ and the fluid phase viscosity η_f.

Similar relationships can be postulated for the diagonal terms of the stress tensor,

$$\sigma_{xx}^f = q_x(\rho)\sigma_{xx}, \quad \sigma_{zz}^f = q_z(\rho)\sigma_{zz}, \tag{6.38}$$
$$\sigma_{xx}^s = (1 - q_x(\rho))\sigma_{xx}, \quad \sigma_{zz}^s = (1 - q_z(\rho))\sigma_{zz}, \tag{6.39}$$

where the functions $q_{x,z}(\rho)$ can in general differ from $q(\rho)$.

Combining eqns 6.35–6.39, we obtain the constitutive relation in the closed form,

$$\sigma_{\alpha\beta} = p_f \delta_{\alpha\beta}/q_\alpha(\rho) - \eta_f \dot{\gamma}_{\alpha\beta}/q(\rho), \quad \alpha, \beta \in \{x, z\}. \tag{6.40}$$

Finally, to close the system, we have to specify the boundary conditions. In addition to the standard boundary conditions for the granular velocity, we need boundary conditions for the order parameter. While this is a complicated issue in general, a simple but reasonable choice is to take no-flux boundary conditions at free surfaces and smooth walls, and a solid phase condition $\rho = 1$ near sticky or rough walls.[3]

6.2.5 Validation of partial fluidization theory in MD simulations

In order to specify functions $q(\rho)$ and $q_{x,z}(\rho)$, the shape of the free-energy density $F(\rho)$ and the dependence of the fluid phase viscosity η_f on pressure p, molecular dynamics simulations were performed in a small two-dimensional shear cell (Volfson et al., 2003b; Volfson et al., 2003a). About 500 slightly polydisperse grains were placed between two horizontal rough walls (periodic boundary conditions were assumed at the vertical "walls") and sheared by applying opposite forces along the horizontal x-direction to the walls without gravity, but with varied external pressure p applied in the vertical y-direction). At each value of the shear force the system was allowed to equilibrate, and the components of the shear stress and the order parameter were averaged over the whole system. This procedure was performed for multiple values of the external pressure p, and yielded the following approximations for $q(\rho)$ and $q_{x,z}(\rho)$

$$q(\rho) \approx (1-\rho)^{2.5}$$
$$q_x(\rho) \approx (1-\rho)^{1.9}, \quad q_z(\rho) \approx (1-\rho^{1.2})^{1.9}. \tag{6.41}$$

The functions $q(\rho)$ and $q_{x,z}(\rho)$ for different values of P are illustrated by Fig. 6.7, and indeed an excellent collapse of all the data is observed. Remarkably, these curves are not significantly different from a simple linear dependence $q = 1 - \rho$ used in the original version of the model (Aranson and Tsimring, 2001; Aranson and Tsimring, 2002).

---

[3]The generalized, frame-independent three-dimensional representation of partial fluidization theory can be found in Gao et al. (2005).

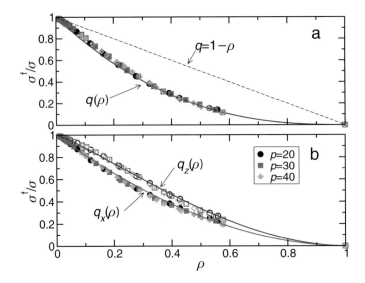

Fig. 6.7 Ratios of the fluid stress components to the corresponding full stress components $\sigma^f_{\alpha\beta}/\sigma_{\alpha\beta}$ vs. ρ for three different pressures p: a–shear stress components, closed symbols σ_{zx}, open symbols σ_{xz}, line is a fit $q(\rho) = (1-\rho)^{2.5}$. b–normal stress components, closed symbols σ_{xx}, open symbols σ_{zz}, lines are the fits $q_x(\rho) = (1-\rho)^{1.9}$, $q_z(\rho) = (1-\rho^{1.2})^{1.9}$. Dashed line indicates function $q(\rho) = 1 - \rho$.

The shape of the free-energy density $\mathcal{F}(\rho, \delta)$ in eqn 6.31 can be estimated from the same simulations of a shear flow in the two-dimensional planar cell geometry (Volfson et al., 2003b). The simulations can be used to find the zeroes of free-energy density derivative $\partial_\rho \mathcal{F}(\rho, \delta) = (\rho - 1)f(\rho, \delta) = 0$. The corresponding values of the order parameter ρ, i.e. fixed points of eqn 6.31 are shown as a function of δ in Fig. 6.8. The overall behavior can be summarized as follows: ramping up the normalized shear rate δ results in the abrupt jump of the order parameter from $\rho = 1$ (static regime) to a small value $\rho \approx 0.2$ (the onset of flow) at $\delta \approx 0.3$ and further decrease of ρ towards zero with the increase of δ. Ramping down the value of δ from a well-fluidized state results in a gradual increase of ρ and a consequent jump from $\rho = \rho_* \approx 0.6$ to $\rho = 1$ (flow arrest) at $\delta \approx 0.25$ (hysteresis). A good fit to the function $f(\rho, \delta)$ is given by the following expressions:

$$f(\rho, \delta) = \begin{cases} \rho^2 - 2\rho\rho_* + \rho_*^2(1-\zeta) & \text{for } \zeta < 1 \\ f = \rho^2 - \rho\rho_*(1+\sqrt{\zeta}) & \text{for } \zeta > 1, \end{cases} \quad (6.42)$$

where $\zeta = A(\delta^2 - \delta_0^2)$ with $A = 15, \delta_0 = 0.25, \rho_* = 0.6$.

The rationale for the expression 6.42 is the following: equation $f(\rho, \delta) = 0$ has two roots $\rho_{1,2} = \rho_*(1 \pm \sqrt{\zeta})$ for $\zeta < 1$ and $\rho_1 = \rho_*(1 + \sqrt{\zeta})$, $\rho_2 = 0$ for $\zeta > 1$. Thus, one sees that one of the roots, $\rho_1 = \rho_*(1 + \sqrt{\zeta})$, remains the same for all ζ. The other root, $\rho_2 = \rho_*(1 - \sqrt{\zeta})$ becomes identical zero for $\zeta > 1$. It prevents unphysical negative values of the order parameter for large values of δ.

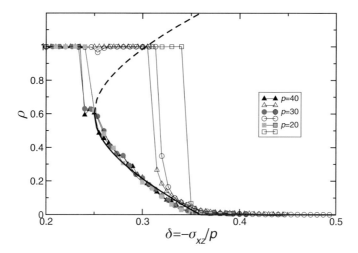

Fig. 6.8 Bifurcation diagram for a planar shear cell showing the order parameter ρ in the steady-state regime vs. control parameter δ obtained from simulations for different values of applied pressure p, reproduced from Volfson et al. (2003b). Open symbols correspond to ramping up external driving force, and closed symbols correspond to ramping down. The thick line shows the fit to eqn 6.42 (its solid part indicates the stable branch, and the dashed part the unstable branch).

In the context of viscosity of the fluid phase, the simulations suggest the following scaling of the fluid part of the stress tensor σ_{xz}^{f} vs. shear strain rate $\dot{\gamma}$ for sufficiently large $\dot{\gamma}$:

$$\sigma_{xz}^{f} = a_0 \sqrt{p}\dot{\gamma}, \tag{6.43}$$

where constant $a_0 \approx 2$ depends on the materials properties such as restitution coefficient, etc. This implies that the fluid component of the partially fluidized granular flow exhibits a Newtonian rheology with the fluid phase viscosity $\eta_f \sim \sqrt{p}$. Remarkably, expression 6.43 has a direct connection with the rheology of dense granular flows discussed in Section 6.1.2. Indeed, the corresponding "fluid friction coefficient" is $\sigma_{xz}^{f}/p = a_0 \dot{\gamma}/\sqrt{p} = a_0 I$, where I is the same inertial parameter as in eqn 6.2. Therefore, well-fluidized flows (small ρ) would show the same rheology as discussed in previous sections, including Bagnold-type flow profiles and a similar flow rule. For very small shear rates, however, due to grain compressibility, our simulations show deviations from the scaling given by eqn 6.43, $\sigma_{xz}^{f} \sim p^{1/4}\dot{\gamma}$.

6.2.6 Thin granular layers: a partial fluidization approach

When a layer of granular material flows over a rough topography, the grains close to the bottom are trapped by the topography and remain largely immobile. In this case, one cannot assume that the layer is uniformly fluidized, and therefore even if the layer is thin, the Saint-Venant approach becomes more difficult to implement. However, the problem can be treated within the framework of the partial fluidization theory outlined above. In this section we focus on the dynamics of thin granular layers on

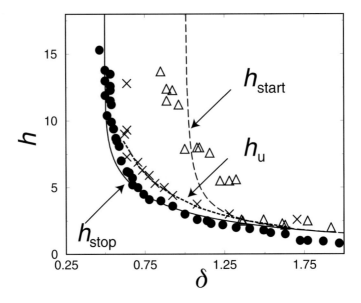

Fig. 6.9 Comparison of theoretical and experimental phase diagrams. Lines are computed using the partial fluidization theory, symbols depict rescaled experimental data from Daerr and Douady (1999). Solid line and ● indicate the range of existence of avalanches and steady-state flows (curve h_{stop}); dashed line and △ (curve h_{start}) correspond to the linear stability boundary of the static granular layer of constant depth; short-dashed line and × denote the boundary between wedge and balloon avalanches (curve h_c) for parameter values $\beta = 3.15, \alpha = 0.025$ of eqns 6.58 and 6.59. Experiential data are scaled by the typical grain size $d \approx 200$ μm, from Aranson and Tsimring (2001), Aranson and Tsimring (2002).

rough inclined planes adopting the simplified formulation of the partial fluidization theory, namely with the linear constitutive relation for stresses ($q(\rho) = 1 - \rho$), fixed viscosity of the fluid phase η_f and the quartic form of the free-energy density (eqn 6.33). Generalization of this model will be discussed later in this chapter.

We consider an initially flat layer of grains of thickness h on a sticky flat surface tilted by angle φ to the horizon. We again introduce a Cartesian coordinate frame aligned with the inclined plane, with the z-axis normal to the plane, and the x-axis oriented downhill. For thin granular layers on inclined planes, the momentum balance equation (eqn 3.12) and the order–parameter evolution equation (eqn 6.31) can be significantly simplified.

In the case of the stationary shear flow in a flat layer, the force balance conditions yields

$$\partial_z \sigma_{zz} + \partial_x \sigma_{xz} = -g \cos \varphi, \quad \partial_z \sigma_{xz} + \partial_x \sigma_{xx} = g \sin \varphi. \tag{6.44}$$

The solution to eqns 6.44 in the absence of lateral stresses $\sigma_{zz} = \sigma_{zx} = \sigma_{xz} = 0$, is given by eqn 6.3.

First, we can examine the stability of a stationary layer of uniform thickness h on the inclined plane with the slope $\tan\varphi$. Since the static granular layer corresponds to the value of the order parameter $\rho = 1$, the linearized solution to eqn 6.31 can be sought in the following form

$$\rho = 1 - \sum_{n=1}^{\infty} A_n \sin\left(\frac{(2n-1)\pi z}{2h}\right) e^{\lambda_n t}, \qquad (6.45)$$

where $n = 1, 2...$ is the vertical "mode number" and A_n, λ_n are the amplitudes and growth rates of the corresponding perturbations. The solution in this form satisfies the boundary condition $\rho = 1$ at the bottom $z = 0$ and the zero-flux boundary condition $\partial_z \rho = 0$ at the free surface $z = h$. Substitution of the solution 6.45 into eqn 6.31 using the expression for free-energy density (eqn 6.33) yields the following expression for the growth rates

$$\lambda_n = \delta - 1 - \frac{(2n-1)^2 \pi^2}{4h^2}. \qquad (6.46)$$

The lowest-order perturbation ($n = 1$) has the largest growth rate

$$\lambda_1 = \delta - 1 - \pi^2/4h^2, \qquad (6.47)$$

and, therefore, is the most "dangerous" one. The neutral curve $\lambda_1 = 0$ gives the relation between the control parameter δ and the critical layer thickness h

$$h_{\text{start}} = \frac{\pi}{2\sqrt{\delta - 1}}. \qquad (6.48)$$

As follows from eqn 6.48 the layer is stable with respect to the small perturbations below $\delta < 1$, which was interpreted above as corresponding to the *static angle of repose*. Remarkably, from eqn 6.48 it follows that the layer can be stable even for slopes above the static repose angle, as long as the thickness h is below the critical thickness h_{start}. This stabilization is due to the proximity of the sticky bottom that favors the solid phase $\rho = 1$. Above the neutral curve the static layer is linearly unstable. The stability condition (eqn 6.48) is shown by the dashed line in Fig. 6.9.

The linear stability does not guarantee, however, the stability with respect to finite–amplitude perturbations. This additional non-linear stability criterion can be obtained from the stationary flow existence condition, i.e. the solutions with $\rho(z) \neq 1$ and non-zero flow velocity $v_x \neq 0$. This flow regime exists only if the thickness of the layers h exceeds a certain threshold value. This threshold thickness of the stationary flow can be naturally interpreted as the "stop height" h_{stop} introduced by Pouliquen (1999) for the chute flow experiments, see Section 6.1.1.

The value of h_{stop} can be computed in the following way. The stationary equation 6.31 can be integrated once after multiplying it by $\partial_z \rho$:

$$\frac{1}{2}(\partial_z \rho)^2 - F(\rho) = \frac{\rho_z^2}{2} - \frac{\rho^4}{4} + \frac{(\delta+1)\rho^3}{3} - \frac{\delta\rho^2}{2} = 2c(\rho_0) = \text{const}, \qquad (6.49)$$

where $c(\rho_0) = \rho_0^4/2 - 2(\delta+1)\rho_0^3/3 + \delta\rho_0^2$ is the integration constant fixed by the value of the order parameter $\rho = \rho_0$ at free surface $z = h$. Expressing from eqn 6.49 the depth

z vs. ρ, after another integration one obtains the implicit relation for the profile of the order parameter $z(\rho)$. Correspondingly, one obtains the following relation between height h and the value ρ_0 of the order parameter at the free surface $z = h$ in the stationary flow regime:

$$h = \int_{\rho_0}^{1} \frac{d\rho}{\sqrt{\frac{\rho^4}{2} - \frac{2(\delta+1)\rho^3}{3} + \delta\rho^2 - c(\rho_0)}}. \tag{6.50}$$

The solution to eqn 6.50 exists if the layer thickness h exceeds a certain critical value h_{stop}. Thus, the value of h_{stop} can be obtained through minimization of integral 6.50 with respect to ρ_0:

$$h_{\text{stop}} = \min_{\rho_0}(h) = \min_{\rho_0} \int_{\rho_0}^{1} \frac{d\rho}{\sqrt{\frac{\rho^4}{2} - \frac{2(\delta+1)\rho^3}{3} + \delta\rho^2 - c(\rho_0)}}. \tag{6.51}$$

This integral can be calculated analytically for $\delta \to \infty$ and $\delta \to 1/2$. For large δ, the critical solution to the stationary equation 6.31 has a form $\rho = 1 + A\cos(kz)$ with $A \ll 1$ and $k = (\delta - 1)^{1/2}$, and therefore, $h_{\text{stop}} \to h_{\text{start}}$. For $\delta \to 1/2$, an asymptotic evaluation of eqn 6.51 gives $h_{\text{stop}} = -\sqrt{2}\log(\delta - 1/2) + \text{const}$. This expression agrees qualitatively with the empirical formula $\tan\varphi - \mu_0 \sim \exp[-h_{\text{stop}}/h_0]$ proposed by Daerr and Douady (1999) and Pouliquen (1999). In the general case, the integral condition (eqn 6.51) can be evaluated numerically.

The neutral stability curve $h_{\text{start}}(\delta)$ and the critical line $h_{\text{stop}}(\delta)$ are shown in Fig. 6.9. They separate the parameter plane (δ, h) into three regions. At $h < h_{\text{stop}}(\delta)$, the trivial static equilibrium $\rho = 1$ is the only stationary solution of eqn 6.31 for the chosen boundary conditions. Within the region $h_{\text{stop}}(\delta) < h < h_{\text{start}}(\delta)$ the static equilibrium state coexists with the stationary flow. For $h > h_{\text{start}}(\delta)$, the static regime is linearly unstable, and the only stable regime is that of the granular flow. These findings agree qualitatively and even quantitatively with rescaled experimental results (Daerr and Douady, 1999; Pouliquen, 1999).

Now we turn to the description of spatially non-uniform flows in thin granular layers using partial fluidization theory. Let us first derive the relation between the layer thickness h and the flux of granular material along the inclined plane. We again assume a no-slip condition for velocity $\mathbf{v} = 0$ at the bottom $z = 0$. At the free surface $z = h$ we assume that the normal derivative of velocity is zero, $\partial_n \mathbf{v} = 0$. We also consider a quasi-two-dimensional situation when the flow parameters depend on both x and y "horizontal" coordinates (x-axis is again directed downhill, and y is perpendicular to that and the normal upward direction z). Then, the kinematic boundary condition on the free surface for an incompressible medium (compare eqn 6.9) can be expressed in the form of the mass conservation law

$$\partial_t h = -\partial_x J_x - \partial_y J_y, \tag{6.52}$$

where

$$J_{x,y} = \int_0^h v_{x,y} dz \tag{6.53}$$

are inplane components of the flux of the granular material $\mathbf{J} = (J_x, J_y)$ (compare eqn 6.17). In a typical situation of chute flows, the downhill velocity v_x is often much larger than the orthogonal y-component v_y, so the mass conservation law in the leading order can be simply expressed as

$$\partial_t h = -\partial_x J_x. \tag{6.54}$$

The velocity v_x is determined from the constitutive relation (eqn 6.40)

$$\eta_f \partial_z v_x = q(\rho)\sigma_{xz} = (1-\rho)g\sin\varphi(h-z). \tag{6.55}$$

Substitution of eqn 6.55 into eqn 6.53 after partial integration taking into account the no-slip boundary condition $v_x = 0$ at $z = 0$ yields the relation between the flux J, the local slope angle φ and the order parameter ρ:

$$J_x = \frac{g\sin\varphi}{\eta_f} \int_0^h (1-\rho)(h-z)^2 dz. \tag{6.56}$$

A similar procedure can be used to derive the transverse flux component J_y (which is important, for example, for understanding of the fingering instability of an avalanche front). Note that in the fully fluidized state ($\rho \to 0$) the constitutive relation with fixed fluid viscosity $\eta_f = $ const gives rise to a quadratic vertical velocity profile for the steady-state flow, $v_x(z) \sim h^2 - (h-z)^2$. However, a more accurate approximation for the fluid viscosity $\eta_f \sim \sqrt{p}$, eqn 6.43, yields the Bagnold-type velocity profile (eqn 6.5) consistent with experimental data.

Now, we simplify the equation for the order parameter (eqn 6.31) assuming the slope angle is close to the critical value given by the condition (eqn 6.48), i.e. to the static angle of repose for a given layer depth h. If the depth h is not too large, we can assume the simple sinusoidal profile of the order parameter ρ across the layer dictated by the structure of the most unstable perturbation (see Eq. (6.45)) and can seek the solution in the form

$$\rho(x,y,z,t) = 1 - A(x,y,t)\sin\left(\frac{\pi z}{2h}\right), \tag{6.57}$$

where $A(x,y,t) \ll 1$ is a slowly varying function of inplane coordinates x,y and time t. Substituting anzatz 6.57 into the order-parameter equation 6.31 and applying a standard orthogonalization procedure with respect to the transverse mode $\sin(\pi z/2h)$, after some algebra we obtain the following equation for $A(x,y,t)$.[4]

$$\partial_t A = \left(\delta - 1 - \frac{\pi^2}{4h^2}\right)A + \nabla_\perp^2 A + \frac{8(2-\delta)}{3\pi}A^2 - \frac{3}{4}A^3. \tag{6.58}$$

Here, $\nabla_\perp^2 = \partial_x^2 + \partial_y^2$.

---

[4] Formally, this equation is rigorous only for $\delta \approx 2$ and $\delta - \delta_c \ll 1$ since quadratic (A^2) and cubic (A^3) terms have different orders of smallness in the expansion procedure.

The equation for the layer thickness h yields in the leading order

$$\partial_t h = \nabla \mathbf{J} = -\alpha \partial_x (h^3 A) + O(\partial_x^2 h, \partial_y^2 h). \tag{6.59}$$

The effective transport coefficient α depends on the flow rheological parameters as:

$$\alpha \approx 0.12 \frac{g\tau l^2 \sin\varphi}{\eta_f}. \tag{6.60}$$

The order-parameter relaxation length l and the time τ appear in the expression for α due to the renormalization of length and time according to eqn 6.32.

Since eqn 6.59 in the leading order only contains the first-order derivative with respect to x, it leads to discontinuities (Whitham, 1974), and, therefore, needs regularization. Until now we have not made a distinction between the local slope of the bottom and the free surface. What in fact determines the stress tensor and hence the control parameter δ is the slope of the free surface. We will use φ for the local slope of the free surface, and denote by $\bar{\varphi}$ the slope of the underlying (flat) bottom. The local surface slope angle φ depends on the depth gradient ∇h: $\varphi \approx \bar{\varphi} - \partial_x h + \ldots$. Substituting this expression into eqn 6.59 and including also the y-component of flux J_y, we obtain in the linear order in $\partial_x^2 h, \partial_y^2 h$:

$$\partial_t h = -\alpha \partial_x (h^3 A) + \frac{\alpha}{\tan \bar{\varphi}} \nabla \left(h^3 A \nabla h \right). \tag{6.61}$$

Additional corrections proportional to $\partial_x h$ arise also in eqn 6.58 due to the dependence of the control parameter δ on $\partial_x h$. Expanding δ in $\partial_x h$, in linear order we derive (see eqn 6.34)

$$\delta \approx \delta_0 - \beta \partial_x h, \tag{6.62}$$

where $\delta_0 = (\tan^2 \bar{\varphi} - \tan^2 \varphi_0)/(\tan^2 \varphi_1 - \tan^2 \varphi_0)$ and the constant $\beta = 2\tan\bar{\varphi}$ $/(\tan^2 \varphi_1 - \tan^2 \varphi_0)/ \cos(\bar{\varphi})^2 \approx 1.5 - 3$ depending on the value of $\bar{\varphi}$, compare eqn 6.34.

The last term in eqn 6.61 is due to the change of the local slope of the free surface φ and leads to the saturation of the avalanche front shape (without it the front would become infinitely steep in a finite time), see the next section. This term is also important for large wave number cutoff of the long-wave instability observed by Forterre and Pouliquen (2003) and the avalanche front transverse instability (Aranson et al., 2006).

In conclusion of this section, we point out that eqns 6.58 and 6.61 derived here are complementary to more traditional Saint-Venant-type models of thin granular flows in situations when the flow is near a fluidization threshold, and the continuous transition between rolling and static phases is an essential feature of the dynamics.

6.3 Avalanches in thin granular layers

Granular avalanches occur spontaneously when the slope of the granular material exceeds a certain angle (static angle of repose) or they can be produced at smaller

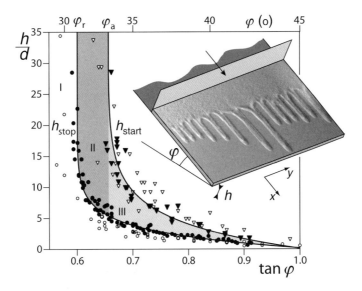

Fig. 6.10 Stability diagram for one-dimensional avalanches: h_{stop} is the thickness of the sediment left after an avalanche for a given angle φ, in air (●) and in water (○); $h_{start}(\varphi)$ is the maximum stable height of sediment, in air (▼) and in water (▽). In region I, an avalanche cannot propagate down the slope: the perturbation decays after the driving stops. Avalanches triggered in region II are stable, while they exhibit a transverse instability in region III: solitary erosion waves are observed when starting from the stable height h_{stop}. The inset shows the experimental setup and a typical pattern arising as a result of the transverse instability, from Malloggi et al. (2006).

angles (in the bistable regime) by applying a finite initial perturbation. Laboratory studies of avalanches are often carried out in a chute geometry when a layer of grains is titled at a certain fixed angle φ. Both quasi-one-dimensional avalanches initiated from a horizontally uniform perturbation (Daerr, 2001; Malloggi et al., 2006) and two-dimensional avalanches developing from a point-like initial perturbation (Daerr and Douady, 1999) have been studied experimentally. Avalanches demonstrate a rich variety of behaviors, including downhill and uphill expansion, solitary waves, transverse front instability, etc. In this section we discuss experimental and theoretical analysis of both one- and two-dimensional avalanches.

6.3.1 Solitary avalanches and transverse instability: experimental results

The most detailed experimental analysis of one-dimensional avalanches in thin granular layers has been recently performed by Malloggi et al. (2006). In order to initiate avalanches in a reproducible manner a special "bulldozer" technique was used to impose a controlled initial perturbation uniform along a direction perpendicular to the slope direction. The experiments were performed in air with sand (dry avalanches) and under water with fine powder (wet avalanches).

The asymptotic behavior of the avalanche depends on the inclination angle φ and the initial layer depth h. Avalanches do not propagate in region I in Fig. 6.10. The avalanche front appears to be stable in the range of angles $\varphi_r < \varphi < \varphi_a$ independent of the layer depth, region II in Figure 6.10. It was observed that after initial acceleration the avalanche eventually converges to a certain solitary wave shape shown in Fig. 6.11. Furthermore, this solitary wave is found to be quite insensitive to the details of the avalanche preparation. For each value of the inclination angle φ there is a family of solitary avalanches: the speed of the front depends on the mass of granular matter carried by the avalanche, we will discuss this phenomenon in detail later.

For $\varphi > \varphi_a$ within the range of bistability $h_{\text{stop}} < h < h_{\text{start}}$ (region III), the steady-state solitary avalanches are transversally unstable. The instability results in the formation of transverse undulations with a certain characteristic wavelength. Then, a sequence of fusion events leads to the gradual increase of the spatial modulation length scale through coarsening. Finally, a fingering pattern is formed. At this final stage, the flowing zones are disconnected from one another, so that the wavelength remains constant.

In order to elucidate the nature of the transverse front instability, Malloggi et al. (2006) employed the following technique: in addition to the experiments started from a flat granular bed, a series of experiments starting from a modulated initial condition were conducted, see Fig. 6.12. The modulation at a given wavelength was simply produced by imprinting on the sediment surface regularly spaced thin grooves. It was found that the modes forced by this technique always decay in region II, but in region III, the front undulations amplify exponentially for a wide band of modes. Non-linear effects become visible when the amplitude of undulations exceeds a certain critical value (of the order of 1 cm for dry avalanches). This technique allows for a direct measurement of the linear growth rate λ of a given transverse mode as a function of the mode wave number k, Fig. 6.12(c).

The transverse instability was found for both underwater and dry avalanches. However, no instability has been detected for monodisperse glass beads. A similar fingering instability was observed in earlier experiments by Pouliquen et al. (1997) in a somewhat different system with a rigid (non-erodible) substrate, see Fig. 2.12. In those experiments, fingering was also present for rough polydisperse materials and absent for monodisperse glass beads. These observations led Pouliquen et al. (1997) to conclude that the instability was caused by the size segregation in the course of front propagation. In fact, size segregation was detected near the front, with larger and rougher beads localized between the fingers. However, one cannot exclude the possibility that the size segregation is not the cause, but the consequence of the fingering instability, which in turn is caused by other, dynamical mechanisms; one such mechanism is discussed in the end of the next section. The absence of fingering instability for monodisperse glass beads can then be attributed to a very different rheology of those as compared with rough sands and similar granular materials (see Section 6.1).

6.3.2 One-dimensional avalanches: theoretical description

Let us now turn to the theoretical description of one-dimensional avalanches in the framework of the partial fluidization theory above for thin layers, so we ignore the

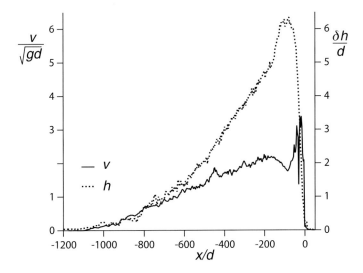

Fig. 6.11 A typical solitary avalanche profile: height $\delta h = h - h_{\text{stop}}$ rescaled by the grain size d (dotted line) and the surface velocity profile v rescaled by typical velocity $\sqrt{gd}$ (solid line). The experiment is performed in air for inclination angle $\varphi = 32°$, $h_{\text{stop}} = 2.3$ mm $= 7.8$ d, region II of Fig. 6.10, from Malloggi et al. (2006).

Fig. 6.12 Flowing part of solitary waves visualized by image difference (air, particles size $d = 300$ µm, inclination angle $\varphi = 35$ deg, time interval 1.1 s, starting (a) from a flat bed, or (b) from a bed with imposed depth modulation with the wavelength of 6.5 cm. (c) Linear growth rate of transverse perturbations λ as a function of the wave number k. The solid line is the best fit to $\lambda \sim k(1 - a_0 k)$, $a_0 = $ const, from Malloggi et al. (2006).

y dependence in the set of equations 6.58 and 6.61 for the amplitude of the order parameter $A(x,t)$ and thickness of the fluidized layer $h(x,t)$. Numerical analysis of eqns 6.58 and 6.61 in large enough domains confirms the existence of avalanches in a certain range of the initial thickness h and the parameter δ related to the inclination angle φ. For small values of h and δ corresponding to the region left of the curve h_{stop} in Fig. 6.9, all perturbations eventually decay and motion stops. In the range of angles and heights restricted by curves h_{stop} and h_{start} the bistable behavior is

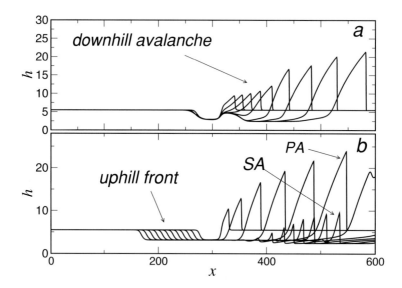

Fig. 6.13 One-dimensional solutions to eqns 6.58 and 6.61 illustrating the evolution of localized perturbation introduced at the location $x = 250$ for the initial layer thickness $h(x) = h_0 = 5.5$ and for two different values of the chute inclination characterized by parameter δ. (a) $\delta = 1.02$, propagation of a downhill avalanche for 10 consecutive moments of time with time interval $\Delta t = 10.0$; (b) higher inclination $\delta = 1.07$, perturbations grow both downhill and uphill, and small secondary avalanches (SA) emerge following the passage of a larger primary (PA) avalanche. Other model parameters: transport coefficient $\alpha = 0.05$ and $\beta = 0.25$, from Aranson and Tsimring (2002).

observed: small perturbation decay, whereas finite-amplitude perturbations develop into avalanches similar to those shown in Fig. 6.13. Above the curve h_{start}, a static granular layer is linearly unstable with respect to arbitrary small perturbations of the height h or the fluidization amplitude A, leading to eventual fluidization of the entire system.

In the bistable range a strong enough localized perturbation triggers an avalanche. Depending on the values of h and δ, two different types of avalanches are observed, similar to the experiment. For small values of h the avalanche propagates strictly downhill, see Fig. 6.13(a). Eventually, the downhill avalanche assumes a "universal" wedge-like shape. In the course of the downhill motion the amplitude of the avalanche and the avalanche speed grow due to the accumulation of granular matter involved in the motion. Correspondingly, the depth of the layer left *behind* the avalanche, h^-, becomes smaller than the initial thickness h. The value of the residual layer h^- is close to the value of h_{stop}. However, if the numerical solution to eqns 6.58 and 6.61

is performed in a periodic domain, the amplitude and the speed of the avalanche approaches certain asymptotic values after the avalanche circles the entire integration domain many times: the avalanche "collects" all the material that is above h_{stop}. Thus, using periodic boundary conditions one can easily obtain steady-state avalanches propagating with constant speed and without change of amplitude, as in the experiment by Malloggi et al. (2006) where the avalanches were triggered at $h \to h_{\text{stop}}$.

For larger h (or δ) avalanches expand both downhill and uphill (we call them *expanding avalanches*), see example in Fig. 6.13(b). Eventually, the whole granular layer (both below and above the initial perturbation) is involved in the downhill flow. Interestingly, in addition to the primary avalanche triggered by the initial localized perturbation, secondary avalanches may occur in the wake of the primary avalanche (Fig. 6.13). These secondary avalanches are also seen in experiments. The "pure" downhill avalanches appear to exist only in a relatively narrow range of h and δ, see Fig. 6.9. The curve $h_u(\delta)$ separating expanding and downhill avalanches can be obtained from the numerical solutions to eqns 6.61 and 6.58. The results, shown in Fig. 6.9, exhibit reasonable agreement with the experimental data by Daerr and Douady (1999).

The steady-state solitary avalanche solution propagates with velocity V without shape deformation:

$$A(x,t) = \bar{A}(x - Vt), \quad h(x,t) = \bar{h}(x - Vt). \tag{6.63}$$

Thus, in the coordinate system co-moving with the velocity V, steady-state equations 6.61 and 6.58 assume the form:

$$-V\partial_x \bar{h} = -\alpha \partial_x(\bar{h}^3 \bar{A}) + \frac{\alpha}{\tan\bar{\varphi}} \nabla \left(\bar{h}^3 \bar{A} \nabla \bar{h}\right) \tag{6.64}$$

$$-V\partial_x \bar{A} = \left(\delta - 1 - \frac{\pi^2}{4\bar{h}^2}\right)\bar{A} + \nabla^2 \bar{A} + \frac{8(2-\delta)}{3\pi}\bar{A}^2 - \frac{3}{4}\bar{A}^3. \tag{6.65}$$

In order to find solutions to eqns 6.64 and 6.65, we can solve numerically the one-dimensional equations 6.61 and 6.58 in a large domain with periodic boundary conditions. Asymptotically, the solution from a single initial "bump" converges on an asymptotically stable localized solution that decays to the "unperturbed" depth h_0 at $x \to \pm\infty$ (Fig. 6.14). Calculations reveal the existence of a one-parameter family of these localized solutions that can be parameterized by the value of the asymptotic layer thickness h_0 or the "trapped mass" m carried by the soliton, i.e. the area between h and h_0.[5] The main features of these localized solutions are: (a) the velocity of the avalanche V is an increasing function of m, see Fig. 6.15(a); (b) the family of admissible solutions for a solitary wave terminates at $m = m_c$ and $V = V_c = V(m_c)$; (c) the critical mass m_c decreases with α. The dependence of V on m is qualitatively consistent with experimental data, see Fig. 6.15(a). The shape of the solutions is sensitive to the value of the transport coefficient α: for large α the solution has a well-pronounced

---

[5]Values m and h_0 (as well as all other parameters characterizing the avalanche) are uniquely related to each other within this one-parameter family, however the mass rapidly diverges with the increase of h_0 away from h_{stop}, so h_0 has to be kept very close to h_{stop} in order to remain within the domain of validity of eqns 6.61 and 6.58.

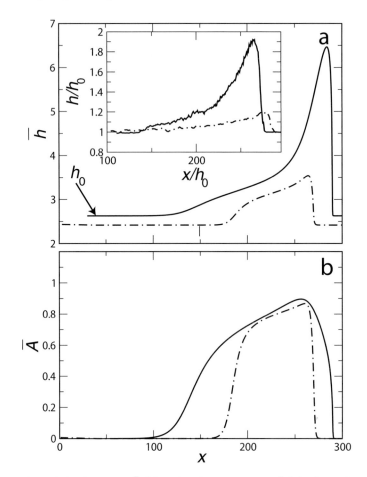

Fig. 6.14 Profiles of displacement $\bar{h}$ (a) and order parameter $\bar{A}$ (b) of steady-state solitary avalanches at two different sets of parameters: $m = 147.7$, $\alpha = 0.08, \beta = 1, \delta = 1$, avalanche velocity is $V = 2.72$ (solid line); $\alpha = 0.025, \delta = 1.15, m = 62, V = 0.86$ (dot-dashed line). Inset: Typical experimental (Malloggi et al., 2006) height profiles for avalanches in 300 μm sand, $\bar{\varphi} = 32.3°$ (solid line) and 500 μm glass beads, $\bar{\varphi} = 22°$ (dot-dashed line), from Aranson et al. (2006).

shock-wave-like shape (Fig. 6.14) with the height of the crest $h_{\max}$ several times larger than h_0. The relative height of the crest decreases for $\alpha \to 0$. The results are qualitatively consistent with the shape of sand (large α) and glass bead ($\alpha \to 0$) avalanches, see the inset of Fig. 6.14 showing representative experimental profiles.

The stability of the one-dimensional solitary avalanche with respect to *transverse* perturbations can be analyzed numerically. Applying small perturbation to h, A in the form $\sim \exp[\lambda(k)t + iky]$, we can linearize eqns 6.61 and 6.58 near the one-dimensional solution (eqn 6.63) and obtain the growth rate of linear perturbations $\lambda(k)$, see Fig. 6.15. The growth rate of the perturbations is positive at sufficiently small k (long-

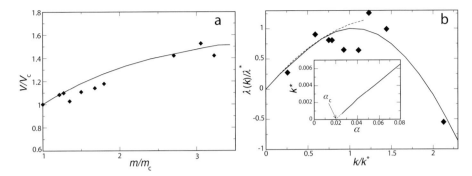

Fig. 6.15 (a) the solitary wave speed V vs. trapped mass m (solid line), diamonds depict the experimental data for sand avalanches, $\bar{\varphi} = 32.3°$, from (Malloggi et al., 2006); (b) the growth rate of transverse modulations $\lambda(k)$ vs. modulation wavenumber k for $\delta = 1.15$, $\alpha = 0.08$ and $m = 102$, k is scaled by k^*, and the growth rate by $\lambda^* = \lambda(k^*)$. Solid line: $\lambda(k)$ obtained by the numerical stability analysis of the one-dimensional solution (eqn 6.63). Dashed line is the solution of eqn 6.67. Symbols depict experimental data for sand avalanches. Inset: optimal wave number of k^* vs. α for $\delta = 1.15$

wave perturbations) and turns negative at large k. It reaches a maximum at a certain $k = k^*$. These theoretical results can be directly compared with the experimental results obtained by the forced modulation technique, see Fig.6.12. Despite a strong scatter related to intrinsic technical difficulties of the forced modulation approach, the experimental data is consistent with these theoretical results. The inset to Fig. 6.15(b) shows the dependence of the optimal wave number k^* vs. transport coefficient α, obtained by the numerical linear stability analysis of the solitary avalanche solution. According to these calculations, the transversal instability only occurs for α above some α_c.

For long-wave perturbations the mechanism of *transverse instability* can be studied analytically by considering the solitary avalanche solution with slowly varying position $x_0(y,t)$:

$$A(x,t) = \bar{A}(x - x_0(t,y)), \quad h(x,t) = \bar{h}(x - x_0(t,y)). \tag{6.66}$$

The results of the asymptotic analysis can be summarized as follows (technical details can be found in Aranson et al. (2006)). The growth rate of small perturbations $\lambda(k)$ for $k \to 0$ is of the form

$$\lambda = |k|\sqrt{\zeta_1 V_m} - (1 + \zeta_2)k^2/2 + O(k^3), \tag{6.67}$$

where $V_m = \partial V/\partial m$ is the derivative of avalanche velocity with respect to the mass of the avalanche, and constants $\zeta_{1,2}$ are evaluated using the unperturbed one-dimensional solution (eqn 6.63) of the steady-state problem eqns 6.64 and 6.65,

$$\zeta_1 = \frac{\alpha}{\tan\bar{\varphi}} \int_{-\infty}^{\infty} \left(\bar{A}\bar{h}^3 \partial_x \bar{h}\right) dx, \quad \zeta_2 = \frac{\alpha}{\tan\bar{\varphi}} \int_{-\infty}^{\infty} \left(\bar{A}\bar{h}^3 \partial_m \bar{h}\right) dx.$$

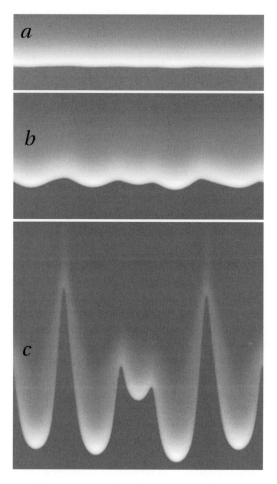

Fig. 6.16 Development of the transversal instability. Graycoded images of the granular layer height $h(x, y, t)$ (white corresponds to larger h), showing a slightly perturbed front at moments of time $t = 100$ (a), front with well-developed modulations at $t = 300$ (b), and the fingering state $t = 500$ (c). Domain size is 600 units in x-direction and 450 units in y-direction, the number of mesh points was up to 1200×600, only part of the domain in the x-direction is shown. Parameters: $\delta = 1.16$, $\alpha = 0.14$, $\beta = 2$ and initial height $h_0 = 2.285$.

Figure 6.15 shows that eqn 6.67 gives the correct description of the linear stage of the transverse instability for small k. Qualitatively, the transverse instability of a planar front is caused by the following mechanism: since the velocity of the avalanche increases with its mass, a local increase of avalanche mass results in the increase of its velocity and, consequently, "bulging" of the front. Since the bulge moves forward faster than its tails and leaves a trough behind it, grains flow towards the bulge, further draining and slowing down the trailing regions.

The evolution of the avalanche front beyond the initial linear instability regime

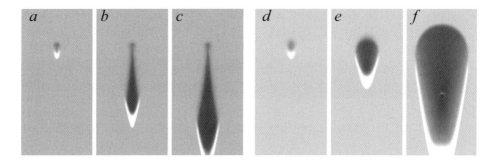

Fig. 6.17 Sequence of images demonstrating the evolution of a wedge avalanche (a–c) and balloon avalanche (d–f) obtained from a numerical solution of eqns 6.58 and 6.61, from Aranson and Tsimring (2001), Aranson and Tsimring (2002).

can be studied by the two-dimensional numerical analysis of eqns 6.58 and 6.61 in a rectangular domain with periodic boundary conditions in the x- and y-directions. In one such simulation, a narrow strip of height $h = 4.285$ along the y-direction was deposited on a flat layer with height $h = 2.285$. To trigger the transverse instability, a small noise was added to the initial conditions. The initial conditions rapidly developed into a quasi-one-dimensional solution described by eqn 6.63. Due to the periodicity in the x-direction, the solitary avalanche could pass through the integration domain several times, allowing an analysis to be performed in a relatively small domain in the x-direction. The transverse modulation of the avalanche leading front was observed after about 100 units of time for the parameters of Fig. 6.16. The modulation initially grew in amplitude with the typical wave number k close k^*, eventually coarsened and led to the formation of large-scale finger structures, consistent with the experimentally observed shapes, see Fig. 6.10. No saturation of finger length was found either numerically or experimentally.

Fingering patterns observed in the course of the instability of granular avalanche fronts are remarkably similar to those in thin films on inclined surfaces, both with clear and particle-laden fluids (Troian et al., 1989; Zhou et al., 2005). The instability occurs in the large surface tension limit. However, despite the visual similarity the physical mechanism leading to this fingering instability is rather different: in fluid films the instability is driven (and stabilized) by the surface tension, whereas in dry or underwater granular flow the surface tension plays no role.

6.3.3 Two-dimensional avalanches in thin granular layers

Daerr and Douady (1999) conducted experiments with two-dimensional avalanches in a thin layer of grains on a sticky (velvet) inclined plane; see Fig. 2.11. As in the one-dimensional case, in sufficiently thin layers (h slightly above h_{stop}) avalanches expand only downhill from the point of initial perturbations. In the two-dimensional case they eventually acquire the wedge-like shape with an opening angle that increases with the layer thickness and/or inclination angle. In thicker and/or steeper layers, the region of flowing sand expands both downhill and uphill from the initial perturbation (the

152 *Patterns in gravity-driven granular flows*

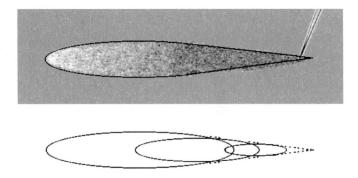

Fig. 6.18 Top: Total area overrun by the wedge avalanche (solid line), compared with experimental image from (Daerr and Douady, 1999). Bottom: superimposition of avalanche boundaries given by eqn 6.69 for three different moments of time, from Rajchenbach (2002).

grains themselves of course only flow downhill). The avalanche region in this case has the shape resembling a growing balloon; see Fig. 2.11. The rear front of the balloon-like avalanche propagates uphill with a velocity roughly one half of the downhill velocity of the head front, and the velocity of the head is also two times larger than the depth-averaged flow velocity (Rajchenbach, 2001; Rajchenbach, 2003). The regions of existence of the two types of avalanches are superimposed on the stability diagram discussed earlier (see Fig. 6.9): between the dashed and the solid lines the layer exhibits a bistable behavior: a finite perturbation can trigger an avalanche, otherwise the layer remains stable. The line with ×- symbols indicates the transition between wedge and balloon avalanches.

The existence of non-trivial avalanche shapes calls for their theoretical description. In fact, both wedge-like and balloon-like structures are reproduced by the partial fluidization theory. Direct simulations of two-dimensional equations 6.61 and 6.58 for the same parameter values but different inclination angles yield the structure and dynamics of avalanches closely resembling the experimental observations, compare Figs. 6.17 and 2.11. The line in the phase diagram (h, δ) separating the two classes of avalanches, also agrees well with experimental observations, see the dotted line in Fig. 6.9.

An elegant description of the avalanche shapes was proposed by Rajchenbach (2002) on the basis of kinematic ideas similar to the Huygens principle in the theory of diffraction (see, e.g. Born and Wolf (1959)). Rajchenbach (2002) proposed that the front of the avalanche creates a sequence of point-like sources of perturbations along the track of the avalanche that then expand. For balloon-like avalanches these expanding regions are assumed to have the form of circles. The "earlier" circles are evidently bigger than the "younger" ones. The shape of the avalanche at the moment of time t is defined as the envelope of these circles (labelled for each time τ, $(0 < \tau < t)$). The shape of a balloon-like avalanche is given by the following equation:

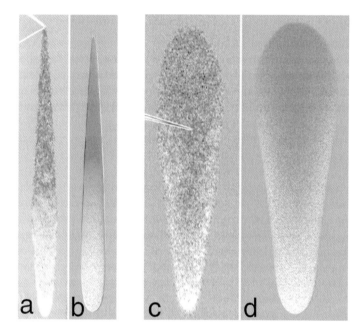

Fig. 6.19 Wedge-like (a,b) and balloon-like (c,d) avalanches obtained from numerical solutions to eqns 6.29. Panels (a) and (c) feature experimental data, and (b), (d) show numerical results. Experimental conditions shown in panel (a) correspond to the smaller chute inclination angle φ compared to that in panel (c). The pin in the experimental images (a),(c) corresponds to the location of the initial perturbation, from Douady et al. (2002)

$$x^2 + \left(y - 2\bar{v}t + \frac{5}{2}\bar{v}\tau\right)^2 = \left(\frac{1}{2}\bar{v}\tau\right)^2, \quad 0 < \tau < t, \tag{6.68}$$

where $\bar{v}$ is the free parameter corresponding to the velocity of the rear front. For the wedge avalanches the shape is reproduced if one assumes that the expanding regions are ellipses with different velocities of downhill and perpendicular expansion, then the envelope is given by

$$\left(\frac{\bar{v}x}{2v_\perp}\right)^2 + \left(y - \frac{3}{2}\bar{v}t\right)^2 = \left(\frac{1}{2}\bar{v}t\right)^2. \tag{6.69}$$

Here, $v_\perp$ is the perpendicular velocity. While these heuristic relations agree rather well with experimental observation; see Fig. 6.18, their connection to the actual dynamics of granular flows remains to be understood.

Douady et al. (2002) applied the modified BCRE equations 6.28 and 6.29 to the description of two-dimensional avalanches. Despite the narrowness of the granular layer, they assumed a two-phase structure with a thin layer of rolling grains flowing over another thin layer layer of static grains. By fitting the empirical functions $\bar{\Gamma}$ and $\mu(z_R)$ they were able to reproduce the shape of the avalanches remarkably well, see Fig. 6.19.

6.3.4 Avalanches in deep granular layers: relation between BCRE and partial fluidization theories

The similarity in the two-dimensional avalanche solutions obtained above within the partial fluidization theory and the modified BCRE models may seem surprising, however there is an underlying reason for it. Indeed, partial fluidization theory describes a continuous transition from fluidized to static grains, however, this transition in certain cases can be approximated by a sharp interface (fluidization front). Then, the partial fluidization theory can be used to compute the position of the fluidization front, and the resulting equations can be cast in the form similar to that of the modified BCRE theory.

To demonstrate this, we choose the Cartesian coordinate system aligned with the unperturbed free surface with the origin at the surface, the z-axis pointing upwards normal to the surface, and the x-axis pointing downhill along the surface. For smooth horizontal variations of the flow, the local vertical profile of the the order parameter ρ can be approximated by the following dependence at $-\infty < z < 0$,

$$\rho = 1 - \left(\tanh[(z+z_R)/\sqrt{8}] - \tanh[(z-z_R)/\sqrt{8}]\right)/2. \tag{6.70}$$

Here, the slowly varying parameter z_R plays the role of the depth of the fluidized layer. This expression is very close to the exact front solution in the infinite domain,

$$\rho = \frac{1}{2}\left(1 \pm \tanh[(z-z_R)/\sqrt{8}]\right), \tag{6.71}$$

if $z_R \gg 1$ and $\delta \to 1/2$ and differs from it only in the vicinity of the free surface $z = 0$, where it is corrected in order to satisfy the no-flux boundary condition $\partial_z \rho = 0$.[6]

Let us introduce the new variable $\bar{z}$ characterizing the entire amount of the "fluid phase" in the system

$$\bar{z} = \int_{-\infty}^{0} (1-\rho)dz. \tag{6.72}$$

It is easy to check by direct integration that for the anzatz 6.70 $\bar{z} = z_R$. We show below that eqn 6.31 can be reduced to the equation for the "thickness" of the fluidized region z_R. Such a description is accurate in two important limits: $\bar{z} \gg 1$ and $\bar{z} \ll 1$. For the intermediate values of $\bar{z}$ the above approximation for the order parameter (eqn 6.70) gives a smooth interpolation between these two limits. Numerical solutions to corresponding equations indicate that qualitative features are not sensitive to the specific choice of interpolation since the solution tends to "avoid" the intermediate area (qualitatively similar results can be obtained using piece-linear approximations).

For simplicity we use here the quartic from of the free energy (eqn 6.33), the linear function $q(\rho) = 1 - \rho$, and the Newtonian constitutive equation (eqn 6.35). As we discussed already in Section 6.2.1, the specific flow rheologies only change numerical $O(1)$ factors in the corresponding equations for mean flow velocity, while the form of these equations remains unchanged.

---

[6]Note that for simplicity we assume the layer to be infinitely deep and do not take into account the effect of the bottom. In order to generalize anzatz 6.70 to finite depth layers, one needs to add to eqn 6.70 an extra term $\sim \tanh[(z+h(x))/\sqrt{8}]$ to satisfy the boundary condition at the base.

After integration of eqn 6.31 we obtain

$$\partial_t z_R = \partial_x^2 z_R + \int_{-\infty}^0 \rho(1-\rho)(\delta-\rho)\mathrm{d}z + \int_{-\infty}^0 (v_x\partial_x\rho + v_z\partial_z\rho)\mathrm{d}z. \tag{6.73}$$

The horizontal velocity profile $v_x(z)$ is found from the constitutive relation (eqn 6.35) using the expression for the shear stress, compare eqn 6.55,

$$v_x = -\alpha_0 \int_{-\infty}^z (1-\rho)z'\mathrm{d}z' \tag{6.74}$$

and

$$v_z = -\partial_x \int_{-\infty}^z \mathrm{d}z' v_x = \partial_x z_R \alpha_0 \int_{-\infty}^z \mathrm{d}z' \int_{-\infty}^{z'} \mathrm{d}\zeta \zeta \partial_\zeta (1-\rho). \tag{6.75}$$

Here, $\alpha_0 = g\sin\varphi/\eta_f$. Now, substituting eqns 6.70, 6.74, and 6.75 into eqn 6.73, after some algebra we derive the evolution equation for the position of fluidized layer z_R,

$$\partial_t z_R = \partial_x^2 z_R + K(z_R) - \alpha_0 G(z_R)\partial_x z_R. \tag{6.76}$$

Function $K(z_R)$ can be found in the closed form

$$K = \frac{6}{\sqrt{2}(s-1)} + \frac{2\delta-1}{\sqrt{2}} - \frac{2z_R}{s-1}\left(\frac{3}{s-1} + \delta + 1\right), \tag{6.77}$$

with $s = \exp(\sqrt{2}z_R)$. Function $K(z_R)$ has the following asymptotic behaviors

$$K(z_R) = \begin{cases} (\delta-1)z_R & \text{for } z_R \ll 1 \\ \sqrt{2}\left(\delta-\frac{1}{2}\right) & \text{for } z_R \gg 1. \end{cases} \tag{6.78}$$

This expression shows that the depth of the fluidized layer increases if $\delta > 1$ for small z_R, and at smaller $\delta > 1/2$ for large z_R. This is yet another indication of the bistability of the granular layer. At small z_R eqn 6.76 complies with the behavior of eqn 6.31 linearized near $\rho = 1$, and for large z_R eqn 6.76 gives the asymptotically correct result for the velocity of the front between fluidized and solid states at $\delta \to 1/2$.

Function $G(z_R)$ can only be found in an integral form. However, asymptotic values of $G(z_R)$ can be found for large and small z_R,

$$G(z_R) = \begin{cases} \frac{12-\pi^2}{3\sqrt{2}}z_R \approx 0.5021 z_R & \text{for } z_R \ll 1 \\ \frac{\pi^2}{3} \approx 3.29 & \text{for } z_R \gg 1. \end{cases} \tag{6.79}$$

We can interpolate these two limits using a single function

$$G(z_R) = \frac{\pi^2}{3}\tanh\left(\frac{12-\pi^2}{\pi^2\sqrt{2}}z_R\right). \tag{6.80}$$

Equation 6.73 has to be solved together with the equation for the height of the free surface $h(x,t)$. The latter can be derived from the mass conservation (eqn 6.54) with the expression for flux given by eqn 6.56. Substituting $\rho(z)$ from eqn 6.70, we obtain

$$\frac{\partial h}{\partial t} = -\frac{\alpha_0}{3}\partial_x f(z_R),\qquad(6.81)$$

where

$$f(z_R) = \begin{cases} 2\pi^2 z_R & \text{for } z_R \ll 1 \\ z_R^3 & \text{for } z_R \gg 1. \end{cases}\qquad(6.82)$$

Again, we interpolate between these limits with a single function, $f(z_R) = z_R(z_R^2 + 2\pi^2)$. Then, using the relation 6.62 between the height gradient $\partial_x h$ and the control parameter, $\delta \approx \delta_0 - \beta\partial_x h$, the height profile becomes coupled to the dynamics of the thickness of the fluidized layer (eqn 6.76).

Equations 6.76 and 6.81 give a simplified description of two-dimensional flows in deep inclined layers or sandpiles. It is interesting to point out the similarities and differences between this simplified model and the BCRE theory and its modifications (Bouchaud et al., 1994; Bouchaud et al., 1995; Boutreux et al., 1998; Douady et al., 1999; Mehta, 1994; Khakhar et al., 2001; Aradian et al., 2002; Douady et al., 2002), see Section 6.2.2. The BCRE theory also operates with two variables, the thickness of the immobile fraction z_S and the thickness of the rolling (flowing) fraction z_R, eqns 6.20 and 6.21 or 6.27. The original BCRE equations were later modified by Boutreux et al. (1998) for large values of z_R by replacing the "instability term" $\gamma(\tan\varphi - \tan\varphi_c)z_R$ by the "saturation term" v_{up} for $(\tan\varphi - \tan\varphi_c)z_R > z_0$, $z_0 \gg 1$. One may notice that eqn 6.76 is very similar to the second BCRE equation (eqn 6.27) however, with one important difference. From our derivation it directly follows that the value of the critical angle φ_c must be different for small and large z_R, whereas in eqn 6.27 that value is the same. This important distinction of our model gives rise to the hysteretic behavior of the fluidization transition that is missing in the original BCRE model and its modification (Boutreux et al., 1998). On the other hand, using an *ad hoc* non-monotonous basal friction function $\mu(z_R)$ in eqn 6.29 allowed Douady et al. (2002) to incorporate hysteresis in the BCRE model. Presumably, the specific form of this function can be deduced from the partial fluidization theory, however, it has not yet been done.

We performed numerical simulations of eqns 6.76 and 6.81 and found that in the bistable range ($0.5 < \delta < 1$) small localized perturbations decay, and large enough perturbations trigger avalanches. Figure 6.20 shows the development of the avalanche from a localized perturbation. The avalanche propagates both uphill and downhill. In fact we find this behavior for all values of inclination within the bistable regime. This observation is consistent with our conclusion from previous sections that the domain of existence of wedge avalanches shrinks with the increase of layer thickness and is in the agreement with the experiment of Daerr and Douady (1999).

6.4 Longitudinal instability of granular flow along a rough inclined plane

The steady-state regime of the granular flow along a rough inclined plane exists only in a certain range of the plane inclinations and flow rates. For sufficiently high inclination angles steady-state flow typically does not exist; instead, an accelerating flow with the velocity linearly increasing along the plane ensues. For smaller angles the steady-state

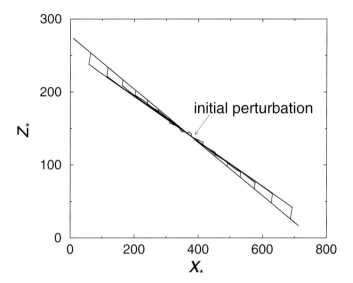

Fig. 6.20 Evolution of the free surface profile during an avalanche within a simplified two-phase description (eqns 6.76 and 6.81) for $\delta_0 = 0.75, \alpha_0 = 0.2, \beta = 3.15$. In the wake of the avalanche the normalized slope of the free surface δ is reduced and approaches the equilibrium value $1/2$. Note the "true" horizontal and vertical variables (x_*, z_*) are related to our original Cartesian variables (x, y) via a rotation by the angle φ.

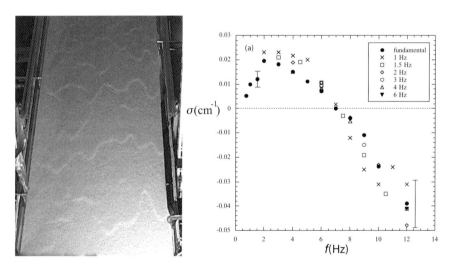

Fig. 6.21 Long-wave instability observed in flow of sand down a rough inclined plane. Left panel: a snapshot of the chute surface illustrating the wave trains. Right panel: Spatial growth rate σ of the long-wave instability as a function of the forcing frequency f, from (Forterre and Pouliquen, 2003).

flow exists only for not too low total flux values (controlled, for example, by the intake rate of grains at the inlet). A decrease in the flux typically results in the onset of a longitudinal instability and modulation waves. Even for smaller values of the intake rate the chute discharges occur in the form of a periodic sequence of avalanches.

Forterre and Pouliquen (2003) presented an experimental study of the long-surface-wave instability developing in granular flows on a rough inclined plane, Fig. 6.21. This instability (sometimes called Kapitza instability of roll waves) was known from previous studies (Savage, 1979; Davies and Foye, 1991), however, no accurate characterization of the instability had been performed. Forterre and Pouliquen (2003) measured the threshold and the dispersion relation of the instability by imposing controlled periodic perturbations at the inlet and measuring their evolution downstream; see Fig. 6.21. The spatial growth rate σ of small perturbations vs. the perturbation frequency f shows features of a long-wave instability: the growth rate σ vanishes at $f = 0$, peaks at a certain optimal frequency (about 3 Hz) and becomes negative for higher frequencies (perturbations decay exponentially downstream).

This long-wave instability, at least on a qualitative level, can be understood in the framework of the depth-averaged Saint-Venant-type theory presented in Section 6.2.1. The corresponding mass and momentum conservation equations assume the form (compare eqns 6.17 and 6.19)

$$\partial_t h + \partial_x (\langle V \rangle h) = 0 \tag{6.83}$$
$$\partial_t (\langle V \rangle h) + \alpha_1 \partial_x (\langle V \rangle^2 h) = (\tan \varphi - \mu(\langle V \rangle, h) - \partial_x h) \, gh \cos \varphi,$$

where h is the local thickness of the flow (for a flat bottom $z_R \equiv h$), φ is the chute inclination angle, $\langle V \rangle$ is the depth-averaged flow velocity, $\mu(\langle V \rangle, h)$ is the bottom friction coefficient (see Fig. 6.5), $\alpha_1 \sim O(1)$ is a constant determined by the velocity profile within the layer. Equations 6.83 admit the steady-state flow solution, $\langle V \rangle = V_0 = \text{const}, h = h_0 = \text{const}$. The depth-averaged steady-state flow velocity V_0 is in fact determined by the flux of grains at the inlet of the chute, and the steady-state layer depth h is given by the slope $\tan \varphi$ through the solution to the equation

$$\mu(V_0, h_0) = \tan \varphi. \tag{6.84}$$

The Saint-Venant equations can be rewritten in dimensionless variables

$$x \to x/h_0, \ t \to t V_0 / h_0, \ \bar{h} = h/h_0, \ \bar{V} = \langle V \rangle / V_0. \tag{6.85}$$

Then, the steady state is simply $\bar{h} = 1, \bar{V} = 1$. In order to examine the stability of this solution with respect to small perturbations, we substitute the ansatz $\bar{h} = 1 + h_1, \bar{V} = 1 + V_1$ into eqn 6.83 and retain only terms linear in h_1, V_1:

$$\partial_t h_1 + \partial_x V_1 + \partial_x h_1 = 0$$
$$\text{Fr}^2 (\partial_t V_1 + (\alpha_1 - 1) \partial_x h_1 + (2\alpha_1 - 1) V_1) = -a V_1 - b h_1 - \partial_x h_1, \tag{6.86}$$

where $a = \partial_{\bar{V}} \mu, b = \partial_{\bar{h}} \mu$, and

$$\text{Fr} = \frac{V_0}{\sqrt{gh_0 \cos \varphi}} \tag{6.87}$$

is the dimensionless steady-state flow velocity (the Froude number). Seeking the solution to linear eqns 6.86 in the form $h_1, V_1 \sim \exp[\lambda(k)t + ikx]$, we obtain the following dispersion relation

$$\lambda^2 + 2i\alpha_1 k\lambda + \mathrm{Fr}^{-2}\left[i(a-b)k - a\lambda\right] + \left(\mathrm{Fr}^{-2} - \alpha_1\right)k^2 = 0. \tag{6.88}$$

The uniform flow is unstable if the following condition is fulfilled

$$1 - \frac{b}{a} > \alpha_1 + \sqrt{\alpha_1(\alpha_1 - 1) + \mathrm{Fr}^{-2}}, \tag{6.89}$$

(see for details Forterre and Pouliquen (2003)). Since $b/a = -\partial \bar{V}/\partial \bar{h}$, this instability condition can be rewritten as

$$c_0 > c_g \tag{6.90}$$

where $c_0 = 1 + \partial \bar{V}/\partial \bar{h}$ is the dimensionless velocity of kinematic wave (compare with the first eqn 6.84 in rescaled variables), and $c_g = \alpha_1 + \sqrt{\alpha_1(\alpha_1 - 1) + \mathrm{Fr}^{-2}}$ is the velocity of a "gravity wave" propagating downstream. Therefore, the stability criterion states that flow becomes unstable when the kinematic velocity is larger than the velocity of gravity waves. This theory correctly predicts the existence of the stability threshold. However, eqn 6.88 does not account for the cutoff of the instability for large values of k. Indeed, for $k \to \infty$ one finds that $\lambda \sim \sqrt{\alpha_1 - \mathrm{Fr}^{-2}} k \to \infty$. In the subsequent publication Forterre (2006) used a full linear stability analysis of the local mass and momentum balance equations supplemented by the constitutive relations for stresses in the form of eqn 6.6 to compute additional higher-order linear terms $\sim \partial_x^2 h, \partial_x^2 V$ in the Saint-Venant equations 6.83. With this modification the set of Saint-Venant equations describes the cutoff of the longitudinal instability for large k.

The long-wave surface instability can also be analyzed in the framework of the partial fluidization theory. Linearizing eqns 6.61 and 6.58 near the steady flowing solution $A = A_0 + \tilde{a}\exp[\lambda t + ikx], h = h_0 + \tilde{h}\exp[\lambda t + ikx]$, after some algebra one can immediately obtain that the growth rate of the linear perturbations λ is positive only in a band restricted from above by some critical wave number and only in the vicinity of h_{stop}; see Fig. 6.22. The shape of the dispersion curve $\lambda(k)$ is typical for a generic long-wave instability, i.e. $\lambda \to 0$ for $k \to 0$ and $\lambda \to -\infty$ at $k \to \infty$. With the increase of the layer depth h and correspondingly the flux of granular matter, the instability disappears, in agreement with experiments.

The non-linear saturation of the instability results in the development of a sequence of avalanches, which is generally non-periodic; see Fig. 6.23. The structure shows slow coarsening due to merging of the avalanches: larger avalanches move faster and eventually overcome and absorb smaller ones.

6.5 Pattern-forming instabilities in rotating cylinders

Granular media in rotating horizontal cylinders (drums) often show behavior similar to chute flows. Four main flow regimes are shown in Fig. 6.24. For very small rotation rates (as defined by the Froude number, which for the rotating drum geometry is

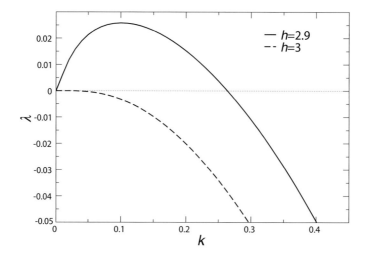

Fig. 6.22 The growth rate $\lambda(k)$ of small perturbations against the uniform steady flow solution with constant $h = h_0$ and $A = A_0$ derived from eqns 6.58 and 6.61 for $\alpha = 0.025, \beta = 2, \delta = 1.1$ and two values of h_0. Instability occurs at $h = 2.9$ that is near $h = h_{\text{stop}}(\delta)$ (see Fig. 6.9) and disappears with increase of h.

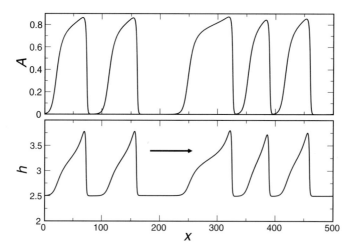

Fig. 6.23 Typical profiles of the height h and the amplitude of the order parameter A in the regime of long–surface wave instability for $\beta = 2, \alpha = 0.025, \delta = 1.1$. Starting from generic initial conditions $h = h_0, A = $ const plus small noise, a sequence of avalanches develops.

$\text{Fr} = \Omega^2 R/g$, where Ω is the angular velocity of drum rotation and R its radius), well separated in time avalanches occur when the slope of the free surface exceeds a certain critical angle φ_d thereby reducing this angle to a smaller angle φ_s (Rajchenbach, 1990; Jaeger et al., 1989; Tegzes et al., 2002; Tegzes et al., 2003). These critical angles have the same meaning and are close numerically to the angles $\varphi_{\text{stop,start}}$ in the chute flow

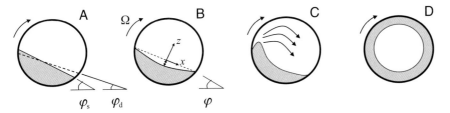

Fig. 6.24 Sketches of different flow regimes in a rotating cylinder with increasing Froude number: (a) avalanching/slumping regime, (b) rolling/cascading regime (c) cataracting regime, (d) centrifugal regime.

geometry with a similar material, see Section 6.1.1. The difference between angles φ_d and φ_s is usually a few degrees. At an intermediate rotation speed, a continuous flow of grains emerges instead of discrete avalanches through a hysteretic transition, see Fig. 6.25, similar to the transition from periodic avalanches to steady flow in the chute geometry at large rates of grain deposition (Lemieux and Durian, 2000). In the bulk, the granular material rotates almost as a solid body with some internal slipping. As moving grains reach the free surface they slide down within a thin near-surface layer; Fig. 6.24(a) (see below Section 7.3). At low rotation speed the surface has a nearly flat shape; its average slope defines the dynamic angle of repose φ_d. For higher rotation speeds the surface assumes a characteristic S-shape (so-called cataracting), Fig. 6.24(c), (Zik et al., 1994; Ottino and Khakhar, 2000). Eventually, for a very high rotation rates Ω particle inertia becomes dominant, and a transition to centrifuging at Froude number Fr = 1 is observed, see Fig. 6.24(d).

The S-shape of the free surface of grains in the drum at intermediate rotation rates can be deduced from the following simple argument. Let us assume that the bulk of the drum is in a solid-body rotation, and a shear flow occurs only within a thin layer of width z_R small compared to the drum radius R, see Fig. 6.24(b). The flow on the surface starts when the tangential projection of the gravity force $f_g = -gz_R \sin\varphi$ exceeds the frictional force $f_{fric} = \mu p$, where μ is the static friction coefficient, and $p = -gz_R \cos\varphi$ is the pressure within the flowing layer. The flux of grains J can be obtained from the integration of the equation for the velocity tangential to the surface in the flowing layer, $J = \int_{-z_R}^{0} v_x(z)dz$. Velocity v_x is given by the constitutive relation for the Newtonian shear stress $\eta\partial_z v_x = f_g - f_{fric} = -gz_R \sin\varphi + \mu gz_R \cos\varphi$ with the no-slip condition $v_x = 0$ at $z = -z_R$. Here, we assume for simplicity the particle density $\nu = 1$. Simple algebra yields

$$J = -\frac{gz_R^3}{3\eta}\cos\varphi\,(\tan\varphi - \mu). \qquad (6.91)$$

Due to the solid-body rotation of grains inside the drum, the mass conservation dictates that at an arbitrary point on the surface the flux

$$J = \frac{\Omega}{2}(R^2 - r^2), \qquad (6.92)$$

where r is the distance from the center of the drum to the point at surface. Equating eqns 6.91 and 6.92, one obtains an equation relating the free surface angle φ and the

162 *Patterns in gravity-driven granular flows*

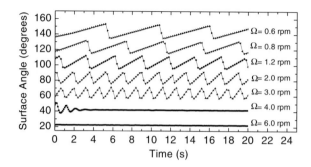

Fig. 6.25 The mean surface angle as a function of time at various rotation rates Ω. The data were taken using large $d = 0.9$ mm beads (different curves are offset by $20°$ for visual clarity), from Tegzes *et al.* (2003)

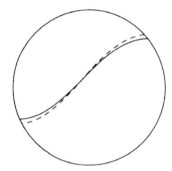

Fig. 6.26 *S*-shaped profile of the free surface in a drum rotating in the anticlockwise direction obtained from eqn 6.93 for 60% filling and the friction coefficient $\mu = 0$, from Zik *et al.* (1994)

distance from the drum center r,

$$-\frac{gz_R^3}{3\eta}\cos\varphi\,(\tan\varphi - \mu) = \frac{\Omega}{2}(R^2 - r^2). \tag{6.93}$$

However, eqn 6.93 contains an unknown variable z_R that is in general a function of the velocity and pressure. In order to exclude z_R, Zik *et al.* (1994) made a simplifying assumption that the flow stops when the pressure reaches some "equilibrium" value p_0, i.e. $z_R = -p_0/g\cos\varphi$. With this assumption, eqn 6.93 gives the free surface profile in a parametric form. Plotting solutions to this equation indeed yields *S*-shaped profiles, see Fig. 6.26.

Various models addressing the nature of the transition from discrete avalanches to the continuum flow have been suggested, see, e.g., Benza *et al.* (1993). The simplest phenomenological description capable of capturing this transition was proposed by Linz and Hänggi (1995). It is based on a system of two ordinary differential equations for the mean flow velocity $\bar{v}$ and the mean free surface angle $\bar{\varphi}$

$$\dot{\bar{v}} = g\left[\sin\bar{\varphi} - \mu(\bar{v})\cos\bar{\varphi}\right]\chi(\bar{\varphi},\bar{v}),$$
$$\dot{\bar{\varphi}} = \Omega - a\bar{v}, \qquad (6.94)$$

where Ω is the rotation frequency of the drum, $\mu(\bar{v}) = b_0 + b_2\bar{v}^2$ is the velocity dependent friction coefficient, $\chi(\bar{\varphi},\bar{v})$ is some phenomenological cutoff function guaranteeing that the system stays at rest for $\bar{v} = 0, \bar{\varphi} < \varphi_s$, and a, b_0, b_2 are parameters of the model. Despite the simplicity, the model yields a qualitatively correct transition from discrete avalanches to continuous flow with the increase of rotation rate Ω, and also suggests a logarithmic relaxation law of the free surface angle in the presence of external vibrations (in the experiment by Jaeger et al. (1989) periodic shaking of the drum was induced by a loudspeaker mechanically coupled to the system).

The transition from avalanches to flow also naturally arises in the framework of the partial fluidization theory (Section 6.2.4). From that general formulation one can derive a system of coupled equations for the parameter δ (which is related to the derivative of surface local slope φ as $\delta = \delta_0 - \beta\partial_x h$, and the width of the fluidized layer z_R (cf. eqns 6.76 and 6.81):

$$\partial_t z_R = \partial_s^2 z_R + K(z_R, \delta) - \alpha_0 G(z_R)\partial_s z_R,$$
$$\partial_t \delta = \bar{\Omega} + \partial_s^2 J, \qquad (6.95)$$

where $-s_0 < s < s_0$ is the coordinate along the slope of the granular surface inside the drum, s_0 is the endpoint position of the free surface determined by the drum filling fraction; the dimensionless frequency $\bar{\Omega} = \tau\Omega\beta$, τ is the characteristic time in eqn 6.31; $J = \beta f(z_R)$ is the downhill flux of grains, $\bar{v}$ is the averaged velocity in the flowing layer, and functions K, G, and f are given by eqns 6.78, 6.79, and 6.82, see (Aranson and Tsimring, 2002). Note that this model again bears a resemblance to the BCRE-type models of surface granular flows that were applied to rotating drums by Khakhar et al. (1997b) and Makse (1999).

Equations 6.95 exhibit stick–slip-type oscillations of the surface angle for slow rotation rates and a hysteretic transition to a steady flow for larger rates. In the steady-state regime after integration of the second of equations 6.95 with no-flux boundary conditions at the drum wall $J(s = \pm s_0) = 0$ the following expression for the flux J as a function of the arclength s is obtained:

$$J \sim \bar{\omega}(s_0^2 - s^2). \qquad (6.96)$$

Since $J \sim f(z_R) \sim z_R^3$, eqn 6.96 yields the following scaling for the width of the flowing layer z_R vs. rotation frequency: $z_R \sim \bar{\omega}^{1/3}$ and arclength position s

$$z_R(s) \sim \bar{\omega}^{1/3}\left(s_0^2 - s_2\right)^{1/3}. \qquad (6.97)$$

The width of the rolling layer z_R in turn determines the surface velocity v_x according to eqn 6.74: $v_x = \alpha_0 \int (1-\rho)z' dz' \sim z_R^2$. Thus, one obtains that the surface velocity at the middle point scales as $v_x \sim z_R^2 \sim \omega^{2/3}$. This scaling is consistent with measurements by Tegzes et al. (2003).

A further simplification of eqns 6.95 can be achieved by integrating them over the arclength s from $-s_0$ to s_0. Then, the model is reduced to a system of two coupled

164 *Patterns in gravity-driven granular flows*

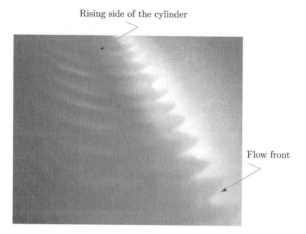

Fig. 6.27 Granular fingers in a long horizontal drum rotating with the angular speed 2 rev/s at the filling fraction 1.8%, from Shen (2002)

ordinary differential equations for the averaged drum angle $\langle \delta \rangle$ and the averaged flow thickness $\langle z_R \rangle$ similar to the model of Linz and Hänggi (1995).

Under certain conditions (relatively low volume filling fraction of grains and large rotational speed) granular flows in long rotating drums exhibit fingering instability that is similar to that observed in chute flows (Malloggi et al., 2006), see Fig. 6.27. The similarity between fingering in rotating drums and chute flows suggests that the same underlying mechanism described in Section 6.3.2 may be at play here.

6.6 Self-organized criticality and statistics of granular avalanches

It is well known that in real sandpiles avalanches can vary widely in size. The broad distribution of scales in real avalanches stimulated Bak et al. (1987) to introduce a "sandpile" cellular automaton as a paradigm model for the *self-organized criticality* (SOC), the phenomenon that occurs in slowly driven non-equilibrium systems when they asymptotically reach a critical state characterized by a power-law distribution of event sizes. The set of rules that constitute the sandpile model is very simple. In a two-dimensional form, the cellular automaton model is formulated for an integer variable $z_{i,j}$ (local height) on a square lattice. Unit size "grains" are dropped one by one on a lattice in random places and form vertical stacks. If the height difference between neighboring stacks exceeds a certain threshold K_0, the values of $z_{i,j}$ are updated synchronously according the following rules:

$$z_{i,j} = z_{i,j} - 4,$$
$$z_{i\pm 1,j} = z_{i\pm 1,j} + 1,$$
$$z_{i,j\pm 1} = z_{i,j\pm 1} + 1, \tag{6.98}$$

which describe hops of four grains from the higher stack to the neighboring ones. This event may trigger an "avalanche" of subsequent hops until the sandpile returns to the stable state. After that, another grain is dropped and the relaxation process repeats. The size of an avalanche is determined by the number of grains set into motion by adding a single grain to a sandpile. This model asymptotically reaches a critical state in which the mean "angle" (difference in heights among neighbors) is equal to K_0, and the avalanches have a universal power-law distribution of sizes, $P(S) \propto S^{-\zeta}$, where the avalanche size S is defined as the total number of hops after a deposition of a single particle, an $\zeta \approx 1.5$.

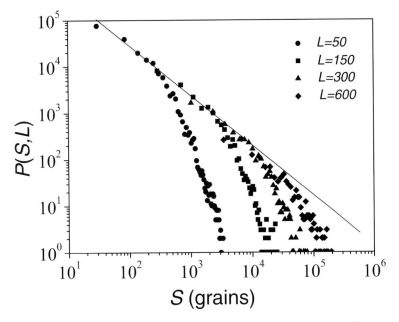

Fig. 6.28 The unscaled size distributions P of avalanche sizes S from different subsets of experiments on a three-dimensional pile of rice corresponding to different sizes of the field of view $L = 50, 150, 300,$ and 600 mm. Approximate power-law behavior with an exponent of $\zeta = 1.20(5)$ is observed in a finite range of avalanche sizes S, from Aegerter et al. (2003)

Despite the tremendous popularity of the SOC idea (the original paper Bak et al. (1987) has been cited more then 3000 times), the relevance of this model and its generalizations to avalanches in real sandpiles is still a matter of debate. Some experiments, for example, with glass beads poured on a conical sandpile (Costello et al., 2003) or with long rice (Aegerter et al., 2003) do show the power-law distribution of avalanche sizes (see Fig. 6.28) however, the scaling exponent varies significantly, and other experiments do not find the power-law statistics of avalanches (Jaeger et al., 1989; Rajchenbach, 2000; Lemieux and Durian, 2000). This should not be entirely surprising since real sandpiles are quite different from the idealized model of Bak et al. (1987). Most importantly, the sandpile model by Bak et al. (1987) is defined via a single "angle of repose" parameter K_0, and so its asymptotic behavior has the properties of the critical state for a second-order phase transition. As we have seen above, real sandpiles are characterized by two angles of repose and thus exhibit features of a first-order phase transition. While the difference in these angles may be small, it can prove significant for the extreme statistics that indicate the presence of SOC. Furthermore, reliable measurements of these statistics are very hard, since they require a large-scale experimental setup and long-term data acquisition. Thus, in real physical experiments one might expect significant finite-size effects influencing the statistics.

Interestingly, rice piles were also observed to demonstrate roughening dynamics of their surface as the distribution of active sites in the self-organized critical state shows a self-affine structure with the fractal exponent $d_B = 1.85$ (Aegerter et al., 2003; Aegerter et al., 2004). This is consistent with the theoretically predicted mapping between self-organized criticality and roughening observed, for example, in the Kardar–Parizi–Zhang model (Paczuski and Boettcher, 1996).

7
Patterns in granular segregation

One of the most fascinating features of heterogeneous granular materials (i.e. consisting of grains of different sizes, shapes, or other physical characteristics) is their tendency to segregate under external agitation rather than to mix, as one would expect from naïve entropy considerations. This property is widespread in Nature (see, e.g. Iverson (1997)) and has important technological implications, Cooke et al. (1976). Almost any type of mechanical excitation leads to segregation of grains with different mechanical properties. Granular segregation has been found in granular convection (Knight et al., 1993), hopper flows (Makse et al., 1997b; Gray and Hutter, 1997; Samadani et al., 1999; Samadani and Kudrolli, 2001), flows in rotating drums (Zik et al., 1994; Hill, 1997; Choo et al., 1997), etc. Segregation among large and small particles due to shaking has been termed the "Brazil-nut effect" (Williams, 1963; Rosato et al., 1987) after the well-known fact of everyday life that in a jar of mixed nuts the biggest nuts (and Brazil nuts are often the biggest) are usually found at the top, see Fig. 7. In addition, flows of granular mixtures can separate spontaneously through electrification (e.g. due to rubbing) or magnetization (Mehrotra et al., 2007).

Granular segregation was known and used for millennia by peasants as a way to separate chaff from grains and soil from potatoes. It is widely used to sort materials in the mineral processing industries. On the other hand, in many applications (especially in pharmacology) segregation is extremely undesirable. Often, separation of grains produces interesting patterns. For example, if a binary mixture of particles that differ both in size *and* in shape is poured down on a flat surface, a heap that consists of thin alternating layers of separated particles is formed (Gray and Hutter, 1997; Makse et al., 1997b; Makse and Herrmann, 1998). Rotating of mixtures of grains with different sizes in long drums produces well-separated bands of pure monodisperse particles (Zik et al., 1994; Hill, 1997; Choo et al., 1997; Chicarro et al., 1997).

7.1 Basic mechanisms of granular segregation

Despite the obvious importance of segregation and the need to understand and control it, the basic mechanisms are still the subject of active research (Ottino and Khakhar, 2000; Kudrolli, 2004; Aranson and Tsimring, 2006b). Segregation of small and large particles can occur even in systems at thermal equilibrium (Asakura and Oosawa, 1958), driven by a purely entropic mechanism called the *depletion force*. Since the excluded volume for small particles around large ones becomes smaller when large grains clump together, the separated state possesses higher entropy. This effect induces attraction among large particles. The entropic segregation has been extensively

168 *Patterns in granular segregation*

Fig. 7.1 Illustration of the Brazil-nut: after shaking, the biggest Brazil nuts rise on top of smaller peanuts.

studied for colloid–polymer mixtures and other binary fluids (Asakura and Oosawa, 1958; Lebowitz, 1964; van Duijneveldt *et al.*, 1993). In the context of granular matter the effect of the depletion force was demonstrated in experiments and simulations with vertically vibrated mixtures of spheres by Melby *et al.* (2007). However, granular systems are driven and strongly dissipative, and equilibrium thermodynamic arguments can only be applied qualitatively. Furthermore, the granular segregation is more widespread than would be dictated by thermodynamics, because there are many other mechanisms that may lead to segregation in real granular materials. In fact, practically any variation in mechanical properties of particles (size, shape, density, surface roughness, etc.) may lead to their segregation.

7.1.1 Entropic mechanism of segregation

Segregation of granular matter can be a result of purely statistical (entropic) effects since in many cases the segregated state corresponds to a higher entropy than a mixed state. This can be understood based on the following qualitative argument (Srebro and Levine, 2003). Imagine two sets of grains that are identical to each other except for the value of friction, which is larger for set 1 than for set 2. Clearly, more frictional grains 1 can have all the same static configurations as grains 2 plus some additional configurations not available for set 2. Thus, the entropy of the first system S_1 is greater than that of the second one. This suggests that in a mixture of the two sets of grains, a segregated state may attain a higher entropy than a mixed state.

Srebro and Levine (2003) computed the free energy of the mixture of N identical spherical grains among which a fraction f is of type 1 and the rest is of type 2 that differ by their frictional properties, as characterized by three friction coefficients $\mu_{11}, \mu_{12}, \mu_{22}$ among particles 1, between particles 1 and 2, and among particles of type 2, respectively. In the mean-field approximation similar to the one described in Chapter 3, one can compute the partition function for the mixture, and assuming that the compactivity of the system X is known, find the free energy

$$\mathcal{F} = XN[f \ln f + (1-f)\ln(1-f) + 2f(1-f)S(X) \\ - \ln X - fS_{11}(X) - (1-f)S_{22}(X)], \tag{7.1}$$

where $S_{ij} = \ln\left(e^{-s_{\min}/X} - e^{-s_{ij}/X}\right)$, $s_{\min}$ is the minimum size of a Voronoi cell for one particle (it corresponds to a hexagonal packing of disks in two dimensions and fcc packing of spheres in three dimensions), s_{ij} is the maximum size of a Voronoi cell of particle of type i surrounded by particles of type j, and $S = (S_{11} + S_{22})/2 - S_{12}$. The equilibrium concentration f can be found by solving the equation

$$\partial_f \mathcal{F} = XN \left[\ln\left(\frac{f}{1-f}\right) \ln(1-f) + 2(1-2f)S(X)\right] = 0, \tag{7.2}$$

or $2f - 1 = \tanh[S(X)(2f-1)]$. For small $S(X) < 1$, the only solution of this equation is $f = 1/2$, which corresponds to a fully mixed state, whereas for $S(X) > 1$ two solutions with $f > 1/2$ and $f < 1/2$ exist, which corresponds to a segregated solution in a mixture. The function $S(X)$ can be rewritten as

$$S(X) = \ln \frac{\sqrt{(1 - e^{-\Delta s_{11}/X})(1 - e^{-\Delta s_{22}/X})}}{1 - e^{-\Delta s_{12}/X}}, \tag{7.3}$$

where $\Delta s_{ij} = s_{\min} - s_{ij}$. For $X \to 0$, $S(X) \to 0$, which corresponds to no segregation. Indeed, for small compactivity, the probability to attain a state with a Voronoi cell bigger than $s_{\min}$, is small, and so all particles are highly compacted, and the difference in friction coefficients is irrelevant. At large X, function $S(X)$ approaches a limiting value $S_\infty = \ln\left(\sqrt{\Delta s_{11} \Delta s_{22}}/\Delta s_{12}\right)$, and if this value is greater than 1, segregation occurs. For large enough X, large Voronoi cells are more likely, and it is in those "arched" states that the friction difference plays a key role.

Thus, the frictional entropy-driven segregation of static grains depends crucially on the values of Δs_{ij}. The values of s_{ij} can be computed using minimalistic models of one particle of type i resting on two (in two dimensions) or three (in three dimensions) particles of type j. The resulting phase diagram indicating the regions in the (μ_{11}, μ_{22}) plane where segregation may occur, is shown in Fig. 7.2.

As we have seen in Chapter 3, compactivity is difficult to measure experimentally, however, one can relate compactivity with the volume fraction either theoretically (Srebro and Levine, 2003) or experimentally (Nowak et al., 1998; Schröter et al., 2005). Then, the phase diagram, Fig. 7.2, can lead to quantitative predictions on the segregation as a function of packing fraction of the material, but such experimental verification is still lacking. However, a direct comparison of this model with a vertical

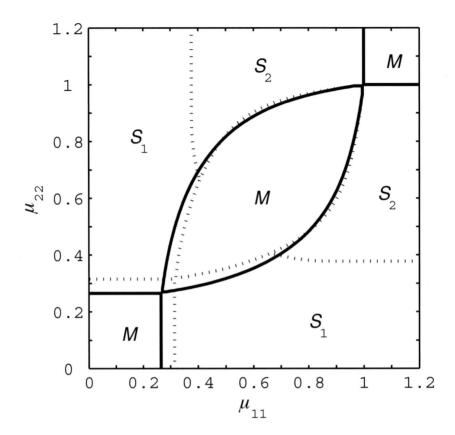

Fig. 7.2 Phase diagram in the friction coefficients (μ_{11}, μ_{22}) plane assuming that $\mu_{12} = \min(\mu_{11}, \mu_{22})$ for two dimensions (solid lines) and three dimensions (dotted lines). M indicates regions of mixing ($f = 0.5$) at any compactivity, S_1 regions correspond to grains segregated at any compactivity, and in S_2 grains are segregated only at sufficiently high compactivity, from Srebro and Levine (2003)

shaking experiment is difficult since there are many other factors, most importantly gravity and inertia, that come into play. Pohlman et al. (2006) investigated experimentally the role of surface roughness of spherical grains on their segregation properties. Surprisingly, they found that while variations in roughness do affect the repose angle of a sandpile, they do not lead to a noticeable segregation when the mixture was placed in a rotating drum (see more details below). This may cast some doubt on the validity of the theoretical picture presented above, however, one should realize that the dynamic conditions of the rotating-drum experiment are qualitatively different from the type of static segregation described here.

7.1.2 Kinetic sieving

A dynamic mechanism most often invoked to describe granular segregation is called *kinetic sieving* (Williams, 1976; Rosato et al., 1987; Savage and Lun, 1988). According to this mechanism, in agitated granular matter, voids between grains are constantly being created, and smaller and/or heavier particles are more likely to fall into them due to gravity, and they end up lower compared with large and/or lighter particles. Rosato et al. (1987) illustrated this mechanism by a Monte-Carlo simulation of a periodically shaken two-dimensional box filled with monodisperse disks and one large intruder near the bottom. A shake consisted in lifting the whole assembly upwards uniformly to the top of the box and letting it fall down and relax. After each shake, a new local equilibrium configuration was found by following a sequence of down- and sideways moves of individual particles leading to the decrease of the overall potential energy. In this process, particles move with respect to each other, and voids may occur. A void that opens beneath the large intruder is likely to be filled by smaller particles before the large particle returns to its initial position, and therefore it would end up higher than it started from. A sequence of such events leads to a gradual rise of the intruder to the surface.

Savage and Lun (1988) developed a semiquantitative theory of kinetic sieving in a different situation when a mixture of spherical grains of equal masses but different diameters d_1, d_2 is agitated by a sheared motion. They approximated a shear flow by a discrete set of layers of thickness $d \approx d_{1,2}$ sliding past each other in the horizontal x-direction with a relative velocity v_r. Obviously, the sliding velocity is related to the shear strain rate dv_x/dz, $v_r = d\, dv_x/dz$. Let us consider two neighboring layers, A and B sliding past each other in a shear flow. Assigning equal probability to all possible distributions of particles and voids within a layer, in the limit of a large number of particles one can compute the probability of finding a void with a certain diameter d_v,

$$p(E) = (\bar{E} - E_m)^{-1} \exp\left[-\frac{E - E_m}{\bar{E} - E_m}\right], \tag{7.4}$$

where $E = d_v/d$ is the void diameter normalized by the mean particle diameter d, $\bar{E}$ is the mean void diameter, and E_m is the minimum possible void diameter (for close packing of spheres $E_m = 0.1547$).

Using this expression one can compute the number of small (1) and large (2) particles from layer A per unit area per unit time that fall into voids in layer B,

$$N_{1,2} = n_v \frac{n_{1,2}}{n_1 + n_2} n_p v_r d [E_1 + \bar{E} - E_m + 1] \exp\left[-\frac{E_{1,2} - E_m}{\bar{E} - E_m}\right]. \tag{7.5}$$

Here, $n_{1,2}$ are the number densities of particles of types 1 and 2, n_p is the total number of particles in a layer per unit area, and $E_{1,2} = d_{1,2}/d$, and n_v is the density of voids.

The partial percolation velocities $q_{1,2}$, in the continuum limit, can be found from the mass conservation:

$$\nu_{1,2} q_{1,2} = -m_{1,2} N_{1,2}, \tag{7.6}$$

where $\nu_{1,2}$ are the partial densities of particles $1, 2$. Since both of these percolation velocities are directed *down*, there must exist a counterflow of particles that balances this percolation in the stationary regime. On the experimental grounds, Savage and Lun (1988) assumed that particles have a finite probability of being "squeezed" from a bottom layer upwards, and this probability is independent of the particle size. This assumption leads to the following *net* mass fluxes in the z-direction

$$\nu_1 q_1^N = -\frac{\nu_2}{\nu} m_1 N_1 + \frac{\nu_1}{\nu} m_2 N_2$$
$$\nu_2 q_2^N = \frac{\nu_2}{\nu} m_1 N_1 - \frac{\nu_1}{\nu} m_2 N_2. \qquad (7.7)$$

In the limit of small concentration of smaller (1) particles their net flux normalized by the flow strain rate dv_x/dz and the particle diameter d_2, reads

$$\tilde{q}_1^N = \frac{4}{\pi} \frac{M}{N} \frac{k_1^2}{1 + (\bar{E}^2/k_2)(M/N)} \left[(2 + \bar{E} - E_m) \exp\left(-\frac{1 - E_m}{\bar{E} - E_m}\right) \right.$$
$$\left. -(\zeta + \bar{E} - E_m + 1) \exp\left(-\frac{\zeta - E_m}{\bar{E} - E_m}\right) \right] \qquad (7.8)$$

Here N is the total particle density, M is the total density of voids, $\zeta = d_1/d_2$ is the ratio of particle diameters, and $k_{1,2}$ are $O(1)$ numerical constants depending on the packing of grains inside flowing layers.

Savage and Lun (1988) compared formula 7.8 for the percolation fluxes with experimental data by Bridgwater et al. (1978) and Cooke and Bridgwater (1979) and found good agreement, see Fig. 7.3. Subsequently, the theoretical concepts of Savage and Lun (1988) have been confirmed in direct molecular dynamics simulations by Hirshfeld and Rapaport (1997), who showed that Lennard–Jones particles of different sizes in a two-dimensional chute flow exhibit stratification such that large particles drift to the top of the flowing layer. The situation becomes more complex if other parameters of grains vary as well, such as particle density or roughness.

One should notice that the expressions for the percolation fluxes (eqn 7.7) do not contain gravity acceleration g. Effectively, this constitutes a limit of infinitely large gravity when a small particle falls into a void whenever the latter is big enough to fit the former. However, in a real situation gravity is evidently a factor along with particle inertia determining the rate of percolation. A more realistic model of kinetic sieving in shallow chute flows taking into account these effects explicitly was proposed by Gray and Thornton (2005). This model yields the following simple expressions for the net percolation velocities of small (1) and large (2) particles

$$q_1^N = S_r \phi_2$$
$$q_2^N = -S_r (1 - \phi_2), \qquad (7.9)$$

where ϕ_2 is the volume fraction of the large particles, and S_r is the "segregation number", which is proportional to the ratio of the gravity acceleration to the product of the downhill velocity and the flow thickness. It also depends on the material properties of the particles such as interparticle friction and the size ratio. These equations imply that large particles percolate upward as long as the volume fraction of small particles ϕ is non-zero, and conversely, the small particles percolate downward if the volume concentration of large particles $1-\phi$ is non-zero. Using these expressions combined with the mass balance equation, Gray et al. (2006) obtained a number of non-stationary

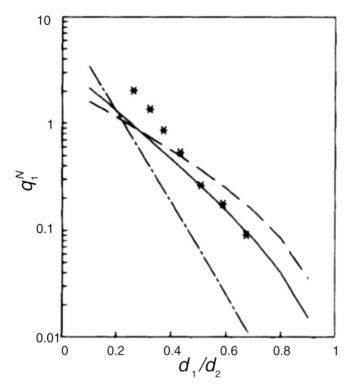

Fig. 7.3 Comparison of theoretical net percolation velocity of small particles q_1^N (eqn 7.8) (lines) with experimental data of Bridgwater et al. (1978), Cooke and Bridgwater (1979) (symbols). Lines are obtained for $M/N = 1, \bar{E} = 0.414, k_1 = 1, k_2 = 0.63$ (dashed), $M/N = 1.2, \bar{E} = 0.33, k_1 = 1, k_2 = 0.59$ (solid), $M/N = 1.48, \bar{E} = 0.25, k_1 = 1, k_2 = 0.55$ (dot-dashed), from Savage and Lun (1988).

solutions describing segregation in shallow flows. In particular, these solutions describe formation of concentration interfaces ("shocks") that develop in shallow chute flows downstream from an inlet feeding uniformly mixed granular material. In the segregated solution, small particles percolate to the bottom, and large particles flow above them (so-called "inverse grading").

A heuristic theoretical model of granular segregation in a binary mixture of grains with different densities ν_1 and ν_2 due to a variant of kinetic sieving can be developed from the mass balance equation at equilibrium:

$$\frac{d}{dz}\left(-D\phi\frac{df}{dz} + J_z\right) = 0, \qquad (7.10)$$

where D is the mass diffusion coefficient, $f = \phi_1/\phi$ is the relative volume fraction of the heavier particles, $\phi = \phi_1 + \phi_2$ is the total volume fraction of the granular material, and J_z is the segregation flux. Assuming that the segregation velocity is proportional to the buoyancy force (Khakhar et al., 1997a), one obtains

$$v_z = -C(\nu_1 - \bar{\nu})g\cos\varphi, \tag{7.11}$$

where the average density $\bar{\nu} = (\nu_1\phi_1 + \nu_2\phi_2)/(\phi_1 + \phi_2)$ and C is a function of particle properties. Substituting v_z from eqn 7.11 and using expression for $\bar{\nu}$ we obtain the segregation flux

$$J_z = v_z\phi_1 = -K_s\phi[1 - rf(1-f)], \tag{7.12}$$

where $K_s = C\nu_1 g\cos\varphi$ is the characteristic segregation speed, $r = \nu_2/\nu_1$ and $f = \phi_2/\phi_1$.

Kinetic sieving is only one of many mechanisms that can play a role in the Brazil-nut effect and related segregation phenomena. Kinetic sieving can be assisted by other factors such as arches of smaller particles supporting large "intruders" (Duran et al., 1993; Dippel and Luding, 1995). Another complicating factor is a large-scale convection flow in a vibrated granular bed, which is usually caused by non-linearity of the sidewall friction: grains drift upwards near the walls and downward in the middle of a cavity. As pointed out by Shinbrot and Muzzio (1998), the motion of particles is also affected by inertia, and therefore the amplitude of driving and the density of particles may play a major role. In particular, Shinbrot and Muzzio (1998) and Breu et al. (2003) demonstrated the so-called *reverse* Brazil-nut effect at large amplitude of driving, when a large but light particle actually sinks to the bottom whereas a heavy particle of the same size rises to the top, see Fig. 2.7. Some of these other mechanisms are discussed below.

7.1.3 Convection

Knight et al. (1993) proposed a different explanation for the Brazil-nut effect based on the granular convection due to friction with sidewalls (see also Knight et al. (1996), Cooke et al. (1996)). During upward motion of the vibrated container grains are compacted, and frictional forces extend deep into the bulk of material. During the downward motion, the granular layer is dilated, and frictional forces only penetrate a narrow boundary layer near the walls. A periodic sequence of upward and downward displacements of the container leads to the convection pattern in which grains move upward near the center and downward near the wall. Such convective motion in a pile of poppy seeds was investigated by non-invasive high-resolution magnetic-resonance imaging (MRI) by Ehrichs et al. (1995), see Fig. 7.4. The stripes seen in Fig. 7.4 carry detailed information on the vertical velocity profile: the velocity is maximal at the center and decreases in the vicinity of the container walls.

In order to study size segregation, Knight et al. (1993) placed a large tracer sphere (diameter 19 mm) in a container filled with granular material of significantly smaller size (2 mm), and subjected it to vertical tapping with frequency 1 Hz, see Fig. 7.5. Some of the grains at the same depth as the large tracer bead were dyed. In the course of tapping, both small grains and the large bead rose to the top together. After that, the large bead remained on the surface, whereas the small grains moved to the wall and submerged. This only occurs if the larger grains are large enough compared with the width of the near-wall boundary layer, in the opposite case both small and large particles are swept by the downward convection flow and no Brazil-nut effect occurs.

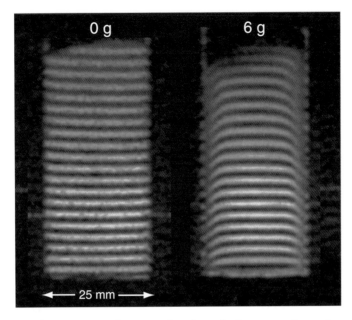

Fig. 7.4 (left panel) MRI image of a 2-mm slice through the center of container, no vibration, bright stripes indicate tagged particles; (right panel) MRI image after a single shake of peak acceleration $\Gamma = 6$ g. Stripe deformations correspond to a plug-like vertical velocity profile, courtesy of Sidney Nagel and Heinrich Jaeger.

7.1.4 Granular segregation in dilute systems

Segregation of granular materials occurs not only in dense but also in dilute systems when the mechanisms described above are presumably not applicable. However, in dilute regimes other segregation mechanisms come into play. These mechanisms can be understood within the kinetic theory of granular gases.

Granular thermal diffusion. Due to inelastic collisions, the granular temperature often varies significantly across a sample. For example, in vertically shaken beds, granular temperature is usually high near the vibrating wall but then it decays away from it. Sometimes, a minimum of the granular temperature is observed at a certain distance from a thermal wall (Feitosa and Menon, 2002; Huan et al., 2004). The gradient of temperature leads to a thermal diffusion of grains, but in granular mixtures the diffusion rates depend on the particle mass and size. Hsiau and Hunt (1996) computed coefficients of thermal diffusion in a binary mixture using Enskog theory and arrived at the conclusion that large particles accumulate in "cold" regions, and small particles accumulate in the "hot" regions. This can explain the prevalence of the normal Brazil-nut effect since in most cases the cold region is near the surface and the hot region is near the vibrating bottom.

Condensation. Granular segregation can also occur due to the differential condensation of grains in the gravity field. Hong et al. (1999) showed that in a monodisperse system, there exist a critical temperature T_c at which grains condense at the bottom of

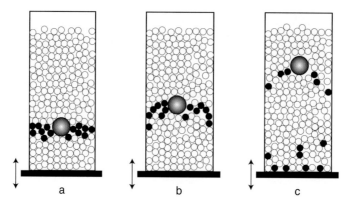

Fig. 7.5 Sketch of the granular convection experiment by Knight *et al.* (1993): (a) initial configuration before the start of the tapping experiment; (b) after several taps, grains near the wall start to move down along the wall; (c) after more taps, the bulk grains start to move upward, and the large tracer bead moves together with them.

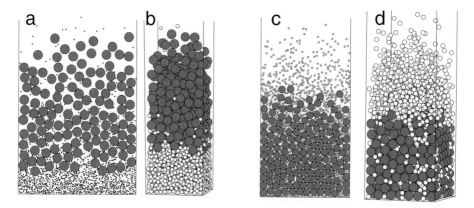

Fig. 7.6 Normal and reverse Brazil-nut effect in molecular dynamics simulations of binary granular mixtures: (a) two-dimensional system with $d_2/d_1 = 8, m_2/m_1 = 4$; (b) three-dimensional system with $d_2/d_1 = 2, m_2/m_1 = 2$; (c) two-dimensional system with $d_2/d_1 = m_2/m_1 = 4$; (d) three-dimensional system with $d_2/d_1 = 2, m_2/m_1 = 6$, from Hong *et al.* (2001).

a cavity. Indeed, if particles had zero size, at a fixed temperature T the density would follow the exponential barometric profile $\nu = \nu_0 \exp[-mgz/T]$ where m is the particle mass. The density at the bottom depends on the total mass of the gas per unit bottom area $M = \int_0^\infty \nu(z)dz$ yielding $\nu_0 = Mmg/T$. So, as the temperature goes down, the density at the bottom would grow indefinitely. Of course, in a gas of hard spheres of diameter d, the density cannot exceed the close-packing density $\nu_c \sim m/d^2$. This leads to a critical temperature $T_c \sim Md^2g$. In a binary system, one can similarly introduce two critical temperatures $T_{c1,c2}$ for the two monodisperse components of the mixture,

$T_{c1} \sim m_1 d_1, T_{c2} \sim m_2 d_2$. If $T_{c1} > T_{c2}$, then for granular temperature $T_{c2} < T < T_{c1}$, particles of type 1 will condense at the bottom, while particles of type 2 will not (assuming that the components of granular gas weakly interact with each other). This difference in vertical profiles of grain density leads to the granular segregation. If two types of particles differ only in size and have the same density, then larger particles have higher critical temperature, and therefore should condense first. Thus, at an intermediate temperature, one would expect to observe the reverse Brazil-nut effect, i.e. condensation of large grains near the bottom. At certain values of parameters this effect is indeed observed in molecular dynamics simulations (see Fig. 7.6), but often this effect is masked by stronger kinetic sieving (percolation) mechanism that acts in the opposite direction. Hong et al. (2001) proposed an approximate criterion for the condensation to overcome percolation in the form

$$\left(\frac{d_2}{d_1}\right)^{D-1} < \frac{m_2}{m_1}, \tag{7.13}$$

where D is the dimension of space.

Kinetic interaction. The limitation of the condensation model proposed by Hong et al. (2001) is that it does not take into account interaction between large and small species. Furthermore, it only predicts segregation in a certain range of temperatures when large particles condense and small ones are still in the dilute regime. In fact, if the interaction of species is taken into account, the segregation of binary mixtures in a gravity field occurs even when both phases are in the dilute state. Jenkins and Yoon (2002) derived the expressions for the relative hydrodynamic velocity of the two species starting from the Enskog theory with an additional interaction term in the momentum balance equation that describes momentum transport between the two species (Jenkins and Mancini, 1987). This interaction leads to a relative diffusion flux even in the absence of a temperature gradient, thus, this mechanism of segregation is fundamentally different from the thermal diffusion described above. It is based on the competition between the particle inertia (through the ratio of their masses) and the geometry as it enters through the finite-density corrections to the partial pressures. In the dilute limit, this segregation mechanism is independent of the grain sizes, and only dependent on the grain masses.

For the case when larger-diameter spheres (type 2) are dilute and smaller-diameter spheres (type 1) are dense, the segregation condition can be written in the form (Jenkins and Mancini, 1987)

$$\psi \equiv \frac{m_2 d_1^3}{m_1 d_2^3} = \frac{m_2}{m_1}\zeta^3 = \frac{1}{24}(\zeta+1)(\zeta+2)(\zeta+3), \tag{7.14}$$

where $\zeta = d_1/d_2$ is the ratio of particle diameters, $m_{1,2}$ are particle masses, and ψ is the density ratio. The condensation condition (eqn 7.13) can also be written in this notation as $\psi = \zeta$. The comparison between the two conditions is illustrated in Fig. 7.7. For the values of ψ above the solid line, the large spheres rise; for values of ψ below it, large spheres fall. As it follows from Fig. 7.7, while the mechanism of kinetic interaction of Jenkins and Yoon (2002) is significantly different from the condensation

178 Patterns in granular segregation

proposed by Hong et al. (2001), the criteria for the segregation to occur are quite similar.

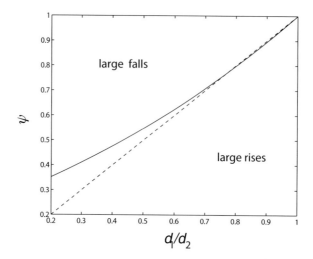

Fig. 7.7 Illustration of segregation condition (eqn 7.14) (solid line) and the condensation condition (eqn 7.13) (dashed line) in the plane ζ, ψ, from Jenkins and Yoon (2002).

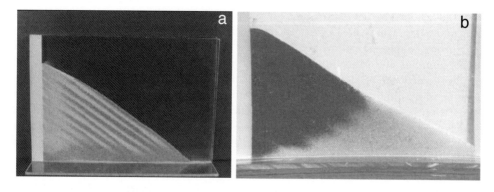

Fig. 7.8 Side view of granular stratification and segregation in a flow down a two-dimensional sandpile between two glass walls: (a) mixture of small rounded grains (white glass beads) and large rough grains (black sugar); (b) large round grains (glass spheres) and small rough particles (black sand). From Makse et al. (1998).

7.1.5 Experimental tests

The large number of proposed mechanisms of granular segregation call for experimental validation. However, unambiguous experimental conclusions are difficult to reach. Various mechanisms may compete and thus mask the presence of each other. Many system parameters have to be explored, and in different parameter regimes different mechanisms may become prevalent. Breu et al. (2003) performed a number of experiments on vertical shaking of binary mixtures of various materials, and they observed that depending on the mass ratio or the amplitude of driving, a bidisperse system of large and small particles may exhibit either normal or reverse Brazil-nut effect (see Fig. 2.7). To date, the most detailed experimental study of granular segregation in vertically vibrated containers was carried out by Schröter et al. (2006). They eliminated the mechanisms of segregation based on the mass and friction differences by using small and large particles of identical material, and systematically varied other parameters of the system: the ratio of particle sizes, amplitude and frequency of driving, volume fraction of large and small particles, and the total height of the granular layer. Schröter et al. (2006) also augmented their experimental study by molecular dynamics simulations of an identical system. This comparison allowed probing of the role of granular convection by turning it "off" in the numerical system via replacing frictional boundary conditions on sidewalls by periodic boundary conditions. The results of this extensive work appear to favor the kinetic sieving mechanism as the most prevalent for the Brazil-nut effect at small amplitudes (less than one large particle diameter) and high frequencies (80 Hz) of driving. At low frequencies (15 Hz) the normal Brazil-nut effect persists, but it evidently is caused mainly by granular convection, because in parallel numerical simulations with periodic boundary conditions, no noticeable segregation occurs at these conditions. For strong shaking in shallow layers, the thermal diffusion in a granular temperature gradient is the most likely candidate for the mechanism of reverse Brazil-nut effect. While this study did not find the signatures of other proposed mechanisms of segregation such as condensation or kinetic interaction, one cannot exclude that these mechanisms may become dominant in different parameter regimes, for example for significantly different densities of particles.

7.2 Granular stratification

Granular stratification is a particular form of segregation that produces layered structures similar to that shown in Fig. 7.8(a). Typically, it occurs when a binary mixture of particles with different physical properties is involved in an intermittent near-surface flow. Such flows may occur when sand is slowly poured down on a sandpile (Gray and Hutter, 1997; Makse and Herrmann, 1998; Koeppe et al., 1998) or when a binary mixture is slowly rotated in a drum, see Fig. 2.15. The primary mechanism of stratification is related to the avalanches acting as kinetic sieves (Savage and Lun, 1988; Savage, 1993; Gray and Thornton, 2005). During an avalanche, voids are continuously being created within the flowing near-surface layer, and small particles are more likely to fall into them. This creates a downward flux of smaller particles that is compensated by the upward flux of larger particles in order to maintain a zero total particle flux across the flowing layer. Thus, the avalanche has the form of a rolling double layer where large particles roll above smaller particles. If the avalanche reaches the bottom of the sandpile, as in laboratory experiments (Makse and Herrmann, 1998), it forms a kink that stops the oncoming flow of grains and propagates upward to the top of the hill. As it moves up, it leaves behind a double layer of small and large grains, see Fig. 7.9. Once the kink reaches the top of the sandpile, a new avalanche forms, and

180 Patterns in granular segregation

the process repeats. This process may occur only if the two types of grains differ both in size and in roughness, such that larger grains have additionally larger roughness resulting in a larger angle of repose. Indeed, in the opposite case, smaller grains get trapped in the upper part of the sandpile, and only larger grains reach the bottom, so instead of stratification, a large-scale segregation occurs, see Fig. 7.8(b).

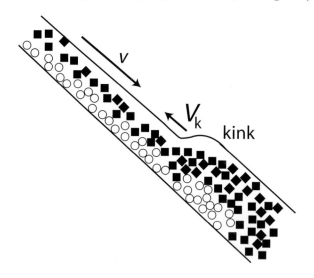

Fig. 7.9 Sketch of a double-layer formation in a chute flow.

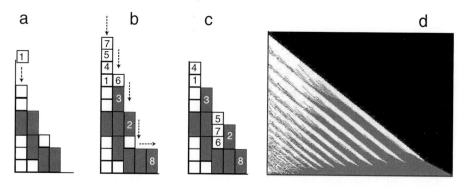

Fig. 7.10 Cellular automata model of granular stratification: (a–c) relaxation rules; (d) a typical stratification pattern obtained in numerical simulations. From Makse et al. (1997b)

Makse et al. (1997b), Makse et al. (1997a) proposed a cellular automata model that generalized the classical sandpile model by Bak et al. (1987) (see Section 6.6). In this model, a sandpile is built on a lattice, and rectangular grains have identical horizontal sizes but different heights (see Fig. 7.10(a)–(c)). Grains are released at the top of the heap sequentially, and they are allowed to roll down the slope. A particle would become rolling if the local slope (defined as the height difference between neighboring columns) exceeds a certain number corresponding to the angle of repose in a real sandpile. To account for differences in grain properties, four different angles of repose $\varphi_{\alpha\beta}$ are introduced for grains of type α rolling on a substrate of type β ($\alpha, \beta \in \{1, 2\}$, where 1 and 2 again stand for small and large grains, respectively). Normally, $\varphi_{21} < \varphi_{12}$ because of the geometry (small grains tend to get trapped by large grains), and one-component angles of repose usually lie within this range, $\varphi_{21} < \varphi_{11}, \varphi_{22} < \varphi_{12}$. However, the ratio of $\varphi_{11}, \varphi_{22}$ depends on the relative roughness of the grains. For $\varphi_{21} < \varphi_{11} < \varphi_{22} < \varphi_{12}$ (large grains are rougher), the model yields stratification in agreement with experiment (Fig. 7.10(d)). If, on the other hand, $\varphi_{22} < \varphi_{11}$ (which corresponds to smaller grains being rougher), the model yields only large-scale segregation: large particles collect at the bottom of the sandpile.

This discrete cellular automata model can also be recast in the form of continuum equations for a two-component near-surface granular flow (Boutreux and de Gennes, 1996; Makse et al., 1997a) that generalize the single-species (BCRE) model of monodisperse near-surface granular flows (Bouchaud, 1994) (see Section 6.2.2):

$$\partial_t z_{R\alpha} = -v_\alpha \partial_x z_{R\alpha} + \Xi_\alpha, \tag{7.15}$$

$$\phi_\alpha \partial_t z_S = -\Xi_\alpha, \tag{7.16}$$

where $z_{R\alpha}(x, t), v_\alpha$ are the thickness and velocity of rolling grains of type α ($\alpha = 1$ for small particles and 2 for large particles), ϕ_α is the concentration of *static* grains of type α near the surface, $z_S(x, t)$ is the instantaneous profile of the "substrate" of the sandpile, and Ξ_α characterizes the rate of conversion of rolling grains of type α into the substrate of static grains. The total thickness of the rolling layer $z_R(x, t) = z_{R1}(x, t) + z_{R2}(x, t)$. It is natural to assume that for thin flowing layers, the conversion rates Ξ_α depend linearly on the partial thicknesses of flowing layers of two types of grains via a 2×2 "collision" matrix $\hat{M}$,

$$\Xi_\alpha = \sum_{\beta=1}^{2} M_{\alpha\beta} z_{R\beta}. \tag{7.17}$$

The collision matrix $\hat{M}$ itself can be a complicated function of the surface slope angle φ and relative concentrations of two species ϕ_1 and ϕ_2. Boutreux et al. (1999) deduced the canonical form of the interaction matrix based on a microscopic picture of grain–grain interactions leading to the exchanges between flowing and static layers (see Fig. 7.11). Figure 7.11(a) depicts the process of sending a static grain of type 1 into rolling by a collision with a rolling grain of the same type (autoamplification). For a thin rolling layer, the corresponding rate of exchange is linearly proportional to z_{R1} and ϕ_1, i.e. $\Xi_1^a = a_1(\varphi)\phi_1 z_{R1}$. The proportionality coefficient $a_1(\varphi)$ is a function of the local slope angle φ. For three other processes shown in Fig. 7.11, the interaction rates are $\Xi_1^b = -b_1(\varphi)\phi_1 z_{R1}$ (Fig. 7.11(b), autocapture), $\Xi_1^c = c_1(\varphi)\phi_2 z_{R1}$ (Fig. 7.11(c), cross-amplification), $\Xi_1^d = -d_1(\varphi)\phi_2 z_{R1}$ (Fig. 7.11(d), cross-capture) and similar for Ξ_2. For simplicity, one can assume that autocapture and cross-capture occur with the same rate, $b_\alpha = d_\alpha$, then the collision matrix takes the form

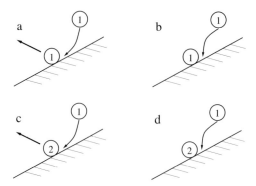

Fig. 7.11 Four types of grain–grain collisions between a rolling grain of type 1 and static grains of types 1 or 2: (a) autoamplification, (b) autocapture; (c) cross-amplification; (d) cross-capture.

$$\hat{M} = \begin{pmatrix} a_1(\varphi)\phi_1 - b_1(\varphi)\phi_2 & c_2(\varphi)\phi_1 \\ c_1(\varphi)\phi_2 & a_2(\varphi)\phi_2 - b_2(\varphi)\phi_1. \end{pmatrix} \quad (7.18)$$

Functions $a_\alpha, b_\alpha, c_\alpha$ near the angle of repose are assumed piecewise linear:

$$\begin{aligned} a_\alpha(\varphi) &= \gamma_a K(\varphi - \varphi_\alpha) \\ b_\alpha(\varphi) &= \gamma_b K(\varphi_\alpha - \varphi) \\ c_\alpha(\varphi) &= \gamma_c K(\varphi - \varphi_\beta), \end{aligned} \quad (7.19)$$

where $K(x) = x$ for $x > 0$ and $K(x) = 0$ for $x < 0$. The tangents of the repose angles φ_α, $\alpha = 1, 2$ are functions of the composition of the static substrate $\phi_{1,2}$. If index 1 corresponds to smaller grains for definiteness, then $\varphi_1(\phi_2) > \varphi_2(\phi_2)$ for any concentration of larger particles ϕ_2. We can assume for simplicity that the difference $\psi = \varphi_1(\phi_2) - \varphi_2(\phi_2)$ is independent on ϕ_2. Parameter ψ determines the difference in size between grains 1 and 2. If particles 1 and 2 have the same shape, then evidently $\varphi_1(\phi_2 = 0) = \varphi_2(\phi_2 = 1)$ because of the scale invariance of the repose angle. But if particles have different shape, than $\varphi_{11} \equiv \varphi_1(\phi_2 = 0)$ and $\varphi_{22} \equiv \varphi_2(\phi_2 = 1)$ are different. If grains 1 are more rounded than grains 2, $\Delta\varphi = \varphi_{22} - \varphi_{11} > 0$, and in the opposite case, $\Delta\varphi < 0$. These two cases are sketched in Fig. 7.12. With the collision matrix completely specified, it is relatively straightforward to find a stationary solution of the governing equations (eqns 7.16) (see Makse et al. (1997b) for details). This solution is divided into two regions: $\varphi_2(\phi_2) < \varphi < \varphi_1(\phi_2)$ (region A) and $\varphi < \varphi_2(\phi_2)$ (region B). The first region corresponds to the upper part of the sandpile, and the second region corresponds to the lower part of the sandpile. The profiles of the concentrations and thicknesses of rolling grains of the two types are shown in Fig. 7.13.

The sketches of the steady sandpile profile corresponding to $\Delta\varphi > 0$ and $\Delta\varphi < 0$ are shown in Fig. 7.14. Makse et al. (1997b) analyzed the stability of this steady-state solution and showed that it became unstable if $\Delta\varphi > 0$. The mechanism of the instability can be understood as follows. Suppose a small amount of large grains is

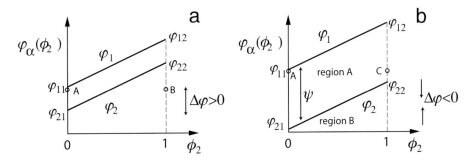

Fig. 7.12 Dependence of the partial angles of repose for two types of rolling grains on the relative concentration of large grains ϕ_2: for two qualitatively different cases: (a) $\Delta\varphi \equiv \varphi_{22} - \varphi_{11} > 0$ and (b) $\Delta\varphi < 0$, from Makse et al. (1997b).

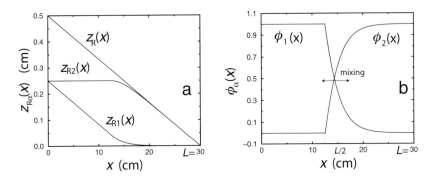

Fig. 7.13 Steady-state solution of the two-component granular flow equations (eqns 7.16): (a) thicknesses of layers of rolling grains $z_{R1,2}(x)$ and $z_R(x) = z_{R1}(x) + z_{R2}(x)$; (b) concentrations $\phi_{1,2}(x)$ of grains 1 and 2 in the static substrate, from Makse et al. (1997b).

embedded near the surface in the top portion of the sandpile where, according to the stationary solution, only small grains should be. Since the amount of large grains is assumed small, it does not change the profile of the sandpile in the first approximation. If a mixture of large and small grains flows over this substrate, one can write down the linear equation for the evolution of the thickness z_{R2} of the flowing layer of large grains

$$\partial_t z_{R2} = -v\partial_x z_{R2} + \gamma(\varphi - \varphi_{22})z_{R2}, \tag{7.20}$$

where $\gamma_1 = \gamma_2 = \gamma$ is assumed. The top part of the layer (Region A) is made of small grains, and the slope $\varphi \approx \varphi_{11}$, so the difference $\varphi - \varphi_{22} \approx \Delta\varphi$. Thus, according to eqn 7.20, z_{R2} evolves as $\exp(\gamma\Delta\varphi t)$. It grows exponentially if $\Delta\varphi > 0$ (Fig. 7.14(a)), and decays exponentially in the opposite case (Fig. 7.14(b)). The exponential growth of z_{R2} corresponds to the large particles being excavated from the substrate, so the small initial perturbation of large particles in the substrate will decay. When z_{R2} is decaying, it means that large particles are being trapped in the upper part of the sandpile, so the initial perturbation in the substrate will grows, and the stationary

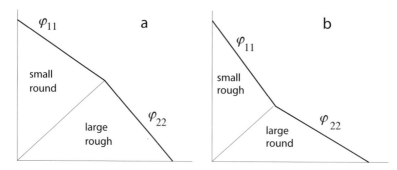

Fig. 7.14 Sketches of sandpile profiles corresponding to the steady-state solution of the two-component granular flow equations (eqn 7.16): (a) $\Delta\varphi > 0$; (b) $\Delta\varphi < 0$.

segregated profile becomes unstable. The result of this instability is the formation of the layered structure similar to shown in Fig. 7.8. This picture agrees both with simulations of the continuum model and experimental data.

Qualitatively similar stratification patterns can be observed in a thin slowly rotating drum that is more than half-filled with a binary mixture (Gray and Hutter, 1997); see Fig. 7.15. Periodic avalanches, occurring in the drum, lead to the formation of strata by the same mechanism as described above.

More recently, Gray and Thornton (2005) proposed another version of the continuum theory for the particle size segregation in shallow granular free-surface flows. The theory, based on the assumption of kinetic sieving of large and small grains in shear avalanche-type flow, is formulated in terms of granular hydrodynamics equations (eqns 3.11 and 3.12) for mass and momentum of each phase (small (1) and large (2) particles). The interaction between the phases is introduced through the additional interaction drag terms $\beta_{s,l}$ in the momentum equations (eqn 3.12):

$$\beta^i = p\nabla f^i - \nu^i C_0(\mathbf{v}^i - \mathbf{v}), i = 1, 2, \quad (7.21)$$

where the factors $f^{1,2}$ determine the proportion of hydrostatic load carried by small and large particles, p is the pressure, $\mathbf{v}^{1,2}$ and $\nu^{1,2}$ are partial velocities and densities of each phase, $C_0 = $ const is the coefficient of interparticle drag. Assuming some functional dependence of the factors $f^{1,2}$ on the corresponding filling fractions $\phi^{1,2}$, the model reproduces certain experimental features, e.g. formation of segregation shocks. The open question is, however, the applicability of this approach to dense granular flows where the validity of hydrodynamics is rather limited (see discussion in Chapter 3).

7.3 Radial segregation patterns in a rotating drum

The most common system in which granular segregation is studied is a rotating drum, or a partially filled horizontal cylinder rotating around its axis. When a polydisperse mixture of grains is rotated in a drum, strong *radial segregation* usually occurs within just a few revolutions. Small, dense and/or rough particles aggregate to the center (*core*) of the drum, large and smooth particles circulate around the core, see Fig.

2.15. The bulk rotates almost as a solid body, and the segregation predominantly occurs within a thin fluidized near-surface layer. Rotating cylinders are widely used in industry for mixing solids or as reactors (kilns) for solid-gas reactions, and have been the subject of extensive engineering studies (Henein et al., 1983; Boateng and Barr, 1996; Davidson et al., 2000). The mechanisms of segregation are the same as the ones discussed above for other granular flows (kinetic sieving, percolation, etc.) and had been fairly well understood in qualitative terms a long time ago (Williams, 1976). In the last decade the focus of research has shifted toward detailed experiments and modeling aimed at the quantitative understanding of fundamental mechanisms controlling the onset and structure of segregation patterns in specific systems (Ottino and Khakhar, 2000; Kudrolli, 2004).

The mechanisms of segregation and resulting structures depend crucially on the flow regime inside the rotating cylinder as defined by the Froude number $\mathrm{Fr} = \Omega R^2/g$ where Ω is the rotation speed, R is the cylinder radius, and g is the acceleration of gravity (see also Section 6.5 for analysis of flows of monodisperse particles in rotating drum).

7.3.1 Segregation in the avalanching regime

In the avalanching regime, well separated in time avalanches occur when the slope of the free surface becomes steeper than the static angle of repose φ_s. Each avalanche takes with it a wedge of grains from the upper half of the drum, and transfers it to the lower half. During this event, the new deposits segregate as was discussed above: small and rough particles get trapped below a layer of larger and smoother particles. Unlike a set of parallel stripes in a sandpile (Fig. 7.8), due to drum rotation these double layers form a Catherine wheel-like pattern with stripes tangent to the core (Fig. 7.15). On a quantitative level, mixing and segregation patterns in slowly rotating drum can be understood in terms of simple mapping of the uphill wedges of granular material (i.e. fraction of the material in the drum above critical angle φ_d) into downhill wedges in the course of avalanche (Metcalfe et al., 1995). This process can also be described by an iterated map of the particles inside the drum, see (Hill et al., 1999).

7.3.2 Segregation in the rolling regime

Radial segregation in rotating drums is a direct consequence of the shear flow that occurs in the drum. At low rotation speeds the flow is confined to the free surface. The bulk exhibits solid-body rotation and simply transfers the particle distribution from the bottom half of the flow to the top half of it. At low rotation speeds, the kinetic sieving mechanism dominates, and small heavy particles sink and form a core surrounded by lighter bigger particles. At large rotation rates, the segregation pattern reverses, and bigger particles assemble at the core, whereas small particles remain on the periphery. Here we will only consider the segregation at slow rotation speeds. The mechanism of segregation can be understood using the same two-component model Makse (1999) as was used above for chute flows. Namely, we assume that the rolling layer of total thickness R is comprised of two components $z_{R1,2}$ associated with particles of types 1 and 2, so $z_R = z_{R1} + z_{R2}$. The equations for the rolling species are (compare to eqns 6.20 and 6.21)

186 *Patterns in granular segregation*

Fig. 7.15 Granular avalanche-induced stratification in a rotating drum observed for low rotation rates: dark and light regions correspond to high concentrations of two different species comprising the mixture; from Gray and Hutter (1997).

$$\partial_t z_{Ri}(x,t) = v(x,t)\partial_x z_{Ri} + \Xi_i, \ i = 1,2, \tag{7.22}$$

and the profile of the static grains $z_S(x,t)$ underlying the rolling layer is described by the equation

$$\phi_i(x,t)(\partial_t z_S - \Omega x) = -\Xi_i, \tag{7.23}$$

where $v(x,t)$ is the velocity of grains in the rolling layer, ϕ_i are the near-surface fractions of static grains, $\phi_1 + \phi_2 = 1$, and Ξ_i are the conversion rates from rolling to static grains given by empirical formulas 7.17–7.19. This model generalizes eqns 7.15 and 7.16 for the case when the static grains surface profile is rotated with the drum with angular velocity Ω and the velocity v is a function of x and t. In fact, the velocity in the rolling layer depends strongly on the normal coordinate z: it is maximal at the free surface and zero at the bottom of the rolling layer $z = -z_R(x,t)$. It is reasonable to assume that the velocity $v(x,t)$ entering the mass conservation equations (7.22) is the height-averaged velocity of the shear flow: $v(x,t) = [z_R(x,t)]^{-1} \int_{-z_R}^{0} v(x,z,t) dz$. Some chute flow experiments (Savage, 1984; Nakagawa et al., 1993; Pouliquen and Gutfraind, 1996) show that the velocity profile in the rolling layer is close to linear with a constant shear rate α (vertical velocity profiles and flow conditions are discussed in detail in Section 6.1): $v(x,z,t) = 2\alpha(z + z_R(x,t))$, then the average velocity v is proportional to the local depth of the rolling layer: $v(x,t) = \alpha z_R(x,t)$. The total flux

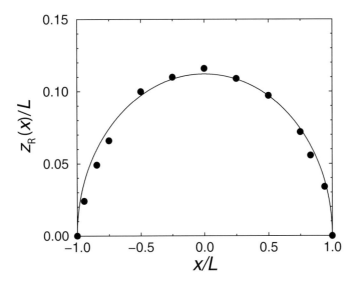

Fig. 7.16 Thickness of the rolling layer $z_R(x)$ from theory (eqn 7.25) with $\alpha = 25s^{-1}$ (line) and experiment (Khakhar et al., 1997b; Khakhar et al., 1997a), from Makse (1999).

of mass through the rolling layer $J = \nu v(x,t) z_R(x,t) = \nu \alpha z_R^2$ can be found from the mass conservation (here ν is the density of granular material),

$$dJ/dx = -\nu \Omega x, \qquad (7.24)$$

since all the mass that is brought to the surface by the sold-body rotation of the bulk has to be "flown away" by the surface flowing layer. From this equation one can easily find the flux as a function of the coordinate x along the free surface (Rajchenbach, 1990), $J = \nu \Omega (L^2 - x^2)$. Substituting $J = \nu \alpha z_R^2$, we obtain the expression for the local thickness of the rolling layer

$$z_R = \sqrt{\Omega(L^2 - x^2)/\alpha}. \qquad (7.25)$$

Figure 7.16 compares expression 7.25 with experimental data (Khakhar et al., 1997b; Khakhar et al., 1997a). A similar expression for the thickness of the rolling layer was obtained by Elperin and Vikhansky (1998) using shear-stress continuity at the interface $z = -z_R$ and assuming that at the interface the Mohr–Coulomb condition is satisfied.[1]

Using eqns 7.22, 7.23, and 7.19 we can now solve for $z_{R1,2}(x)$ in the stationary state. The solution is symmetric with respect to $x = 0$, so we only have to look for a solution at $x < 0$ (upstream from the middle point).

If the two types of grains have similar properties, so all angles of repose $\varphi_{\alpha\beta}$ are close to each other, we can linearize the conversion rates (eqn 7.19)

$$a_i(\varphi) = c_i(\varphi) = C + \gamma[\varphi(x) - \varphi_i(\phi_j)]$$

---

[1] A similar expression can also be obtained from the partial fluidization theory, see Section 6.5 and eqn 6.97. The difference in scaling with x is due to a different assumption on the flow rheology.

$$b_i(\varphi) = C - \gamma[\varphi(x) - \varphi_i(\phi_j)], \qquad (7.26)$$

with

$$\varphi_1(\phi_2) = \varphi_{11} + (\varphi_{12} - \varphi_{11})\phi_2$$
$$\varphi_2(\phi_2) = \varphi_{21} + (\varphi_{12} - \varphi_{11})\phi_2. \qquad (7.27)$$

Note that in this case the difference $\psi = \varphi_1(\phi_2) - \varphi_2(\phi_2) = \varphi_{11} - \varphi_{21} = \varphi_{12} - \varphi_{22}$ is independent of the surface concentration ϕ_2. Small parameter ψ characterizes the difference in properties of the grains that affect their respective repose angles.

Substituting eqns 7.26 and 7.27 into eqns 7.23 and 7.19 we can find the surface profile $\varphi(x) \equiv \partial h/\partial x$ as a function of the profile of the rolling grains,

$$\varphi(x) - \varphi_1(\phi_2) = -\frac{\psi z_{R2}}{z_R}. \qquad (7.28)$$

Now we can substitute it in the stationary equation 7.22 and obtain the solution in a closed form (Makse, 1999)

$$z_{R1}(x) = \frac{z_R(x)}{1 + A(L^2 - x^2)^{-\xi}}, \qquad (7.29)$$

where $\xi = \gamma\psi/C$. The corresponding solution for the concentration profiles is

$$\phi_1(x) = \frac{z_{R1}(x)}{z_R(x)}\left[1 + \xi\left(1 - \frac{z_{R1}(x)}{z_R(x)}\right)\right], \quad \phi_2(x) = 1 - \phi_1(x). \qquad (7.30)$$

As expected, parameter ξ, which is proportional to the difference in the repose angles ψ controls the intensity of segregation. Dashed line in Figure 7.17 shows the solution (eqn 7.30) for a small difference between the species ($\psi \ll 1$). As the degree of difference increases, so does parameter ξ, and eventually the linear approximation that was used to derive it, breaks down. For large enough ψ, particles with smaller repose angle (usually, smaller particles) $\varphi_{12} < \varphi_{21}$ convert to the static phase before reaching the middle of the drum. Then, the smaller static particles will screen the larger rolling particles from the static bulk layer, and so the conversion rate b_2 must vanish. To account for this effect, we have to use the full non-linear form for the conversion functions (eqn 7.19). In this case the solution has to be sought separately in the *inner* region $|x| < x_m$ where $\varphi_2(\phi_2) < \varphi(x) < \varphi_1(\phi_2)$ and in the *outer* region $x_m < |x| < L$ where $\varphi(x) < \varphi_2(\phi_2) < \varphi_1(\phi_2)$. In the inner region $\phi_1 = 1, \phi_2 = 0$, and in the outer region

$$\phi_1(x) = \exp[\gamma\psi(|x| - x_m)/(\sqrt{\Omega a}L)], \quad \phi_2 = 1 - \phi_1. \qquad (7.31)$$

Near the center of the drum we have total segregation, and outside of the total segregation region, the concentration of smaller grains decays exponentially. This solution is also shown in Fig. 7.17. The ratio of the width of the interface to the drum radius $d_s/L = \sqrt{\Omega a}/\gamma\psi$ depends on the rotation speed Ω and material properties of grains α, γ, ψ. The interface becomes sharper for smaller rotation rates and larger ψ,

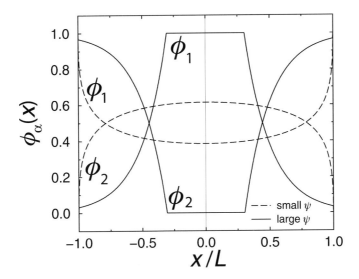

Fig. 7.17 Stationary concentrations of two species $\phi_{1,2}(x)$ in a bulk for small and large ψ, from (Makse, 1999)

which characterizes the difference in particle properties, in agreement with experiments (Clément et al., 1995).

If the rotation rate of the drum is periodically modulated, $\Omega = \Omega_0 + \Omega_1 \sin\omega t$, the core may become unstable and exhibit a multipetal structure as in Fig. 7.18(a). These structures are formed by a non-stationary multicell flow pattern that emerges in the drum as predicted by the theory (Fiedor and Ottino, 2005) in which the thickness and the flow inside the near-surface rolling layer are periodic functions of time. The number n of petals in this pattern corresponds to a number of islands of periodic trajectories surrounding elliptic fixed points in the corresponding Poincaré sections of the flow pattern (Fig. 7.18(b)) that in turn depends on the ratio of the mean angular velocity of drum rotation Ω_0 and the modulation frequency ω, $n = \omega/2\Omega_0$. It is interesting that the Poincaré sections predict chaotic flow patterns in some regions of the bulk flow, which can lead to enhanced mixing of grains.

Multipetal segregation patterns can also be observed at low rotation rates near the transition to the avalanching regime. In this case the near-surface flow becomes periodically modulated even for a constant drum rotation speed (Hill et al., 2004; Zuriguel et al., 2006), and the same mechanism leads to the formation of petals (see Fig. 2.16). Some aspects of segregated multipetal structures in circular and even non-circular drums (tumblers) can be captured by the continuum flow model incorporating collisional diffusion and density-driven segregation (Hill et al., 1999). The model describes the flow in the regime where the flowing layer is steady, thin, and nearly flat, and the rest of the particles is in the solid-body rotation. Surprisingly, under certain conditions, concentration distribution never settles into a steady state and displays persistent chaotic dynamics.

Radial segregation can also be observed in a drum almost completely packed with

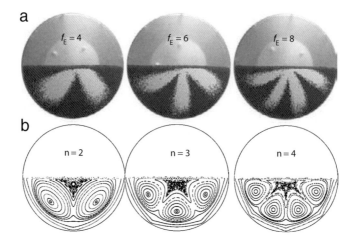

Fig. 7.18 Multipetal segregation patterns in a rotating drum in which the rotation rate is periodically modulated, from Fiedor and Ottino (2005): (a) experimentally observed patterns; (b) theoretical Poincaré sections for different values of $f_E = \omega/\Omega_0$.

polydisperse grains (Nakagawa *et al.*, 1998). Interestingly, at slow rotation rates, i.e. small Froude number, $\mathrm{Fr} = R^2\Omega/g < 1$, small particles accumulate at the core, and for larger rotation rates, the segregation reverses: large particles accumulate near the center of the cylinder. In this case the mechanism of segregation must be different from kinetic sieving in a thin rolling layer because there is no rolling layer in this case. The mechanism here is in fact related to the periodic modulation of gravitational force that acts on particles in the reference frame rotating with the drum (Awazu, 2000). A combination of this periodic force and a centrifugal force due to rotation acts on each particle and leads to a slow percolation flux of small particles along the direction of the force. For large rotation rates, the centrifugal force dominates gravity, and small particles percolate toward the boundary, however, at small Ω gravity dominates, and it pushes small particles toward the center. This simple qualitative picture is confirmed by numerical simulations of a binary mixture of grains in a thin box that is subject to a constant "centrifugal" force acting from the axis to the boundaries and a periodic "gravity" force acting along the z-axis: $F_z = F_0 z + F_1 \cos\Omega t$ (see Fig. 7.19).

7.4 Axial segregation in rotating drums

In long narrow drums with the length significantly exceeding the radius, radial segregation is often followed by the *axial segregation* occurring at much later stages (after several hundred revolutions). As a result of axial segregation, a pattern of well-segregated bands is formed (Zik *et al.*, 1994; Hill and Kakalios, 1994; Hill and Kakalios, 1995) (see Fig. 2.17) which slowly merge and coarsen. At low rotation speeds coarsening saturates at a certain finite bandwidth when discrete avalanches provide granular transport (Frette and Stavans, 1997), and at higher rotation rates in a continuous flow regime axial segregation progresses toward a final state in which all grains are separated in

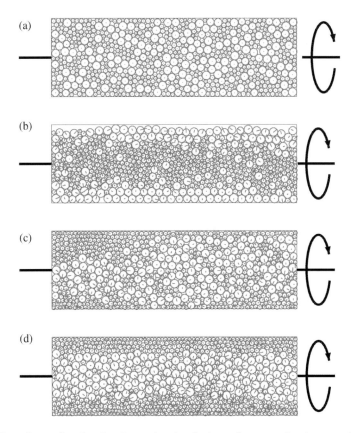

Fig. 7.19 Snapshots of molecular dynamics simulations of segregation in a nearly packed thin box: (a) initially premixed condition; (b) slow rotation, $R\Omega^2 < g$; (c) intermediate rotation $R\Omega^2 \sim g$; (d) fast rotation $R\Omega^2 > g$, from Awazu (2000).

two bands (Zik et al., 1994; Fiedor and Ottino, 2003; Arndt et al., 2005; Finger et al., 2006).

Axial segregation has been known in the engineering community for several decades, apparently it was first observed by Oyama (1939). A representative space-time diagram illustrating evolution and coarsening of bands in a rotating drum is shown in Fig. 7.20. For low rotation rate (Fig. 7.20(a)), no stable band pattern is formed, but smaller beads assemble in irregular non-stationary "clouds". Above a certain rotation speed ($\Omega > 10$ rpm), regular band pattern emerges (Figs. 7.20(b) and (c)). At the initial stage the pattern of more or less equidistant bands slowly coarsens through coalescence in the course of time. Small beads contained in the dissolving bands are distributed between the neighboring bands of small beads in the array.

The decay rate of the number of bands is practically independent of the rotation speed in a certain parameter range. Figure 7.21 shows the number of bands N as a function of the number of rotations for different rotation speeds in a semilogarithmic plot. This dependence is not very sensitive to the rotation speed in the parameter

192 *Patterns in granular segregation*

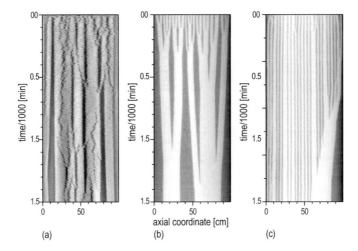

Fig. 7.20 Space-time diagram of the development of segregated bands for a mixture of spherical black and white glass beads with bimodal radius distribution, radius ratio 3 : 1, for three different rotation speeds; (a) $\Omega = 5$ rpm, (b) $\Omega = 15$ rpm, (c) $\Omega = 30$ rpm, from Finger et al. (2006).

range between approximately 10 and 22 rpm; for the 1-m drum it can be fitted with the logarithmic function

$$N(t) = N_0(1 - n\log(\Omega t/2\pi)), \tag{7.32}$$

with fitting constants $N_0 = 55$ and $n = 0.092$.

7.4.1 Mechanisms of axial segregation

Different frictional properties of the grains lead to different *dynamic angles of repose*. The latter is defined as the angle of the slope in the drum corresponding to the continuous-flow regime, however, in real drums the free surface often has a more complicated S-shape (Zik et al., 1994; Elperin and Vikhansky, 1998; Makse, 1999; Orpe and Khakhar, 2001), as was discussed in Section 6.5. It is appealing to suggest that the variations in dynamic angle of repose provide the driving mechanism for axial segregation. This approach was adopted in several models. According to Zik et al. (1994) and Levine (1999), if there is a local increase in concentration of particles with higher dynamic angle of repose, the local slope there will be higher, and that will lead to a local bump near the top of the free surface and a dip near the bottom. As the particles tend to slide along the steepest descent path, more particles with higher angle of repose will accumulate in this location, and an instability will develop. Zik et al. (1994) proposed a phenomenological continuum model of axial segregation based on the conservation law for the relative concentration of the two components ("glass" and "sand"), $c(z,t) = (\phi_1 - \phi_2)$,

$$\partial_t c = -C(\tan\varphi_1 - \tan\varphi_2)\partial_y(1 - c^2)\left\langle (1 + z_x^2)\frac{z_y}{z_x} \right\rangle. \tag{7.33}$$

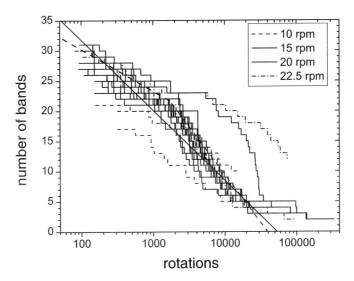

Fig. 7.21 Number of bands N in a half-filled drum as a function of time for different rotation speeds and for intermediate rotation rates $\Omega = 10 - 22.5$ rpm. The data roughly follow a logarithmic decay, with few exceptions for high rotation rates, solid line shows the fit to eqn 7.32, from Finger et al. (2006).

Here, x and y are Cartesian horizontal coordinates across and along the axis of the drum, $z(x, y, t)$ describes the instantaneous free surface inside the drum, ϕ_1, ϕ_2 are normalized concentrations of individual components ($\phi_1 + \phi_2 = 1$), C is a constant related to gravity and effective viscosity of granular material in the flowing layer. The term in angular brackets denotes the axial flux of the glass beads averaged over the cross-section of the drum. The profile of the free surface $z(x, y)$ in turn should depend on $c(y, t)$. If $\langle (1+z_x^2) z_c / z_x \rangle < 0$, linearization of eqn 7.33 leads to the diffusion equation with negative diffusion coefficient that exhibits the segregation instability with growth rate proportional to the square of the wave number. Zik et al. (1994) conclude that the term in angular brackets vanishes for a straight profile $z_x = \text{const}(x)$ due to the assumption that z_y changes sign exactly in the middle of the drum, and, as a result, mixing/demixing processes cancel precisely. However, for the experimentally observed S-shaped profile of the free surface Zik et al. (1994) calculated that the instability condition is satisfied when the drum is more than half full. While experiments show that axial segregation in fact observed even for less than 50% filling ratio, the model gives a good intuitive picture for a possible mechanism of the instability.

7.4.2 Oscillatory segregation

Choo et al. (1997) and Choo et al. (1998) found that at early stages, the small-scale perturbations propagate across the drum in both directions (this was clearly evidenced by the experiments with pre-segregated mixtures), while at later times more long-scale static perturbations take over and lead to the emergence of quasistationary bands of separated grains, see Fig. 7.22. Similar travelling wave patterns and coarsening bands

194 *Patterns in granular segregation*

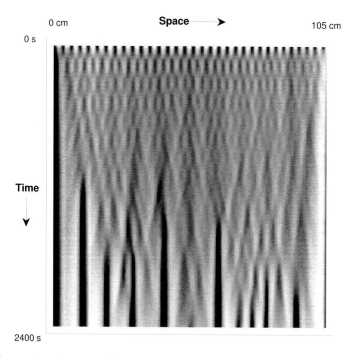

Fig. 7.22 Space-time diagram of the granular surface in a long rotating drum demonstrating oscillatory size segregation. The plot shows the full length of the drum and extends over 2400 s, or 1850 revolutions. Black bands correspond to 45–250 µm black sand and white bands correspond to 300–850 µm table salt, from Choo et al. (1997).

were also observed by Fiedor and Ottino (2003). The slow coarsening process can be accelerated in a drum of a helical shape (Zik et al., 1994). Alternatively, bands can be locked in an axisymmetric drum with the radius modulated along the axis (Zik et al., 1994). Consequent studies (Alexander et al., 2004; Charles et al., 2006) revealed an interesting scaling of the size of the axial bands with the drum diameter and grain size. In particular, segregation shuts off if the ratio of the drum diameter to the grain size is sufficiently small. The band spacing also scales in a non-trivial way with the drum diameter.

In order to account for the oscillatory behavior of axial segregation at the initial stage, Aranson and Tsimring (1999) and Aranson et al. (1999c) generalized the model of Zik et al. (1994). The key assumption was that besides the concentration difference, there is an additional slow variable which is involved in the dynamics. It was conjectured that this slow variable is the instantaneous slope of the granular material (dynamic angle of repose) that unlike model 7.33 is not completely enslaved to the relative concentration c, but obeys its own dynamics. The equations of the model read

$$\partial_t c = -\partial_y(-D\partial_y c + u(c)\partial_y \varphi), \tag{7.34}$$
$$\partial_t \varphi = \alpha(\Omega - \varphi + f(c)) + D_\varphi \partial_y^2 \varphi + \gamma \partial_y^2 c. \tag{7.35}$$

The first term in the rhs of eqn 7.34 describes a diffusion flux (mixing), and the second term describes the differential flux of particles due to the gradient of the dynamic angle of repose. This term is equivalent to the rhs of eqn 7.33 with a particular function $u(c) = G_0(1 - c^2)$. For simplicity, the constant G_0 can be eliminated by rescaling of distance $y \to y/\sqrt{G_0}$. The sign + before this term means that the particles with the larger static angle of repose are driven toward greater dynamic angle of repose. This differential flux gives rise to the segregation instability. Since this segregation flux vanishes with $u(c) \to 0$ (or $|c| \to 1$ that corresponds to pure 1 or 2 states), it provides a saturation mechanism for the segregation instability.

Parameter Ω in eqn 7.35 is the normalized angular velocity of the drum rotation, and $f(c)$ is the static angle of repose, which is an increasing function of the relative concentration (Koeppe et al., 1998), for simplicity it can be assumed linear, $f(c) = F + f_0 c$. The constant F can be eliminated by the substitution $\varphi \to \varphi - F$. The first term in the rhs of eqn 7.35 describes the local dynamics of the angle of repose (Ω increases the angle, and $-\varphi + f(c)$ describes the equilibrating effect of the surface flow), and the term $D_\varphi \partial_y^2 \varphi$ describes axial diffusive relaxation. The last term, $\gamma \partial_y^2 c$, represents the lowest-order non-local contribution from an inhomogeneous distribution of c (the first derivative $\partial_x c$ cannot be present due to reflection symmetry $y \to -y$).[2] This term gives rise to the transient oscillatory dynamics of the binary mixture.

Linear stability analysis of a homogeneous state $c = c_0$; $\varphi_0 = \Omega + f_0 c_0$ reveals that for $u_0 f_0 > \alpha D$ long-wave perturbations are unstable, and if $u_0 \gamma > (D_\varphi - D)^2/4$, short-wave perturbations oscillate and decay (two eigenvalues $\lambda_{1,2}$ are complex conjugate with a negative real part for modulation wave numbers $|k| > k^*$), see Fig. 7.23. This agrees with the general phenomenology observed by Choo et al. (1997) both qualitatively and even quantitatively, Fig. 7.23(b). The results of direct numerical solution of the full model (eqns 7.34 and 7.35) are illustrated in Fig. 7.24. It shows that short-wave initial perturbations decay and give rise to more long-wave non-oscillatory modulation of concentration that eventually leads to well-separated bands. At long times (Fig. 7.24(a)) the bands exhibit slow coarsening with the number of bands decreasing logarithmically with time (see also (Frette and Stavans, 1997; Levitan, 1998; Fiedor and Ottino, 2003)) consistent with experimental observations by Arndt et al. (2005) and Finger et al. (2006), see Fig. 7.21.

7.4.3 Coarsening of bands

The logarithmic scaling for the number of bands (stripes) as a function of time follows from the exponentially weak interaction between interfaces separating different bands (Fraerman et al., 1997; Aranson et al., 1999c) and can be derived from the following qualitative argument. Fronts separating bands of different grains can be found as stationary solutions of eqns 7.33 and 7.35. In an infinite system one can obtain the following relation from stationary equation 7.33:

$$\varphi = \varphi_0 + DG(c), \qquad (7.36)$$

---

[2]In fact, term $\partial_y c$ is possible for the drum with the cross-section increasing along the axis or if the drum is not perfectly levelled.

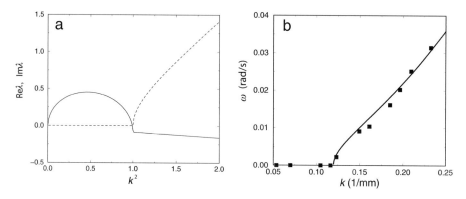

Fig. 7.23 (a) Dispersion relation $\lambda(k)$ for segregation instability, Reλ is shown by the solid line, and Imλ shown by the dashed line. Im$\lambda \neq 0$ for $k > k^* \approx 1$; (b) comparison of the frequency of band oscillations Im(λ) (solid line) with experiment (symbols), from Aranson et al. (1999c).

where φ_0 is an integration constant and function $G(c)$ is of the form $G(c) = \int [u(c)]^{-1} dc = -\frac{1}{2} \log \frac{1-c}{1+c}$. Substituting expression 7.36 into eqn 7.35, we obtain the second-order differential equation for concentration c (symmetric solution corresponds to $\varphi_0 = 0$)

$$\frac{d}{dy}\left[\left(\gamma + \frac{DD_\varphi}{1-c^2}\right)\frac{dc}{dy}\right] + \alpha f_0 c + \frac{\alpha D}{2} \log \frac{1-c}{1+c} = 0. \tag{7.37}$$

This equation indeed possesses an interface solution. Its asymptotic behavior can be found in the limit $D \ll f_0$, when the states on both sides of the interface are well segregated: $c(y \to \pm\infty) \to \pm 1$. Without loss of generality we can fix the position of the interface separating two segregated states at $y = 0$ and choose $c \to 1$ for $y \to \infty$ and $c \to -1$ for $y \to -\infty$. In this limit, one finds from the behavior of linearized eqn 7.37 near $c = 1$: $1 - c \sim \exp[-y/l_0]$ for $y \to \infty$ and $c - 1 \sim \exp[y/l_0]$ for $y \to -\infty$. Here, the characteristic "mixing" length l_0 is of the form

$$l_0 = \sqrt{\frac{D_\varphi}{\alpha} + \frac{4\gamma}{\alpha D} \exp(-2f_0/D)}. \tag{7.38}$$

Slow logarithmic coarsening of the segregated state can then be understood in terms of the weak interaction of these segregated fronts. Since the asymptotic field of the front approaches the equilibrium value of the concentration exponentially $\exp[-x/l_0]$, we expect exponentially large times for front interaction, $T \sim \exp[\Delta/l_0]$, where Δ is the initial distance between fronts. For a multiband configuration as shown in Fig. 7.24(c), the number of fronts N, which is of the order of the average inverse distance between fronts Δ, decreases roughly as a logarithmic function of time $N \approx 1/\Delta \sim 1/(\text{const} + l_0 \log t)$. This dependence is qualitatively consistent with the numerical solution to eqns 7.33 and 7.35, and with experiments (Arndt et al., 2005; Finger et al., 2006).

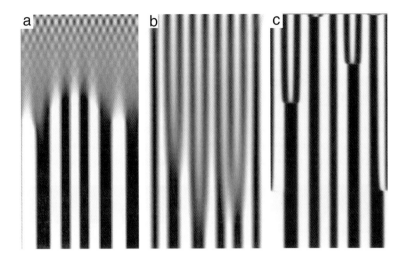

Fig. 7.24 Space-time diagrams demonstrating initial band oscillations and subsequent coarsening. (a) Initial transient, vertical dimension spans time interval $0 < t < 1.25$. Periodic initial conditions with $k > k^*$ excite a decaying standing wave. (b) For the same parameters but for $k < k^*$ periodic initial conditions produce no oscillations (same time interval). (c) Long-term coarsening of bands at higher diffusion constant, time interval $0 < t < 5000$ from Aranson et al. (1999c).

7.4.4 Further experimental findings

While these continuum models of axial segregation showed a good qualitative agreement with the data, recent experimental observations demonstrate that the theoretical understanding of axial segregation is still far from complete (Ottino and Khakhar, 2000). The interpretation of the second slow variable as the local dynamic angle of repose implies that in the unstable mode the slope and concentration modulation should be in phase, whereas in the decaying oscillatory mode, these two fields have to be shifted in phase. Further experiments (Khan et al., 2004) showed that while the inphase relationship in the asymptotic regime holds true, the quadrature phase shift in the transient oscillatory regime is not observed. This led Khan et al. (2004) to hypothesize that some slow variable other than the angle of repose (possibly related to the core dynamics) may be involved in the transient regime.[3] However, so far experiments have failed to identify which second dynamical field is necessary for oscillatory transient dynamics, so it remains an open problem.

One should realize that a direct comparison of simplified phenomenological models similar to eqns 7.33–7.35 with experiments is difficult. These models are based on a number of simplifying assumptions. The most serious of them is the neglect of the radial segregation that typically precedes axial segregation. Radial segregation makes the local concentration of grains an ambiguous quantity. Which concentration

---

[3]Experiments by Khan et al. (2004) however, rule out that the second field is simply related to the width of the submerged core.

actually affects the repose angle: the concentration on the free surface, or the average concentration over the cross-section of the drum, or something else? The important role of the radial segregation in the dynamics of axial segregation was highlighted by the magnetic resonance imaging studies (Hill, 1997) that demonstrated that in fact the bands of larger particles usually had a core of smaller particles. Similarly, Fiedor and Ottino (2003), Khan et al. (2004), and Arndt et al. (2005) showed that small particles formed a modulated shish-kebab-like structure with bands connected by a rod-like core, while large particles formed disconnected rings (see Fig. 7.25). All this suggests that a "unified" three-dimensional theory of radial-axial segregation is needed to explain quantitatively the salient features of the axial segregation.

Fig. 7.25 Image of shish-kebab-like structure of smaller particles in rotating drum (illuminated from below). Smaller particles (shown in dark) form bands between bands of larger particles (disconnected rings, shown in white and gray), from Arndt et al. (2005).

Another recent experimental observation by Khan and Morris (2005) suggested that instead of the normal diffusion assumed in eqns 7.34 and 7.35, a slower subdiffusion of particles in the core takes place, the mean displacement of tracer particles $\langle y \rangle$ scales with time as $\langle y \rangle \sim t^\gamma$ with the scaling exponent γ close to 0.3. The most plausible explanation is that the apparent subdiffusive behavior is in fact a manifestation of a *non-linear* diffusion of concentration that can be described by the following equation

$$\partial_t c = \partial_y D(c) \partial_y c. \tag{7.39}$$

For example, for a generic concentration-dependent diffusion coefficient $D \sim c$, the asymptotic scaling behavior of the concentration $c(y,t)$ is given by the self-similar function $c \sim F(y/t^\alpha)/t^\alpha$ for $t \to \infty$ with the scaling exponent $\alpha = 1/3$ close to 0.3 observed experimentally. The experimentally observed scaling function $F(y/t^\alpha)$ appears to be consistent with eqn 7.39 except for the tails of the distribution where $c \to 0$ and the assumption $D \sim c$ is possibly violated. Normal diffusion behavior corresponding to $D = \text{const}$ and $\alpha = 1/2$ is in strong disagreement with the experiment. Surprisingly, more recent soft particles molecular dynamics simulations by Taberlet and Richard (2006) showed that the self-diffusion of both large and small grains is normal and that the diffusion coefficient is independent of the grain size. While the origin of this discrepancy remains unknown, it possibly originates from the force models used in simulations.

7.4.5 Axial segregation in ternary mixtures

Newey et al. (2004) conducted studies of axial segregation in ternary mixtures of granular materials. It was found that for certain conditions bands of ternary mixtures oscillate axially, which indicates the supercritical *oscillatory instability*. Unlike earlier experiments of Choo et al. (1997), Choo et al. (1998) with binary mixtures, these oscillatory bands in ternary mixtures are *persistent*. Another difference is that while in binary mixtures the oscillations have the form of periodic mixing/demixing, in the ternary mixtures the oscillations are in the form of periodic band displacements. It is likely that the mechanism of band oscillations in ternary mixtures is very different from that of binary mixtures. One possible explanation could be that the third component provides an additional degree of freedom necessary for oscillations. To demonstrate this we can write phenomenological equations for two concentration differences $C_A = c_1 - c_2$ and $C_B = c_2 - c_3$, where $c_{1,2,3}$ are concentrations of individual components. By analogy with eqn 7.34 we write the system of coupled equations for the concentration differences $C_{A,B}$ linearized near the fully mixed state:

$$\partial_t C_A = D_A \partial_y^2 C_A + \xi_A \partial_y^2 C_B,$$
$$\partial_t C_B = D_B \partial_y^2 C_B + \xi_B \partial_y^2 C_A. \tag{7.40}$$

If the cross-diffusion terms have opposite signs, i.e. $\xi_A \xi_B < 0$, the concentrations $C_{A,B}$ will exhibit oscillations in time and in space. Obviously this mechanism is intrinsic to ternary systems and has no counterpart in binary mixtures. However, in experimental situations due to polydispersity often there is no clear distinction between binary and ternary mixtures of grains. Thus, the mechanism described by eqns 7.40 may also have relevance for mixtures with broad size distributions. Experiments by Choo et al. (1998) indeed demonstrated very high sensitivity of the axial segregation, and, especially, oscillatory segregation behavior, to the details of the particle-size distribution.

7.4.6 Molecular dynamics simulations

In parallel with the theoretical studies, molecular dynamics simulations have been performed (Shoichi, 1998; Rapaport, 2002; Taberlet et al., 2004). Simulations allowed researchers to probe the role of material parameters that would be difficult to access in laboratory experiments. In particular, Rapaport (2002) addressed the role of particle–particle and wall–particle friction coefficients separately. It was found that the main role is played by the wall–particle friction: if the friction coefficient between large particles and the wall is greater than that for smaller particles, the axial segregation always occur irrespective of the ratio of particle–particle friction coefficients. However, if the particle–wall coefficients are equal, the segregation may still occur if the friction among large particles is greater than among small particles. Taberlet et al. (2004) studied axial segregation in a system of grains made of identical material differing only in size. The simulations revealed rapid oscillatory motion of bands, which is not necessarily related to the slow band appearance/disappearance observed in experiments (Choo et al., 1997; Choo et al., 1998; Fiedor and Ottino, 2003; Arndt et al., 2005).

In a recent paper Rapaport (2007) studied ternary axial size segregation using a soft-particles simulation technique. The simulations reproduced axial band formation, with medium-size particles tending to be located between alternating bands of big and small particles, as well as coarsening of bands. The axial segregation was also accompanied by partial radial segregation characterized by an inner core region richer in small particles. Overall, the behavior is qualitatively similar to experiments on binary and ternary mixtures.

A different type of discrete element modelling of axial segregation was proposed by Yangagita (1999). This model builds upon the lattice-based sandpile model and replaces a rotating drum by a three-dimensional cubic lattice. Drum rotation is modelled by correlated displacement of particles on the lattice: particles in the back are shifted upward by one position, and the particles at the bottom are shifted to fill the voids. This displacement steepens the slope of the free surface, and once it reaches a critical value, particles slide down according to the rules similar to the sandpile model of Bak et al. (1987) but taking into account different critical slopes for different particles. This model, despite its simplicity, reproduced both radial and axial segregation patterns and therefore elucidated the critical components needed for adequate description of the phenomenon.

7.5 Other examples of granular segregation

As we have seen in the previous sections, granular segregation often originates in near-surface shear granular flows, such as in silos, hoppers, and rotating drums. However, other types of shear granular flows may also lead to segregation. For example, Taylor–Couette flow of granular mixtures between two rotating cylinders leads to the formation of Taylor vortices and then in turn to segregation patterns (Conway et al., 2004).

Pouliquen et al. (1997) observed granular segregation in a thin granular flow on an inclined plane. In this case, segregation apparently occurs as a result of an instability in which the concentration mode is coupled with the hydrodynamic mode. As a result, segregation occurs simultaneously with a fingering instability of the avalanche front, see Fig. 2.12. As implicit evidence of this relation between segregation and fingering instability, Pouliquen et al. (1997) found that a monodisperse granular material does not exhibit fingering instability. However, experiments in rotating drums (Shen, 2002) and avalanches on erodible substrate (Malloggi et al., 2006; Aranson et al., 2006) indicate that fingering instability may occur even in flows of monodisperse granular materials (see Section 6.5). Thus, the segregation could be a consequence rather than the primary cause of the fingering instability.

Interesting segregation patterns emerge in horizontally shaken layers of binary granular mixtures (Mullin, 2000; Mullin, 2002; Reis and Mullin, 2002). After several minutes of horizontal shaking with frequency 12.5 Hz and displacement amplitude 1 mm (which corresponds to the acceleration amplitude normalized by gravity $\Gamma = 0.66$), stripes are formed orthogonal to the direction of shaking. The width of the stripes grows continuously with time as $t^{0.25}$, thus indicating slow coarsening (Fig. 2.8). This power law is consistent with the diffusion-mediated mechanism of stripe merging. Reis and Mullin (2002) concluded from the experimental results that the segregation in this system bears features of a second-order phase transition. Critical slowdown was observed near the onset of segregation. The order parameter is associated with the combined filling fraction C, or the layer *compacity*,

$$C = \frac{N_1 A_1 + N_2 A_2}{S}, \tag{7.41}$$

where $N_{1,2}$ are numbers of particles in each species, $A_{1,2}$ are projected two-dimensional areas of the respective individual particles, and S is the tray area. Ehrhardt et al.

(2005) proposed a simple model to describe the segregation in horizontally vibrated layers. The model is based on a two-dimensional system of hard disks of mass m_α and radius R_α ($\alpha = 1, 2$ denote the species)

$$m_\alpha \dot{\mathbf{v}}_{\alpha i} = -\gamma_i \left(\mathbf{v}_{\alpha i} - \mathbf{v}_{\text{tray}}(t) \right) + \zeta_{\alpha i}(t), \qquad (7.42)$$

where $\mathbf{v}_i$ is the particles velocity $\mathbf{v}_{\text{tray}}(t) = A_0 \sin(\omega t)$ is the oscillating tray velocity, γ provides linear damping, and $\zeta_{\alpha i}$ is white noise acting independently on each disk. The model reproduces the segregation instability and subsequent coarsening of stripes. In more realistic discrete element simulations (Ciamarra et al., 2005) a binary mixture of round disks of identical sizes but two different coefficients of friction with the bottom plate (in fact, velocity-dependent viscous drag was assumed), segregated in alternating bands perpendicular to the oscillation direction irrespective of initial conditions: both a random mixed state and a pre-separated state were used. Using particles of the same size eliminated the thermodynamic "excluded volume" mechanism of segregation. Ciamarra et al. (2005) argued that the pattern-forming mechanism here was related to the dynamical shear instability similar to the Kelvin–Helmholtz instability in ordinary fluids. It was confirmed by a numerical observation of the interfacial instability when two monolayers of grains with different friction constant were placed in contact along a flat interface parallel to the direction of horizontal oscillations. Similar instability is apparently responsible for the ripple formation (Scherer et al., 1999; Stegner and Wesfreid, 1999).

Pooley and Yeomans (2004) proposed a theoretical description of this experiment based on the continuum model for two periodically–driven isothermal ideal gases interacting through frictional force. It was shown analytically that segregated stripes form spontaneously above a critical forcing amplitude. While the model reproduces the segregation instability, apparently it does not exhibit the coarsening of stripes observed in the experiment. Moreover, applicability of the isothermal ideal-gas description to the experimental situation where particles are almost at rest is an open question.

Granular segregation and coarsening in a particularly simple geometry was studied by Aumaître et al. (2001). They investigated the dynamics of a monolayer of grains of two different sizes in a dish involved in a horizontal "swirling" motion. They observed that large particles tend to aggregate near the center of the cavity surrounded by small particles, see Fig. 7.26. The qualitative explanation of this effect follows from simple thermodynamic considerations (see above). Indeed, direct tracing of particle motion showed that the pressure in the area near the large particles is smaller than outside. But small particles do not follow the gradient of pressure and assemble near the center of the cavity because this gradient is counterbalanced by the force from large particles. The reciprocal force acting on the large particles leads to their aggregation near the center of the cavity. Aumaître et al. (2001) proposed a quantitative model of segregation based on the kinetic gas theory and found a satisfactory agreement with experimental data.

As we mentioned earlier, almost any difference in physical properties of grains can lead to segregation in granular matter. Burtally et al. (2002) studied spontaneous separation of vertically vibrated mixtures of particles of similar sizes but different densities (bronze and glass spheres). At low frequencies and at sufficiently large vibrational am-

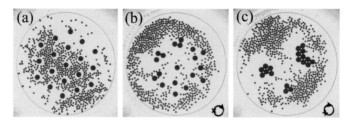

Fig. 7.26 Size segregation in the system of 19 disks and 600 spheres observed in a co-moving frame driven with 1.16 Hz. (a) initial state; (b) After 360 s; (c) final state, after 4896 s. The black circles indicate the motion of the surface in the laboratory frame, from Aumaître et al. (2001).

plitudes, a sharp boundary between the lower layer of glass beads and the upper layer of the heavier bronze spheres was observed. At higher frequencies, bronze particles emerge as a middle layer separating upper and lower glass bead layers. The authors argue that air plays an essential role in the particle separation. A similar conclusion was drawn by Möbius et al. (2001) in experiments with a vertically–vibrated column of grains containing a large "intruder" particle.

Interstitial fluids (e.g. water or oil) may have quantitative and qualitative effects on granular segregation as well. Mixtures of heterogeneous particulates with various fluids are relevant for a variety of chemical technologies. Fiedor and Ottino (2003) and later Arndt et al. (2005) performed detailed experiments on axial segregation in slurries, or bidisperse grain–water mixtures. A mixture of two types of spherical glass beads of two sizes were placed in a water-filled tube at the volume ratio 1:2. The authors found that both the rotation rate and the filling fraction play an important role in band formation. Namely, bands are less likely to form at lower fill levels (20–30%) and slower rotation rates (5–10 rpm). The bands are more likely to appear near the ends of the drum. At higher fill levels and rotation rates, bands form faster, and there are more of them throughout the drum. Fiedor and Ottino (2003), Arndt et al. (2005) also studied the relation between the bands visible on the surface, and the core of small beads, and found that for certain fill levels and rotation speeds, the core remains prominent at all times, while in other cases the core disappears completely between bands of small particles. They also observed an interesting oscillatory instability of interfaces between bands at high rotation speeds. All these phenomena still await theoretical modelling.

8
Granular materials with complex interactions

Until this chapter, our working model of granular matter was an assembly of (nearly) spherical macroscopic objects interacting via short-range contact forces. If grains have essentially non-spherical shape, even these contact forces may lead to novel phenomena and different classes of pattern formation. Furthermore, in many practically important situations a variety of long- and short-range forces may come into play, for example long-range electromagnetic forces due to rubbing and charging of particles, adhesion, interaction with an interstitial fluid, etc. Understanding of the dynamics of granular systems with complex interactions is an intriguing and rapidly developing field. In this chapter we discuss non-trivial pattern-forming effects arising due to the interplay between "traditional" contact forces that are always present in granular systems and other interaction forces.

8.1 Vortices in vibrated layers of granular rods

In Chapter 5 we discussed instabilities and collective motion in mechanically vibrated layers. In most experiments particles had either spherical or irregular but close to spherical shape (as in sand). This shape irregularity did not have a significant effect on pattern formation. However, a strong particle anisotropy may give rise to non-trivial effects. Villarruel et al. (2000) observed the onset of nematic order in packing of long rods in a narrow vertical tube subjected to vertical tapping. Rods initially compactify into a disordered state with predominantly horizontal orientation, but at later times (after thousands of taps) they align vertically, first along the walls, and then throughout the volume of the pipe. The nematic ordering, which occurs spontaneously above the critical filling fraction, can be, at least qualitatively, understood in terms of the excluded volume argument put forward by Onsager (1949).

8.1.1 Onsager theory for rigid rods

Here, we briefly overview the theory of the nematic–isotropic transition in hard-rod liquids pioneered by Onsager (1949). A detailed presentation of this theory can be found in Doi and Edwards (2003). In order to address the effects of particle shape, Onsager (1949) generalized the expression for the free energy $\mathcal{F}$ of a dilute gas of hard spheres of radius r to a dilute solution of very thin rods (large length L to diameter d ratio). This free energy is functionally dependent on the probability distribution function $\Psi(\mathbf{n})$ of rods with respect to their orientation $\mathbf{n}$.

The virial expansion for the free energy of a dilute gas of N hard spheres of number density $\nu = N/V$, (N is the number of particles (spheres), V is the volume) has the form

$$\frac{\mathcal{F}_s}{\nu k_B T} = -1 + \log \nu + \frac{1}{2}\nu B_2 + O(\nu^2), \tag{8.1}$$

where T is the thermodynamic temperature, $\log \nu$ is the entropic contribution, and B_2 is the second virial coefficient (due to the excluded volume) which for hard spheres is a constant equal to $\frac{4}{3}\pi(2r)^3$. According to the Onsager theory, the analogous expression for the free energy of the gas of hard rods is

$$\frac{\mathcal{F}_r}{\nu k_B T} = -1 + \log \nu - S_{or} + \frac{1}{2}\nu B_2 + O(\nu^2), \tag{8.2}$$

where now the second viral coefficient B_2 depends on the distribution of rod orientations $\Psi(\mathbf{n})$,

$$B_2 = \int d\mathbf{n} \int d\mathbf{n}' \tilde{\beta}(\mathbf{n},\mathbf{n}') \Psi(\mathbf{n}) \Psi(\mathbf{n}'), \tag{8.3}$$

and there is an additional entropic contribution S_{or} due to the orientational degree of freedom of the rods:

$$S_{or} = -\int d\mathbf{n} \Psi(\mathbf{n}) \log(\Psi(\mathbf{n})). \tag{8.4}$$

The kernel $\tilde{\beta}(\mathbf{n},\mathbf{n}')$ comprises the excluded volume spanned by two overlapping rods having directions $(\mathbf{n},\mathbf{n}')$. In the three-dimensional case for very long rods ($L/d \to \infty$), one finds $\tilde{\beta}(\mathbf{n},\mathbf{n}') = 2L^2 d|\mathbf{n} \times \mathbf{n}'| = 2L^2 d|\sin\varphi|$

A minimum of the free energy corresponding to the variation of eqn 8.2 with respect to Ψ, yields the following non-linear integral equation

$$\log(\Psi(\mathbf{n})) = C - 2\nu dL^2 \int d\mathbf{n}' |\mathbf{n} \times \mathbf{n}'| \Psi(\mathbf{n}'), \tag{8.5}$$

where the constant C is fixed by the normalization condition $\int d\mathbf{n} \Psi(\mathbf{n}) = 1$. Equation 8.5 can be solved numerically. An approximate solution can be found by substituting into eqn 8.5 an anzatz

$$\Psi(\mathbf{n}) = \frac{\Pi}{4\pi \sinh \Pi} \cosh(\Pi \mathbf{n} \cdot \mathbf{n_0}), \tag{8.6}$$

where $\mathbf{n_0}$ is an arbitrary unit vector, and Π is the (scalar) order parameter characterizing the nematic ordering of rods ($\Pi = 0$ corresponds to the isotropic state, and $\Pi = \infty$ corresponds to a unidirectional distribution along $\pm \mathbf{n_0}$. Substituting eqn 8.6 into eqn 8.5 yields the free energy $\mathcal{F}$ as a function of Π and ν. The result is schematically shown in Fig. 8.1: for small ν there is only one minimum at $\Pi = 0$ (disordered regime), and $\nu > \nu_1$ there is a second minimum at a positive Π corresponding to a nematic state that is, however, still metastable. At some higher ν_c, the first minimum disappears, and the only remaining steady state is nematic. Explicit calculation shows that the critical density ν_c for the nematic–isotropic transition in three dimensions is $\nu_c \approx 4.44(dL^2)^{-1}$.

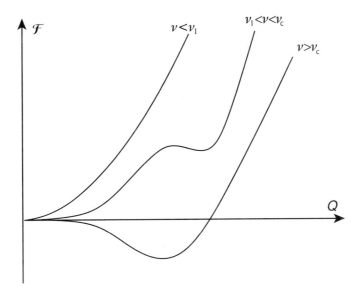

Fig. 8.1 Sketch of the free energy $\mathcal{F}$ as a function of the order parameter Π at various densities of thin rods.

While the Onsager theory has been very successful in application to liquid crystals and polymers, its predictions for granular flows should be taken with caution because it neglects the third- and higher-order viral coefficients that can become important (Frenkel and Eppenga, 1985). Moreover, for finite-aspect-ratio particles the details of the transition and the critical density depend on the shape of the particles. For example, numerical studies indicate that for hard prolate ellipsoids of revolution the isotropic–nematic transition occurs only if the ratio of long to short axis is above two (Frenkel et al., 1984).

8.1.2 Experiments with vibrated rods

One can naïvely expect that a gas of rods subjected to high-frequency excitation would exhibit features similar to the thermodynamic nematic transition as described above. One can argue that due to multiple collisions between the rods and the bottom of a container correlations between impacts should decay on a very short collision time scale, and the effect of the excitation of a submonolayer of granular rods on a vibrated plate should be equivalent to random noise, i.e. the temperature. Indeed, in studies by Villarruel et al. (2000) and follow-up experiments by Galanis et al. (2006), nematic ordering was observed in a wide range of vibration parameters and particles aspect ratios.

However, there was a surprise. Blair et al. (2003) conducted experimental studies of vertically vibrated multilayers of thin rods in a large aspect ratio system (the thickness of the layer of rods, which was of the order of the rod length and hence much greater than the rod diameter, was much smaller than the horizontal system size). For particles with an aspect ratio about 12, at large enough amplitude of vertical acceleration

206 *Granular materials with complex interactions*

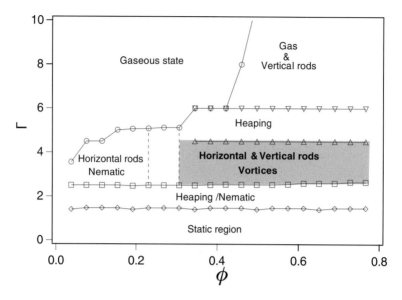

Fig. 8.2 Phase diagram for the system of vertically vibrated rods at a driving frequency 50 Hz. Vortices are observed for sufficiently high filling fraction ϕ and above the critical acceleration Γ, from Blair et al. (2003).

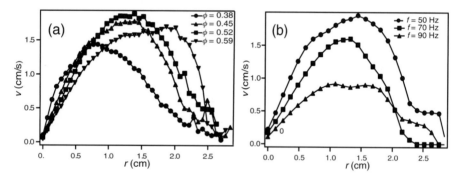

Fig. 8.3 Azimuthal velocity of the vortex v vs. distance from the center r for different parameter values, from Blair et al. (2003).

($\Gamma > 2.2$) rods assume a vertical orientation in the bulk of the system without an apparent influence of the sidewalls. Eventually, most of the rods align themselves vertically in a dense monolayer and begin to synchronously jump on the plate, and then engage in a correlated horizontal motion in the form of rotating vortices; see Figs. 2.19 and 8.2. The vortices exhibit almost solid-body rotation near the core, and then the azimuthal velocity tapers off, Fig. 8.3. The vortices merge in the course of their motion, and eventually form a single vortex.

To identify the nature of the directed motion of tilted rods, in subsequent work by Volfson et al. (2004), a simpler quasione-dimensional system of a single row of tilted

 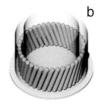 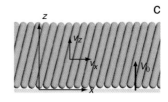

Fig. 8.4 Annulus geometry used in the experiments (a) and simulations (b). For theoretical analysis, a quasione-dimensional system of rods constrained to the $x-z$ plane with periodic boundary conditions along the x-axis (c) is more convenient, from Volfson et al. (2004).

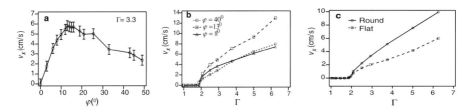

Fig. 8.5 (a) The average horizontal velocity v_x of rods as a function of tilt angle φ. (b) v_x versus Γ for three different tilt angles. (c) The effect of the rod end shape on measured v_x for $\varphi = 28° \pm 2°$. Rods with flat ends are observed to move slower compared to rods with rounded ends. Horizontal motion in a direction opposite to the tilt is observed over a small range of low Γ, from Volfson et al. (2004).

rods in an annulus (Fig. 8.4) was studied, and it was shown directly that the motion of individual rods is a direct consequence of their tilt. The advantage of the quasione-dimensional geometry is that the tilt can be varied systematically simply by changing the number of rods inside the annulus. Effectively, tilted rods bouncing on a frictional surface become self-propelled objects similar to other self-propelled systems for which large-scale coherent motion is often observed (bird flocks, fish schools, chemotactic micro-organism aggregation, etc., see Chapter 9 for details). Of course, here the tilt and collective motion are only maintained because of the dense packing of rods, one rod without contacts with its neighbors would remain nearly horizontal (except for small vibrations) and on average would remain static.[1] The dependencies of the average horizontal rod velocity v_x on the plate acceleration Γ and the tilt of the rods φ are shown in Fig. 8.5. For high plate acceleration, the rods move in the direction of the tilt. However, in a certain range of accelerations and tilt of the rods, they were observed to move in the opposite direction.

Theoretical analysis of collective rods dynamics. Full three-dimensional simulations of a large number of vibrated rods are difficult and time consuming. Volfson et al.

---

[1] This is not entirely true, in certain regimes of vibration even symmetric inelastic objects such as dumbbells, may assume an asymmetric mode of vibration and hence acquire momentum from a frictional substrate. This "self-organized ratchet" motion was investigated experimentally and theoretically in Dorbolo et al. (2005).

208 *Granular materials with complex interactions*

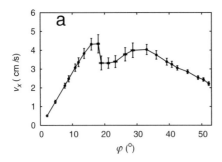 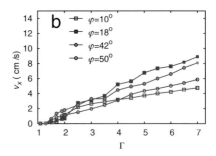

Fig. 8.6 The results of simulations in the annular geometry: average translational velocity of rods as a function of tilt angle φ at $f = 60$Hz, $\Gamma = 3.3$ (a) and the plate acceleration Γ for several tilt angles (b), from Volfson et al. (2004).

(2004) introduced a simplified molecular dynamics algorithm to simulate the dynamics of interacting vibrated rods. This algorithm is based on the standard "soft spheres" molecular dynamics technique, see Section 3.7.1 . The interaction forces between colliding spherocylinders are calculated from the intersection of virtual viscoelastic spheres of diameter d centered at the closest points between the axes of spherocylinders, so it is assumed that the cylinders are in contact whenever virtual spheres are. The normal forces between virtual spheres are computed using the Hertzian spring–dashpot model and the tangential frictional forces are computed by the Cundall-Struck algorithm, see Section 3.7.1. In most simulations the difference between dynamic and static friction coefficients is ignored. The forces arising from the interaction of virtual spheres are then applied to the rods. The motion of rods was obtained as usual by integrating Newton's equations with the forces and torques produced by interactions of a rod with all the neighboring rods, walls of the container, and gravity.

The simulations (see Fig. 8.4(b) and (c)) reproduced the observed phenomenology both qualitatively and quantitatively. By varying three friction coefficients (rod–rod μ_{rr}, rod–bottom μ_{rb}, and rod–sidewall μ_{rs}) separately, the bottom friction was identified as the most important ingredient for occurrence of the horizontal motion. The simulations suggested that the correlated motion can be explained by inelastic frictional contacts between the rods and the vibrated plate. The dependencies of the velocity of horizontal motion v_x on the tilt angle and vibration amplitude are shown in Fig. 8.6. The velocity initially increases with the tilt angle φ, peaks at some optimal angle $\varphi \approx 15°$, and then falls off again, similar to experimental data in Fig. 8.5. Once Γ is above the critical value needed for transport, v_x grows almost linearly with Γ.

Guided by the data from experiments and simulations, Volfson et al. (2004) constructed a contact-mechanical model for the rod dynamics. The starting point of the analysis is the frictional collision of a single rod with an oscillating plate. Despite its apparent simplicity, this problem turns out to be rather difficult. Three friction regimes can be identified: slide, slip–stick, and slip reversal, and during a single collision, a rod can start in one regime (usually, slide) and then move to another regime depending on the conditions (angle, velocity, angular speed) at the start of the collision. As a result of this calculation, the momentum transfer from a plate to the rod during a single collision can be found, which then can be used in a calculation of the average horizontal translation velocity

$$v_x = \frac{(1+e)[\mu(1+3\cos^2\varphi)+3\sin\varphi\cos\varphi]}{2(1-e+6\sin^2\varphi)} V_0, \quad (8.7)$$

where φ is the tilt angle, e is the kinematic restitution coefficient, μ is the friction coefficient, and V_0 is the amplitude of the plate velocity oscillations. This formula is in good agreement with both experiments and simulations.

8.1.3 Phenomenological theory of vortex formation in a rods system

The detailed contact-mechanical theory outlined above sheds light on the nature of the drift of tilted vibrated rods. However, it does not explain the nature of the global pattern formation, and in particular, the formation of coherent vortical structures observed experimentally. The onset and coarsening of vortices can be captured in the framework of a continuum phenomenological theory (Aranson and Tsimring, 2003) based on several key observations: (i) rods drift spontaneously in the direction of tilt; (ii) the drift velocity depends on the tilt angle which in turn is controlled by the degree of packing. The theory is formulated in terms of a local coarse-grained projection of the orientation vector of rods $\boldsymbol{\tau}$ on the bottom plane, or tilt, [2] and the local density ν. The theory assumes that a spontaneous transition to a vertical position occurs if the vibration acceleration exceeds a certain critical value. The transition to the vertical alignment manifests itself as nucleation of high-density islands where vertical rods are packed much denser per unit area than nearly horizontal rods.

Assuming that the motion of rods is overdamped due to the bottom friction, the local horizontal velocity $\mathbf{v} = (v_x, v_y)$ of rods can be written in the form

$$\mathbf{v} = -\nabla p/\mu\nu + \alpha\boldsymbol{\tau} f_0(\tau)/\mu, \quad (8.8)$$

where p is the hydrodynamic pressure and μ is (sliding) friction coefficient. The last term in the rhs is the "driving force" acting on the tilted particle due to vibration, as discussed in the previous section. Function $f_0(\tau)$ characterizes the dependence of the driving force on the tilt magnitude τ. Equation 8.8 combined with the mass conservation law yields

$$\partial_t \nu = -\nabla \cdot (\mathbf{v}\nu) = \mu^{-1} \nabla \cdot (\nabla p - \alpha \boldsymbol{\tau} f_0(\tau)\nu). \quad (8.9)$$

The pressure p is a complicated function of rod density and mean kinetic energy (granular temperature). The dependence of pressure p on physical parameters is not yet available from experiments and simulations. However, in the spirit of models of phase separation and ordering dynamics, such as Cahn–Hilliard theory (see, for example, Bray (1994)), we assume that pressure p can be obtained from the variation of a free-energy functional $\mathcal{F}$ with respect to the density ν, $p = \delta\mathcal{F}/\delta\nu$.

---

[2] Unlike the Onsager's director $\mathbf{n}$, which is invariant with respect to reflection $\mathbf{n} = -\mathbf{n}$, vector $\boldsymbol{\tau}$ has a unique direction associated with the tilt of the rod

We adopt here the generic form of the free energy taking into account the local dynamics and diffusive-type spatial coupling

$$\mathcal{F} = \int\int dxdy \left[\frac{L^2}{2}(\nabla\nu)^2 + f(\nu)\right], \tag{8.10}$$

where L is the length scale related to the rod size. To account for the phase separation, function $f(\nu)$ should have two minima separated by a maximum. We can choose a generic cubic polynomial form of $df/d\nu$. Without loss of generality we can write $df/d\nu = (\nu-\nu_0)[\delta_0 - \delta_1(\nu-\nu_0) + (\nu-\nu_0)^2]$. Since $df/d\nu$ is defined up to an additive constant, we can fix the root ν_0 to be a minimal density for the onset of nematic order, as we discussed in Section 8.1.1. The constants δ_0 and δ_1 depend on the driving acceleration. Experimentally, the dense phase nucleates for large enough vertical acceleration. In our description, a minimum in $df/d\nu$ corresponding to the existence of a dense phase appears for large δ_1. Below, we will associate δ_1 with acceleration Γ and keep δ_0 fixed.

Substituting p from eqn 8.10 into eqn 8.9, after rescaling we obtain the modified Cahn–Hilliard equation (we replaced $\nu - \nu_0$ by ν and used the same notations for the rescaled variables $r \to r/L$, $t \to t/\mu L^2$)

$$\partial_t \nu = -\nabla^2\left(\nabla^2\nu - \nu(\delta_0 - \delta_1\nu + \nu^2)\right) - \alpha\nabla\cdot(\tau f_0(\tau)(\nu+\nu_0)), \tag{8.11}$$

To close the description, an equation for the evolution of the tilt τ is required. For density ν smaller than the maximum packing density, the vertical orientation of rods corresponding to $\tau = 0$ is unstable, as rods spontaneously tilt. We assume that the rate of the instability depends on the rod packing density, so we can write for the local tilt dynamics $\partial_t\tau = f_1(\nu)\tau - |\tau|^2\tau$ with the instability rate $f_1(\nu) = a_0 - a_1\nu$ taken as a liner function of density with some constants $a_{0,1} > 0$, so for $\nu < a_0/a_1$ zero-tilt solution (vertical rods) is unstable. The last term in this equation describes the saturation of the instability, since rods will stabilize at a certain tilt dependent on the local density: the larger the density, the smaller is the equilibrium tilt.

Next, we have to augment the local dynamics by the spatial coupling operator $\hat{D}[\tau]$ describing interactions of neighboring rods in the continuum limit. Since generally the tilt field τ is not necessarily divergence-free ($\nabla\cdot\tau \neq 0$), from general symmetry considerations, in the lowest (second) order, the coupling operator acting on τ takes the form $\hat{D}[\tau] = f_2(\nu)\left(\xi_1\nabla^2\tau + \xi_2\nabla\nabla\cdot\tau\right)$. The coefficients $\xi_{1,2}$ in this expression are analogous to the first and second viscosity in ordinary fluids, Landau and Lifshitz (1987). Function $f_2(\nu)$ describes the decrease of the spatial coupling strength as the rod number density decreases. In a dilute gas phase ($\nu < 0$) the spatial coupling between rods is small, and their tilt becomes large and uncorrelated. Accordingly, we set $f_2 = \nu$, if $\nu > 0$ and $f_2 = 0$ otherwise. Finally, on the symmetry grounds, we should include the simplest term describing coupling between the tilt and the density gradient $\beta\nabla\nu$. Similar terms are also known in the context of liquid crystals and are associated with splay deformations. Summarizing, the evolution of the tilt vector τ is described as follows[3].

---

[3]In the next chapter we will encounter a similar equation derived from microscopic collision rules of interaction between microtubules and molecular motors

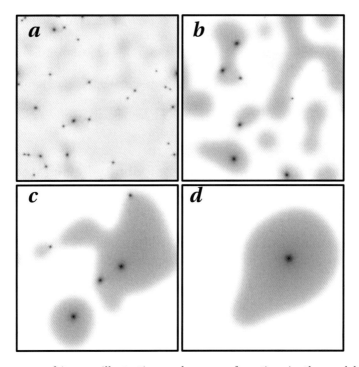

Fig. 8.7 Sequence of images illustrating coalescence of vortices in the model of vibrated rods, the magnitude of the field $|\tau|$ is shown in grayscale (black $|\tau| = 0$, white $|\tau| = 1$), black dots corresponds to vortex cores, from Aranson and Tsimring (2003).

$$\partial_t \tau = f_1(\nu)\tau - |\tau|^2 \tau + f_2(\nu)\left(\xi_1 \nabla^2 \tau + \xi_2 \nabla \nabla \cdot \tau\right) + \beta \nabla \nu. \tag{8.12}$$

For large δ_1 eqn 8.11 exhibits *phase separation* in a certain range of filling fractions defined as $\phi = S^{-1} \int \int \nu \mathrm{d}x \mathrm{d}y$, where S is the system area. Stationary uniform solutions to eqn 8.11 obey

$$\nu(\delta_0 - \delta_1 \nu + \nu^2) = B, \tag{8.13}$$

where the constant B is determined from the mass conservation condition.

In addition to the uniform solutions, eqns 8.11 and 8.12 admit solutions with non-zero topological charge: defects, or vortices. Let us consider a radially symmetric *vortex* solution to eqn 8.12. In this case ν is a function of the polar radius r only, and the tilt τ can be expressed as a complex variable $\hat{\tau} = \tau_x + i\tau_y = \exp(\pm i\theta + i\vartheta_0)w(r)$, where w is the amplitude of the vortex, θ is the polar angle, and ϑ_0 is a constant "phase" characterizing the direction of rod tilt with respect to the center of the vortex. It can be shown that in the limit $\nu = \mathrm{const}$ for $\xi_2 \neq 0$ only the solutions with $\vartheta_0 = \pm \pi/2$ are stable (Aranson and Tsimring, 2003). These solutions correspond to vortices.

In order to follow the evolution of the system from random initial conditions eqns 8.11 and 8.12 were studied numerically. At the initial stage of the evolution many vortices and small dense clusters (islands) are created throughout the domain of integration, see Fig. 8.7. Islands are seen as darker areas because an increase in density ν

results in the decrease of amplitude of the tilt vector $|\tau|$. Some islands trap vortices and are practically immobile, others do not contain vortices and drift in the direction defined by the average orientation of the tilt τ. With time, small islands disappear, and bigger islands grow, see Fig. 8.7(b) and (c). It is interesting to note that due to the tilt-driven drift coarsening occurs much faster than in standard Cahn–Hilliard theory (Bray, 1994). Eventually, one big island with the vortex in the center is formed (Fig. 8.7(d)). This qualitative picture of the phenomenon is in good agreement with experimental observations.

8.2 Swirling motion in monolayers of vibrated elongated particles

As we discussed in the previous section, long cylindrical particles subjected to vertical vibration should exhibit a tendency towards formation of nematic states. Initial experiments supported this observation: nematic-like states were indeed observed. However, more recent studies revealed new phenomena occurring in non-equilibrium systems of vibrated and interacting granular rods that are not captured by a standard thermodynamic description. In particular, it was shown that the resulting patterns depend sensitively on the shape and the physical properties of grains. Narayan et al. (2006) showed that vibrated rice grains of a certain type spontaneously form nematic crystalline-like structures, whereas longer particles with tapered ends exhibit short-range smectic ordering. Cylindrical particles with flat ends with the same aspect ratio instead demonstrate tetratic order, see Fig. 2.20. In several recent experiments, coherent swirling motion of vibrated rod-like objects was discovered (Narayan et al., 2006; Kudrolli et al., 2008; Aranson et al., 2007b).

8.2.1 Swirling of intrinsically polar rods

In a recent experiment, Kudrolli et al. (2008) investigated the dynamics of intrinsically polar rods: while symmetric in shape, the cylinders had an asymmetric weight distribution, so the center-of-mass was displaced from the middle of the cylinder. Such a rod on a vibrated surface typically moves towards the lighter end, thus it becomes an intrinsically self-propelled particle in contrast to symmetric rods that only move in the direction of an externally imposed tilt. When many such rods are placed inside a vibrated closed container, for relatively weak excitation they aggregate at the boundaries, as shown in Fig. 8.8(a). However, when the magnitude of excitation is increased (Fig. 8.8(b)), aggregation at the boundaries is reduced, and coherently moving structures are found in the bulk of the container. In particular, swirling motion of rods can be identified in time-averaged velocity fields as shown in Fig. 8.9(a). This system was simulated using the soft-particle molecular dynamics algorithm described in the previous section. It was further simplified so the details of bouncing of rods on the plate were ignored and it was assumed that the rods are confined to a horizontal plane, and a force acts on each rod along its (horizontal) axis in the direction of the lighter end. This force was assumed to be randomized, with the mean and variance that were fitted using experimental data for a single rod motion. In addition to the driving force, the rods experience velocity-dependent friction with the substrate and inelastic collisions with other rods. This simplification allowed the authors to significantly increase the

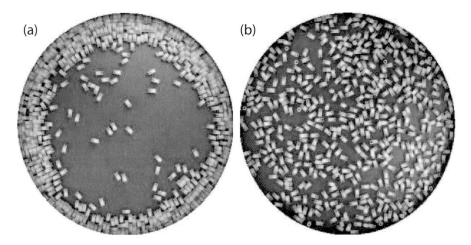

Fig. 8.8 Swarming behavior in the system of vibrated asymmetric rods. (a) Rods migrate and aggregate at the boundaries of a container for modest excitations (number of rods $N = 500$; non-dimensional acceleration $\Gamma = 2$). (b) Homogeneous distribution of rods is observed as the acceleration is increased ($\Gamma = 4$), from Kudrolli et al. (2008).

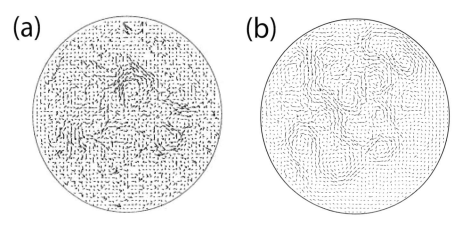

Fig. 8.9 Swirling motion of granular rods. Swirling in time-averaged velocity obtained by computing particle displacements within 5-s time interval. (a) experiment ($\Gamma = 3, N = 900$); (b) numerical simulations, from Kudrolli et al. (2008).

time step and study relatively large systems of up to 5000 rods. Overall, experiments and simulations exhibit a good qualitative agreement, see Figs. 8.9(a) and (b).

8.2.2 Swirling in vibrated apolar rods

Experiments by Narayan et al. (2006) revealed that even apolar rods, under vibration exhibit novel features. For example, some types of rice in a certain range of frequencies and magnitudes of vibration spontaneously form nematic crystalline-like structures, whereas longer particles with tapered ends exhibit short-range smectic ordering. Cylindrical particles with flat ends with the same aspect ratio instead demonstrate tetratic order, see Fig. 2.20. In a certain range of parameters of vibration novel dynamics states, macroscopic swirls, are observed. Nematic states often rotate as a whole in a circular container or exhibit smaller-scale swirls. These swirls are fundamentally different from vortices found by Blair et al. (2003) and Kudrolli et al. (2008): since particles are apolar and on average parallel to the plate, the direction of their motion is not related to the tilt or asymmetric mass distribution.

The mechanism of the swirl formation is related to the instability of an overall rotation of grains around the cavity. In subsequent experiments with vibrated monolayers of elongated particles Aranson et al. (2007b) demonstrated that a small horizontal component of the oscillatory acceleration of the vibrating plate, phase-shifted with respect to the main (vertical) component of acceleration, is the primary source of the drift of particles around the circular container and the subsequent swirl formation.

In these experiment by Aranson et al. (2007b), a wide range of relatively large particles was used: sushi rice (mean length of the order 4 mm, aspect ratio 2–2.5), jasmine rice (mean length about 7 mm, aspect ratio 3.5–4), Basmati rice (mean length about 8 mm, aspect ratio about 6–8), nearly spherical mustard seeds (diameter about 2 mm), monodisperse stainless steel dowel pins (length 4 mm, aspect ratio 4, flat ends), monodisperse steel cylinders tapered to pointed ends (length about 6 mm, aspect ratio 6). For all these particles overall rotation of grains with the angular frequency ω dependent on the frequency f and the amplitude Γ of the plate vibration was found. Almost solid-body rotation was observed for spherical mustard seeds, cylindrical dowel pins, and steel cylinders with tapered ends. In agreement with the studies by Narayan et al. (2006), flat-ended dowel pins exhibited tetratic ordering, while the cylinders with tapered ends showed smectic ordering, similar to that shown in Fig. 2.20. However, for jasmine and sushi rice particles solid-body rotation was accompanied by a significant swirling motion on a scale smaller than the system size, such as seen in Fig. 8.10. The swirls typically show non-stationary behavior and often drift around the container. Well-pronounced large-scale swirling motion occurs only at almost close-packed filling fraction (of the order of 85% area coverage). Practically no swirling was observed at lower filling fractions.

To quantify the collective motion of grains, the velocity field from the sequences of snapshots was extracted using the standard particle image velocimetry technique. Figure 8.10 shows the two-dimensional field of velocity where several recurrent vortices or swirls with the characteristic size about a quarter of the container diameter persisted in the course of an experiment.

The experiments by Aranson et al. (2007b) showed that the angular velocity of the solid-body rotation component of the pattern ω depends strongly on the vibration frequency f and the vertical acceleration Γ_z, see Fig. 8.11. The angular velocity has a pronounced resonance peak at $f \approx 130$ Hz and then changes sign at $f = 134$ Hz. At this frequency we observed a surprising behavior: the angular velocity randomly switches between positive and negative values, as is shown in Figure 8.11, inset (b). The rotation dependence on the vertical acceleration is non-monotonous: the rotational velocity ω initially increases almost linearly with the acceleration Γ_z, then reaches a

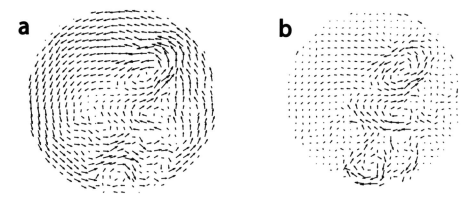

Fig. 8.10 Swirling motion of vibrated jasmine rice: (a) velocity field obtained by the particle-image velocimetry technique of the experimental movie for the jasmine rice at acceleration $2g$ and frequency $f = 142$ Hz; (b) velocity field with overall solid-body rotation subtracted, from Aranson et al. (2007b).

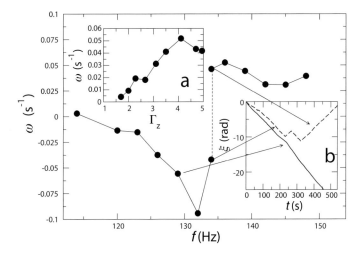

Fig. 8.11 Average angular velocity ω or the solid-body rotation of grains as a function of frequency f for vertical acceleration $\Gamma_z = 3.1$ for jasmine rice grains. Inset a: ω vs. acceleration Γ_z at $f = 142$ Hz. Inset b: angular rotation phase $\xi = \int \omega dt$ vs. time for $f = 129$ Hz (solid line) and $f = 134$ Hz (dashed line), from Aranson et al. (2007b).

maximum at $\Gamma_z \approx 4g$, and finally decreases.

The overall rotation appears to be a bulk effect weakly dependent on the boundary conditions at the lateral wall. To verify this, strips of rough sandpaper or a plastic cable tie with asymmetric teeth were glued to the sidewall, but these did not affect the rotation in the bulk. Furthermore, experiments in a dilute system (10–20 particles) showed them slowly drifting around the cavity as well, which confirms that the mechanism of rotation is due to local coupling between the grains and the substrate.

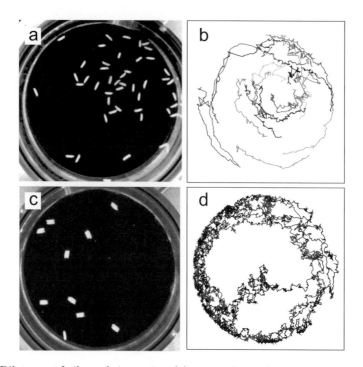

Fig. 8.12 Dilute gas of vibrated rice grains: (a) a typical snapshot of several monomer grains drifting on a vibrated plate at $f = 133$ Hz and $\Gamma_z = 3.5g$; (b) trajectories of several grains extracted from the sequences of snapshots; (c,d) the same for "catamaran" particles in the similar vibration regime, $f = 129$ Hz, $\Gamma_z = 3.5g$, from Aranson *et al.* (2007b).

Several representative long-particle trajectories are depicted in Figs. 8.12(a) and (b).

In order to pinpoint the mechanism of the particle drift, the amplitudes of all three components of the plate acceleration and their relative phases at two locations at the edge of the plate orthogonal with respect to its center (see inset to Fig. 8.13) were simultaneously measured. The measurements showed that the horizontal part of the acceleration was mostly oriented tangentially to the edge of the plate, and in both locations the azimuthal components Γ_a oscillated in phase, which indicates the prevalence of the azimuthal mode of plate vibrations. In a wide range of frequencies the azimuthal component significantly exceeded the radial component Γ_r, see Fig. 8.13.

Furthermore, the measurements show that the angular velocity of particle rotation ω is correlated with the amplitude of azimuthal acceleration Γ_a, see Fig. 8.13. In particular, the peak in angular velocity and the overall rotation of grains changes direction near $f = 130$ Hz where the amplitude of the azimuthal acceleration of the plate has a peak and the phase ϑ_{az} between Γ_z and Γ_a changes rapidly, compare Figs. 8.11 and 8.13. This rapid change of the phase ϑ_{az} is an indication of a plate resonance with respect to the twisting mode of vibration. The exact value of the resonant frequency appears to depend slightly on the amount of material loaded onto the vibrated plate. While it did not change significantly for the rice (total weight of

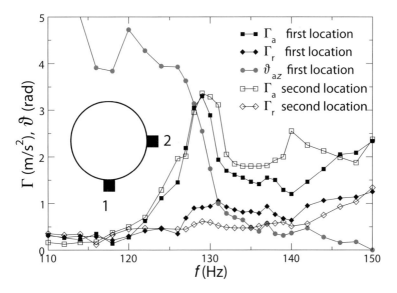

Fig. 8.13 Amplitudes of azimuthal Γ_a (squares) and radial Γ_r (diamonds) components of acceleration vs. frequency f at fixed amplitude of the vertical acceleration $\Gamma_z = 3.5g$. The measurements are performed at two locations at the edge of the plate orthogonal with respect to its center (see sketch in the inset). Line with circles depicts the phase difference ϑ_{az} between vertical Γ_z and azimuthal Γ_a components of acceleration at the first location, from Aranson et al. (2007b).

a monolayer of rice in our experiment was 24 g), for heavy steel pins (total weight of a monolayer of pins was about 200 g) a significant shift of the resonant frequency occurred.

These measurements demonstrated that at least in the experiments by Aranson et al. (2007b) (and likely in similar experiments by Narayan et al. (2006), Narayan et al. (2007)), the overall rotation of grains was a consequence of the imperfection of the vertical shaker system. It showed a strong resonant behavior near a certain vibration frequency. This behavior is common for any mechanical shaker system, however in different experimental setups this component may be smaller or larger, or it may peak at different oscillation frequencies. That possibly explains why different experimental groups observed swirling motion at different experimental conditions.

8.2.3 Friction anisotropy

Experiments with a dilute gas of particles allowed us to determine that the drift of particles in the azimuthal direction is affected by their orientation. Due to the cylindrical geometry of the experiment, the position of a particle can be characterized in polar coordinates by the radius r and polar angle θ, and the orientation of the particle is characterized by the angle φ with respect to the radial direction. Processing about 10^6 data points, we computed velocity distributions for different grain orientations φ. These measurements show that the width of the distributions in the radial and

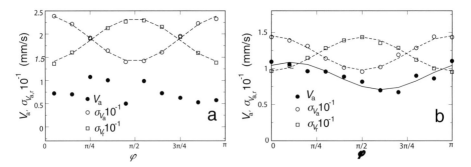

Fig. 8.14 Statistical characteristics of velocity distributions for individual particles (a) and catamarans (b) for a run at $f = 129$ Hz and $\Gamma_z = 3.5$g: Average azimuthal velocity V_a (solid black circles) and standard deviations of V_a and radial velocity V_r, σ_{V_a} and σ_{V_r}, as a function of the angle φ between the particles and the radius-vector from the center of the cavity to the center of the particle. Dashed lines show fits of the data by the sinusoidal functions expected for driven particles with anisotropic friction force in the linear approximation (8.17), (8.18). For individual particles (a) the best fit yields for the standard deviations $\sigma_{V_a} = 19.1 + 4.95\cos(2\varphi)$ and $\sigma_{V_r} = 19.1 - 4.95\cos(2\varphi)$. For "catamaran" particles (b) the fit to experimental data yields $\sigma_{V_a} = 12.1 + 2.33\cos(2\varphi)$, $\sigma_{V_r} = 12.1 - 2.33\cos(2\varphi)$ for the standard deviations and for the average azimuthal velocity $V_a = 0.87 + 0.19\cos(2(\varphi - 0.34))$ correspondingly.

azimuthal directions, and, correspondingly, the standard deviations σ_{V_r}, σ_{V_a} of radial V_r and azimuthal V_a velocities are different. For $\varphi = 0$ (particle is oriented along the radius), standard deviations $\sigma_{V_a} > \sigma_{V_r}$, while for $\varphi = \pi/2$ the relation is opposite. Figure 8.14 shows the dependence of standard deviations σ_{V_r} and σ_{V_a} on the orientation angle φ. This dependence is consistent with a simple model that the grains are driven by a force with non-zero mean that acts predominantly in the azimuthal direction, plus a strong isotropic fluctuating component, and they are damped by an orientation-dependent frictional force ($\mathbf{V}$ is the particle velocity with respect to the container bottom). Ignoring particle inertia, this model can be cast in the following form

$$\hat{\mathbf{B}}\mathbf{V} = \mathbf{F}_0 + \xi(t), \tag{8.14}$$

where $\hat{\mathbf{B}}$ is a 2×2 "friction" matrix, $\xi(t)$ is the white noise modelling the effect of vibration, and $\mathbf{F}_0$ is the driving force. On symmetry grounds, components of the friction matrix $\hat{\mathbf{B}}$ in the first order can be written as

$$B_{ij} = \beta_0 \delta_{ij} + \beta_1 (2n_i n_j - \delta_{ij}), \tag{8.15}$$

where coefficients β_0 and β_1 characterize the isotropic and anisotropic contributions, $\{i,j\} \in \{\|,\perp\}$, and $n_{\|,\perp}$ are projections of the grain director on the direction of particle translation and the orthogonal direction, respectively. In our circular geometry when the driving force is directed azimuthally, particles move predominantly in the azimuthal direction, and $n_\| = \cos\varphi, n_\perp = \sin\varphi$, so the friction tensor can be written as

$$\hat{\mathbf{B}} = \beta_0 \mathbf{I} + \beta_1 \begin{pmatrix} \cos 2\varphi & \sin 2\varphi \\ \sin 2\varphi & -\cos 2\varphi \end{pmatrix}, \quad (8.16)$$

where $\mathbf{I}$ is the identity matrix. For positive β_1, this expression implies that friction is maximal when the particle is translated along itself and minimal when it is translated in the perpendicular direction. We will use this expression below for the description of the dense flow regime. In the dense phase, the stochastic component of the driving force should be strongly suppressed due to confinement by neighboring grains, but the anisotropy of friction still would be a significant factor in the selection of a flowing regime. Using eqns 8.14 and 8.16, for small anisotropy ($\beta_1 \ll \beta_0$) one can obtain the standard deviations of velocity components as functions of the orientation φ:

$$\sigma_{V_a} = \sigma_0 \left(1 - \frac{\beta_1}{\beta_0} \cos(2\varphi)\right), \quad \sigma_{V_r} = \sigma_0 \left(1 + \frac{\beta_1}{\beta_0} \cos(2\varphi)\right), \quad (8.17)$$

which fit well with the experimental data as shown in Fig. 8.14.

The anisotropic friction should also lead to the dependence of the mean azimuthal velocity V_a on the orientation angle φ. Assuming that the driving force F_0 acts in the azimuthal direction, by balancing the driving force and the friction force (eqn 8.14) we obtain

$$V_a = \frac{F_0}{\beta_0}\left(1 - \frac{\beta_1}{\beta_0}\cos(2\varphi)\right) + O(\beta_1^2/\beta_0^2). \quad (8.18)$$

Experiments with individual grains were not able to confirm this dependence, see Fig. 8.14(a), probably due to large velocity fluctuations: the standard deviation of the velocity is an order of magnitude greater than the mean. These large velocity fluctuations are likely related to rolling and bouncing of individual grains shadowing the effect of the anisotropic sliding. In order to reduce this effect and suppress the rolling motion of grains, pairs of particles were glued together to form "catamaran" objects, see Figs. 8.12(c) and (d). The results of data processing for these catamaran particles are shown in Fig. 8.14(b). This data shows the evidence of mean velocity anisotropy consistent with the anisotropic friction model, eqn 8.18. However, it can be fitted with eqn 8.18 only up to a certain phase shift $\Delta\varphi \approx -0.34$. Most likely this phase shift originates from the fact that in the experiment plate vibrations have both azimuthal and linear modes of horizontal displacement. Consequently, the driving force was not purely azimuthal (the ratio of $|\Gamma_r/\Gamma_a| \approx 0.2$), which could skew the dependence of mean azimuthal velocity V_a vs. the orientational angle φ.

8.2.4 Theoretical model of swirl formation

In the hydrodynamic description the nematic ordering of apolar rod-like particles such as rice grains can be characterized by the symmetric traceless alignment tensor $\mathbf{Q}$ related to the nematic director $\mathbf{n} = (\cos\tilde{\varphi}, \sin\tilde{\varphi})$ as follows

$$\mathbf{Q} = \frac{Q}{2}\begin{pmatrix} \cos 2\tilde{\varphi} & \sin 2\tilde{\varphi} \\ \sin 2\tilde{\varphi} & -\cos 2\tilde{\varphi} \end{pmatrix}, \quad (8.19)$$

where Q is the magnitude of the order parameter ($Q = 0$ means total disorder, and $Q = 1$ corresponds to perfect nematic alignment), and $\tilde{\varphi}$ is the mean grain orientation angle with respect to an arbitrary fixed direction within a mesoscopic area.[4] Here, we neglect effects of the smectic ordering, also sometimes observed experimentally, and will stay in the framework of nematodynamics. We are interested in the time scales that are much larger than the period of plate vibrations, so we shall ignore the vertical vibrations of individual grains and only consider two-dimensional inplane transport.

The evolution of the alignment tensor in the vicinity of the equilibrium isotropic–nematic transition can be described in the framework of the de Gennes–Ginzburg–Landau theory (Olmsted and Goldbart, 1992; de Gennes and Prost, 1993). The theory describes the dynamics of tensor $\mathbf{Q}$ in terms of variation of a certain free-energy functional $\mathcal{F}(\mathbf{Q})$. The nematic state corresponds to a minimum of the free-energy functional at non-zero $\mathbf{Q}$, see Fig. 8.1. A similar approach was also applied to the non-equilibrium situation in the context of self-propelled or active particles (Kruse et al., 2004; Simha and Ramaswamy, 2002; Hatwalne et al., 2004). Accordingly, a generic equation describing the evolution of the alignment tensor $\mathbf{Q}$ in two dimensions is of the form (Olmsted and Goldbart, 1992; de Gennes and Prost, 1993)

$$\frac{\partial \mathbf{Q}}{\partial t} + (\mathbf{v}\cdot\nabla)\mathbf{Q} = \epsilon\mathbf{Q} - \frac{1}{2}\mathrm{Tr}(\mathbf{Q}\cdot\mathbf{Q})\mathbf{Q} + D_1\nabla^2\mathbf{Q} + D_2\nabla(\nabla\cdot\mathbf{Q}) + \mathbf{\Omega}\mathbf{Q} - \mathbf{Q}\mathbf{\Omega} + B\kappa^s, \quad (8.20)$$

where $\mathbf{v}$ is the hydrodynamic velocity, $\mathbf{\Omega} = \frac{1}{2}[\nabla\mathbf{v} - \nabla\mathbf{v}^T]$ is the vorticity tensor (we assume that the flow of particles is incompressible), $\kappa^s = \frac{1}{2}[\nabla\mathbf{v} + \nabla\mathbf{v}^T]$ is symmetric-traceless part of the strain rate tensor, $B = \mathrm{const}$ is the coupling constant, $D_{1,2}$ are the corresponding elastic constants, compare with liquid crystals (de Gennes and Prost, 1993), and $\epsilon \sim \nu - \nu_c$ is the parameter controlling the nematic transition that depends on the grain packing density ν, and ν_c is the critical density of the nematic phase transition. Thus, coupling between the coarse-grained (or "hydrodynamic") velocity $\mathbf{v}$ of the particles and the alignment tensor $\mathbf{Q}$ is included through the advection by flow $(\mathbf{v}\cdot\nabla)\mathbf{Q}$ and through rotation of the particle orientation by the vorticity $\mathbf{\Omega}\mathbf{Q} - \mathbf{Q}\mathbf{\Omega}$. Since here we deal with almost rigidly rotating flows (i.e κ^s is small compared to $\mathbf{\Omega}$), we neglect the last term in eqn 8.20.

For the velocity $\mathbf{v}$ we have the following analog of the two-dimensional Navier–Stokes equations

$$\partial_t \mathbf{v} + (\mathbf{v}\cdot\nabla)\mathbf{v} = \eta\nabla^2\mathbf{v} - \nabla p - \mathbf{F}_f(\mathbf{Q},\mathbf{v}) + \mathbf{F}; \quad \nabla\cdot\mathbf{v} = 0, \quad (8.21)$$

where η is the shear viscosity of the granular flow,[5] p is the pressure due to particle motion, $\mathbf{F}_f$ is the velocity-dependent effective bottom friction force, and $\mathbf{F}$ is the anisotropic driving (or conveying) force due to mixed vertical/horizontal vibrations and friction with the bottom of the plate as described in Section 8.2.3.[6] Based on the experiential results for the friction force of catamaran particles (eqn 8.14), it is

---

[4]Note that the order parameter Q is different from the order parameter Π introduced in eqn 8.6 and defined as an inverse typical width of the orientation distribution function $\Psi(\mathbf{n})$.

[5]While liquid crystals in general have anisotropic viscosity tensor, here we neglect for simplicity the anisotropy of the viscosity and consider the case of $\eta = \mathrm{const}$.

[6]Of course driving of rod motion is also caused by frictional interaction between rods and the bottom, so the separation of the total frictional force into $\mathbf{F}_f$ and $\mathbf{F}$ is a matter of convenience

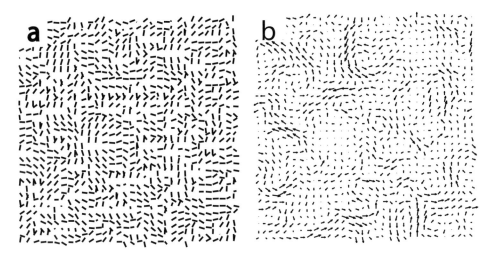

Fig. 8.15 Snapshots illustrating director field **n** (a) and velocity field **v** (b) from numerical simulations of the continuum model (8.25),(8.26). Parameters of simulations are: $\epsilon = 1, \beta_0 = 0.2, \eta = 3, D_1 = 0.8, D_2 = 0.4, F_0\alpha = 1$, and integration is performed in a periodic domain of size 200 × 200 dimensionless units, from Aranson et al. (2007b).

reasonable to assume the following weak dependence of the friction force of the velocity **v** and alignment tensor **Q**

$$\mathbf{F}_f(\mathbf{Q}, \mathbf{v}) = (\beta_0 + \beta_1 \mathbf{Q})\mathbf{v}, \quad \beta_1 \ll \beta_0. \tag{8.22}$$

To simplify calculations one can consider a limiting case of a very large container and introduce a local rectangular Cartesian coordinate system instead of the cylindrical one. Then we can choose the x-direction along the horizontal component of plate acceleration Γ_x coinciding with the direction of the driving force **F**. Since the symmetric traceless tensor **Q** has only two independent components in two dimensions, it is convenient to introduce a *quasivector* of local orientation

$$\boldsymbol{\tau} = (\tau_x, \tau_y) \equiv (Q_{xx}, Q_{xy}) = \frac{Q}{2}(\cos 2\tilde{\varphi}, \sin 2\tilde{\varphi}), \tag{8.23}$$

where the angle $\tilde{\varphi}$ is now between the director and the orientation of the driving force **F**.

We further assume that the hydrodynamic velocity **v** is always close to the uniform translation velocity $\mathbf{v}_0 = \beta_0^{-1} F_0 \mathbf{x}_0$ ($\mathbf{x}_0$ is a unit vector in the x-direction that we choose to coincide with the direction of the mean driving force). Since $\beta_1 \ll \beta_0$, we can rewrite the hydrodynamic equation (eqn 8.21) in the form

$$\partial_t \mathbf{v} + (\mathbf{v} \cdot \nabla)\mathbf{v} = \nu \nabla^2 \mathbf{v} - \nabla p - \beta_0 \mathbf{v} + F_0 + \alpha F_0 \boldsymbol{\tau} + O(\alpha^2), \tag{8.24}$$

where $\alpha = \beta_1/\beta_0$ is a small parameter. Equation 8.20 in the same approximation can be rewritten as

222 Granular materials with complex interactions

$$\frac{\partial \boldsymbol{\tau}}{\partial t} + (\mathbf{v} \cdot \nabla)\boldsymbol{\tau} = \epsilon \boldsymbol{\tau} - |\boldsymbol{\tau}|^2 \boldsymbol{\tau} + D_1 \nabla^2 \boldsymbol{\tau} + D_2 \nabla(\nabla \cdot \boldsymbol{\tau}) + \boldsymbol{\Omega} \times \boldsymbol{\tau}, \qquad (8.25)$$

where $\boldsymbol{\Omega} = (\partial_y v_x - \partial_x v_y)\mathbf{z}_0$ is the vorticity component directed along the vertical coordinate z. Then, in two dimensions one can exclude pressure by applying curl operation to eqn 8.24 and obtain a single equation for the vertical vorticity component Ω:

$$\partial_t \Omega + (\mathbf{v} \cdot \nabla)\Omega = \nu \nabla^2 \Omega - \beta_0 \Omega + \alpha F_0 \left(\partial_y \tau_x - \partial_x \tau_y\right). \qquad (8.26)$$

Equations 8.25 and 8.26 form a closed system of equations describing co-evolution of the alignment tensor $\mathbf{Q}$ and the particle velocity $\mathbf{v}$. These equations appear to be rather generic: we will encounter them in the context of dynamics self-propelled particles in the next chapter. Furthermore, eqn 8.25 is similar to eqn 8.12 for the rod tilt vector $\boldsymbol{\tau}$. Thus, one can argue that due to the presence of a symmetry-breaking driving force $\mathbf{F_0}$ the dynamics of initially apolar grains become similar to those of self-propelled or polar particles.

Uniform transport of particles corresponds to the stationary solution $\tau_x = \tau_0, \tau_y = 0, v_y = 0, v_x = F_0(1 + \alpha \tau_0)/\beta_0, p = \text{const}, |\tau_0| = \sqrt{\epsilon}$. Here, τ_0 is the magnitude of the order parameter characterizing local nematic order ($\tau_0 = 0$ corresponds to a disordered packing, and $|\tau_0| = \sqrt{\epsilon}$ corresponds to an aligned nematic state). Swirls could then emerge if this solution was unstable with respect to spatially periodic modulations.

The linear stability analysis shows that the instability of the uniformly moving grains indeed occurs in a finite range of wave numbers if the driving force F_0 is greater than some critical value. The eigenmode corresponding to the instability has a form of periodic undulations of the local orientation accompanied by a periodic shear. The onset of instability can be obtained in the long-wavelength limit $k \to 0$. The maximum growth rate occurs at a certain wave number k_m which is a function of the model parameters. The selected wave number k_m can be found in the limit of a relatively large value of the speed v_0 or the force F_0 (here we skip rather straightforward but lengthy calculations, the details can be found in Aranson et al. (2007b)):

$$k_m^{3/2} = \frac{\sqrt{2 F_0 \alpha \sqrt{\epsilon}}}{4(D_1 + \eta)}. \qquad (8.27)$$

The length scale $L \sim 1/k_m$ determines the characteristic size of the swirls. With the decrease of the driving F_0, the size of swirls diverges and eventually exceeds the size of container, indicating a transition to solid-body rotation.

To obtain further insights into the dynamics beyond linear stability, we performed a full numerical solution of eqns 8.25 and 8.26 in a periodic domain. A snapshot of a typical simulation in the parameter range corresponding to the linear instability is shown in Fig. 8.15: panel (a) shows the director field $\mathbf{n}$ reconstructed from vector $\boldsymbol{\tau}$ and panel (b) shows the velocity field $\mathbf{v}$. As seen from Fig. 8.15, indeed the system exhibits spontaneous formation of an array of swirls, in qualitative agreement with the experiment.

One of the surprising experimental findings is a very strong sensitivity of swirling to the shape of the particles. For example, no swirling or smectic ordering was observed for monodisperse metal cylindrical particles. Very little swirling was also observed

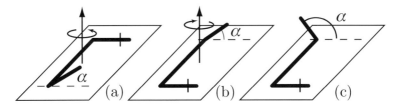

Fig. 8.16 Schematics of a bent-wire chiral particle: central stem and two bendable arms. (a) Chiral wire with positive angle α which the second arm rises out of the plane formed by the stem and the first arm (marked with a vertical line). (b) Mirror image of (a), formed by reflection through a plane perpendicular to the central stem (negative α). The arrows indicate the direction of spin under vibration. (c) Configuration used in the granular gas experiments with $\alpha > -3\pi/4$, from Tsai et al. (2005)

for Basmati rice. While the specific mechanism of this strong sensitivity remains unclear, in the framework of our phenomenological continuum model this effect can be accounted for by variations of effective elastic constants $D_{1,2}$ due to interlocking of particles and formation of the tetratic structures. Tetratic structures possibly possess higher rigidity and resistance to shear, driving the system below the threshold of swirling instability.

8.3 Global rotation in the system of chiral particles

In addition to nematic or smectic ordering originated from the shape anisotropy of particles, some types of particles exhibit chirality when colliding with an oscillating bottom of the container. In this situation, a particle's angular momentum is created by vertical oscillations of the platform. This situation is rather unusual: there are very few systems in which angular momentum can be delivered uniformly throughout the bulk, such as ferrofluids and liquid crystals in a rotating magnetic field and colloidal dispersions of anisotropic particles illuminated with circularly polarized light.

An interesting experiment with macroscopic chiral particles was performed by Tsai et al. (2005). The role of chiral particles was played by bent-wire objects shown in Fig. 8.16 that rotated in a preferred direction under vertical vibration. The direction of rotation is determined by the angle α at which the "second" arm of the wire rises above the plane. This experiment was inspired by rattleback toys or *celtic stones* (elongated objects shaped like the hull of a boat with a curved bottom and a flat top) that have a preferred direction of rotation on a smooth substrate upon shaking.

The experiments demonstrated that angular momenta of individual particles due to collisions were converted into the collective angular momentum of the granular gas of these chiral objects (Fig. 8.17(a)). To quantify these experimental observations, the authors measured the coarse-grained angular velocity of particles' center-of-mass motion relative to the center of the plate $\omega_0(r) = v(r)/r$ in three concentric annuli as a function the number of particles N, see Fig. 8.17(b). The results show that the average angular velocity increases with distance from the center r, and in the outer annulus reaches a value about 7 times greater than that in the middle annulus. Thus,

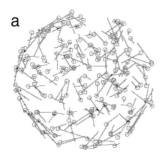 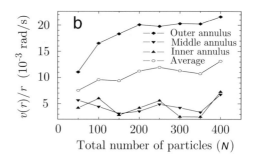

Fig. 8.17 (a) Vectors of total displacements of individual particles accumulated for 26.6 s, with their initial positions marked as circles, for $N = 350$, $f = 60$ Hz, $\Gamma = 2$, arm angle $\alpha = -3\pi/4$, stem length 10 ± 0.2 mm, arm length 4 ± 0.2 mm, and wire diameter 0.91 ± 0.05 mm. (b) Measured center-of-mass angular velocity $v(r)/r$ for $f = 60$ Hz and $\Gamma = 2$ averaged over the three annuli described in the text for different numbers of particles. The angular velocity is greatest in the outer region, from Tsai et al. (2005)

in contrast to solid-body rotation exhibited by elongated particles due to horizontal twisting acceleration considered in Section 8.2, vibrated chiral particles exhibit differential rotation.

The theoretical description of this system was formulated in Tsai et al. (2005) in the framework of three phenomenological equations for the density ν, center-of-mass momentum density $\nu \mathbf{v}$ and the spin angular momentum density $l = \mathcal{I}\Omega$ arising from the rotation of particles around their center-of-mass, Ω is the particle's rotation frequency and $\mathcal{I}$ is the moment of inertia density. Whereas the equations for density and velocity are somewhat similar to those for the vibrated rod system (eqns 8.11 and 8.12), the equation for the spin momentum has no counterpart in the vibrated rod system. On symmetry grounds, it was postulated in the following form,

$$\partial_t l = -\nabla \cdot \mathbf{v} l + \chi - \gamma_\Omega \Omega - \gamma(\Omega - \omega) + D_\Omega \nabla^2 \Omega, \quad (8.28)$$

where χ is the source of the angular rotation (due to chirality of the particles), ω is the collective angular velocity, $\omega = (\nabla \times \mathbf{v})_z$, γ_Ω and γ are dissipative coefficients due to friction and D_Ω is the angular momentum diffusion coefficient. Equation 8.28 can be interpreted as follows: the first term on the right-hand side describes the advection of spin angular momentum and the second (χ) models generation of the angular momentum due to collisions of particles with vibrated bottom. The last three terms describe different types of dissipation due to friction and diffusion.

Equation 8.28 predicts, in agreement with the experiment, the onset of collective rotation of the gas of particles with the averaged angular velocity $\omega_0 = v(r)/r$ resembling experimental observations in Fig. 8.17(b). It is feasible that this model can also exhibit rather non-trivial spatiotemporal dynamics similar to those in the system of vibrated rods. However, due to a small number of particles (about 350) in the experiment these collective regimes have not been explored and remain a challenge. Interesting phenomena should also occur in granular systems composed of particles of opposite chirality, so the overall chirality remains zero. In this situation one may anticipate phase-separation instability leading to chirality sorting.

8.4 Patterns in solid–fluid mixtures

In the previous section we considered complex interactions of grains due to various forms of shape anisotropy of particles, but the interaction forces themselves were collisional, and, therefore, short range. However, in many practical situations long-range forces of different nature may come into play alongside the contact forces. The most typical example of such forces is hydrodynamic interaction due to the presence of interstitial liquid. When particles are completely immersed in fluid, their motion induces hydrodynamic flows that in turn lead to the viscous drag and anisotropic long-range interactions among particles. But even a tiny amount of liquid leads to significant changes in the interactions among particles due to liquid bridges and cohesion among the particles. This can have a profound effect on macroscopic properties of granular assemblies such as the angle of repose, avalanching, ability to segregate (Samadani and Kudrolli, 2000; Samadani and Kudrolli, 2001; Tegzes et al., 2003; Tegzes et al., 2002).

In this section we will discuss mostly the situation of the first kind, when the volume fraction of fluid in the two-phase system is large, and the grains are completely immersed in fluid. This is relevant for many industrial applications, such as, for example, fluidized beds. Fluidized beds have been widely used since German engineer Fritz Winkler invented the first fluidized bed for coal gasification in 1921. Typically, gas or fluid is pumped upwards through a long vertical column (height H is much greater than the width W) containing granular matter. Granular material *fluidizes* when the drag force exerted by the fluid on the granulate exceeds gravity. A uniform fluidization, the most desirable regime for most industrial applications, turns out to be prone to the bubbling instability: bubbles of clear fluid are created spontaneously at the bottom, traverse upwards through the granular layer, and destroy the uniform regime. Instabilities in fluidized beds is still an active area of research in the engineering community, see (Gidaspow, 1994; Jackson, 2000), however, a detailed discussion of these essentially three-dimensional processes goes beyond the scope of this book. Let us just note briefly that similar instabilities occur in shallow fluidized beds (when the aspect ratio $W/H \ll 1$), however, instead of bubbles, oscillatory patterns are formed (Tsimring et al., 1999; Li et al., 2003; Orellana et al., 2005). Furthermore, if the air flow is periodically modulated, the pattern formation in a shallow fluidized bed shows many similarities (however, there are also some differences) with mechanically vibrated layers (Chapter 5), including formation of squares at low frequency of modulation, stripes at higher frequency, and hexagons at higher amplitudes of modulation, see Fig. 8.18.

8.4.1 Vibration of particles immersed in fluid

In Chapters 4 and 5 we discussed the pattern formation in vertically vibrated layers of particles. In most of these experiments the hydrodynamic interaction between particles was not relevant (some experiments were even performed in vacuum to eliminate

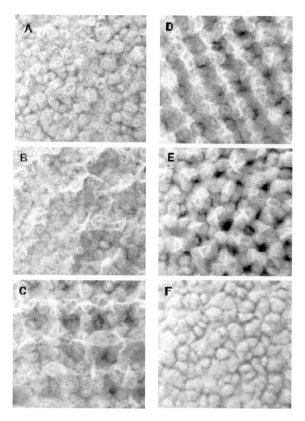

Fig. 8.18 Patterns in a shallow fluidized bed with periodically modulated air flow. Snapshots of the granular layer surface are shown for different flow parameters (amplitude and frequency of air flow modulations): (a) disordered; (b) random patterns; (c) squares; (d) stripes; (e) distorted hexagons; (f) quasicrystalline structures, from Li et al. (2003)

air drag completely) and therefore it was also neglected in theoretical modelling. However, in certain situations long-range hydrodynamic interactions of vibrating particles immersed in fluid can lead to entirely different phenomenology even in the traditional shaking experiment setup.

Voth et al. (2002) and Thomas and Gollub (2004) performed a series of these shaking experiments and discovered a host of novel phenomena occurring as a consequence of both attractive and repulsive interactions between macroscopic particles when they are vibrated in a fluid. A specific pattern is determined by the interplay of short-range repulsive forces due to collisions and long-range attractive hydrodynamic forces. The latter can be varied by tuning the frequency ω and amplitude A of the vibration as illustrated in Fig. 2.22. At a certain amplitude and frequency of oscillations, bound states characterized by a certain equilibrium distance between the particles were observed, see Fig. 2.22(a). They can have a form of crystalline objects, for example hexagons surrounding a particle in the center, see Fig. 2.22(b). Larger numbers of particles (> 7) move chaotically in a bound state with no long-range order, as shown in Figs. 2.22(e)–(f).

Following analysis of Voth et al. (2002), the hydrodynamic interaction among particles can be qualitatively understood by considering the flow produced by the periodic vibration of each individual particle (Batchelor, 1967). Periodic vibration of a spherical particle generates an oscillatory flow around each sphere. If the associated Reynolds number $\text{Re} = \omega A a/\eta \ll 1$ is small, outside of a boundary layer with the thickness $d_{\text{osc}} = \sqrt{\eta/\omega}$ (η is kinematic viscosity of the fluid, a is the radius of the particle), the oscillatory flow can be approximated by a solution of the Stokes equation $\eta \nabla \mathbf{v} - \nabla p = 0$. This solution can be expressed via the hydrodynamic potential Ψ, $\mathbf{v} = -\nabla \Psi$, where $\Psi = -\frac{1}{2} A \omega \sin(\omega t) a^3 \cos\theta / r^2$, (r, θ) are polar coordinates with the origin at the center of the particle and θ is the angle with respect to the vertical z-axis. Correspondingly, at the particle surface, the component of the velocity field parallel to its surface is $\mathbf{v}_\| = \frac{1}{2} A \omega \sin(\omega t) \sin\theta$.

According to Rayleigh (1883), if the magnitude of the oscillatory flow $v_\|(\theta)$ varies along a surface of the particle, a steady secondary flow v_{steady} is generated. The mass conservation then implies that there is also a flow perpendicular to the boundary with magnitude $v_\perp \; d_{\text{osc}} \partial_\| v_\|$. Since this normal flow is not exactly out of phase with $v_\|$, every oscillation cycle transports a finite amount of momentum into the boundary layer. As a result, a non-zero averaged over the period of vibration force imparted on the fluid in the boundary layer and parallel to the boundary and of order of $\nu_f \langle v_\| \partial_\| v_\| \rangle$ appears (here ν_f is the fluid density). This force is balanced by the viscous force $\nu_f \eta v_{\text{steady}} / d_{\text{osc}}^2$, yielding the steady flow $v_{\text{steady}} \sim \langle v_\| \partial_\| v_\| \rangle \sim A^2 \omega / a \sin\theta \cos\theta$ at the edge of the oscillatory boundary layer.

The steady flow v_{steady} is the primary cause of attractive interaction between two particles: it expels the fluid from the poles of each particle, thus generating a perpendicular inflow velocity towards the equator. According to Voth et al. (2002), a good approximation for the inflow velocity at the distance r from the particle center is

$$v(r) = -0.53\sqrt{\omega\eta}\frac{a^2}{r^3}. \tag{8.29}$$

Thus, the corresponding equation of motion $dR/dt = -2v(R)$, $R(t)$ is the distance between the particle centers, implies that the separation between the two spheres should decrease according to the law

$$R(t) = \left(R_0^4 - 4.24 A\sqrt{\omega\eta}a^2 t\right)^{1/4}. \tag{8.30}$$

This estimate is indeed consistent with the measurements of distance between particles as function of time, see Fig. 8.19.

8.4.2 Formation of wind-blown ripples and dunes on the granular surface

Wind-blown sand forms dunes, sand ripples, and beaches of various shapes not just on Earth, but even on Mars, see Fig. 8.20. The first systematic study of airborne (or aeolian) sand transport was conducted by Ralf Bagnold before and during World War II. Bagnold (1956) identified two primary mechanisms of sand transport: *saltation* and

228 *Granular materials with complex interactions*

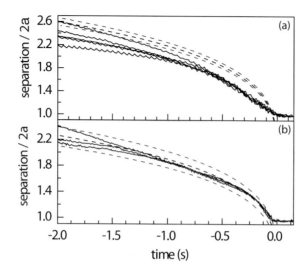

Fig. 8.19 Evolution of the distance between two particles under the action of the attractive flow. The solid curves are different experimental runs, and the central dashed curve shows the theoretical prediction (eqn 8.30). Upper and lower dashed curves show the effect of measurement uncertainty in the amplitude of vertical vibration A. (a) $f = 20$ Hz, $\Gamma = 2.9$, $A = 0.41$ mm, the vertical particle motion is periodic. (b) $f = 50$ Hz, $\Gamma = 4.6$, $A = 0.15$ mm, the vertical motion is chaotic, from Voth *et al.* (2002)

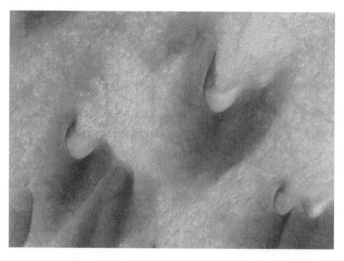

Fig. 8.20 Barchan (crescent) dunes in the Arkhangelsky Crater on Mars, from Parteli *et al.* (2005)

creep. Saltation is the wind-driven bouncing motion of individual grains on the surface, and the creep is a slow gravity-driven relaxation of sand within the near-surface layer. On the basis of laboratory-scale experiments, Bagnold proposed the first empirical relation for the sand flux J driven by the wind shear stress ς:

$$J = C_B \frac{\nu_a}{g} \sqrt{\frac{d}{d_0}} u_*^3, \qquad (8.31)$$

where ν_a is air density, d is the grain diameter, $d_0 = 0.25$ mm is a reference grain size, and $u_* = \sqrt{\varsigma/\nu_a}$ is the wind friction velocity, and C_B is a constant. Later many refinements of eqn 8.31 were proposed; see e.g. Pye and Tsoar (1990).

Since the mechanisms of ripple and dune formation are extremely complex and include phenomena occurring simultaneously on very different time and length scales (from the fraction of a millimeter scale of grain–grain interaction to hundreds of meters length scale of large desert dunes), many simplified phenomenological models were put forward. The basics of ripples and dunes formation can be captured in a simple cellular automata-type model (Nishimori and Ouchi, 1993) that operates with the height of a stack of grains at each lattice site at discrete moments of time n, $h_n(x,y)$. This model incorporates saltation and creep processes in the following way. The full time step includes two substeps. The saltation substep is described as

$$\bar{h}_n(x,y) = h_n(x,y) - j \qquad (8.32)$$
$$\bar{h}_n(x+L(h(x,y)),y) = h_n(x+L(h_n(x,y)),y) + j,$$

where j is the height of grains being transferred from one (coarse-grained) position (x,y) to the other position $(x+L,y)$ (wind is blowing in the positive x-direction), L is the flight length in one saltation that characterizes on a phenomenological level the wind strength. It is assumed that j is fixed and the saltation length L depends on on the local height h_n as

$$L = L_0 + b h_n(x,y), \qquad (8.33)$$

with L_0 characterizing the wind velocity and $b = $ const. The creep substep involves spatial averaging over neighboring sites in order to describe the surface relaxation due to gravity,

$$h_{n+1}(x,y) = \bar{h}_n(x,y) \qquad (8.34)$$
$$+ D \left[\frac{1}{6} \sum_{NN} \bar{h}(x,y) + \frac{1}{12} \sum_{NNN} \bar{h}(x,y) - \bar{h}(x,y) \right],$$

where $\sum_{NN}$ and $\sum_{NNN}$ denote summation over the nearest neighbors and next nearest neighbors, respectively, and $D = $ const is the surface relaxation rate. Simulations of this model reproduced formation of ripples and eventually arrays of barchan (crescent-shaped) dunes; see Fig. 8.21. Nishimori and Ouchi (1993) found that above a certain threshold an almost linear relation holds between the selected wavelength of the dune pattern and the "wind strength" L_0.

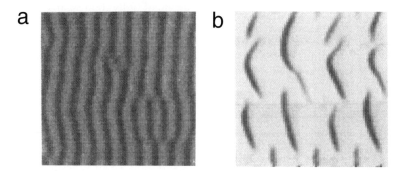

Fig. 8.21 Sand ripple pattern (a) and barchan dunes (b) obtained from simulations of eqns 8.32–8.34, from Nishimori and Ouchi (1993).

In the long-wave limit eqns 8.32–8.34 can be reduced to more traditional continuum models in which the evolution of the sand surface profile h is governed by the mass conservation equation

$$\nu_s \partial_t h = -\nabla \mathbf{J}, \tag{8.35}$$

where ν_s is the density of sand and $\mathbf{J}$ is the sand flux. In order to close eqns 8.31 and 8.35, several authors proposed various phenomenological relations between the wind shear stress at the surface ς and the height h; see, e.g., (Nishimori and Ouchi, 1993; Kroy et al., 2002; Hersen et al., 2004; Prigozhin, 1999; Andreotti et al., 2002; Schwämmle and Herrmann, 2003; Caps and Vandewalle, 2001).

There are many theories combining elements of a two-phase description of dense gravity driven flows considered in Chapter 6 with an aeolian mechanism of grain deposition similar to that of Nishimori and Ouchi (1993). For example, Prigozhin (1999) proposed a system of two equations similar to the BCRE model discussed earlier in Section 6.2.2. As in the BCRE model, one equation describes the evolution of the local height h, while another equation describes the density z_R of particles rolling above the stationary sand bed profile (reptating particles),[7]

$$\begin{aligned}\partial_t h &= \Xi(h, z_R) - f \\ \partial_t z_R &= -\nabla \cdot \mathbf{J} + Q - \Xi(h, z_R)\end{aligned} \tag{8.36}$$

where Ξ (compare eqns 6.21) is the rolling-to-steady sand transition rate, $\mathbf{J} = -\eta_0 R \nabla h$ is the horizontal projection of the flux of rolling particles, $\eta_0 > 0$ is the phenomenological constant grain mobility coefficient. However, there are also additional terms: Q accounts for the influx of falling reptating grains, and f describes the erosion rate of the surface due to the impact of saltating grains. These two quantities are related through a certain "splash function", $p(x, y)$, describing a probability that a grain, ejected by an impact at the position x will land at the position y: $Q(x, t) = \int f(y, t) p(x, y) dy$. In turn, the erosion rate f is postulated to be proportional to the impact intensity, which

---

[7]Since the thickness of the rolling layer z_R is much smaller than the thickness of the static layer underneath z_S, the difference between z_S and $h = z_S + z_R$ can be ignored

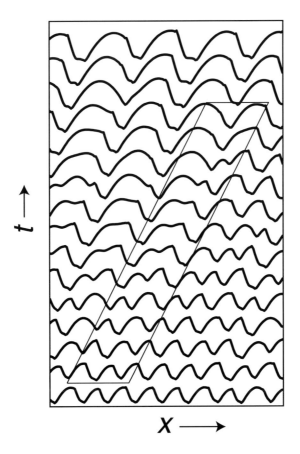

Fig. 8.22 Illustration of interaction and coarsening of one-dimensional dunes, the plot shows height profiles $h(x,t)$ at different moments of time, wind direction is from left to right, from Prigozhin (1999).

depends on the dune surface orientation with respect to the direction of saltation (or wind direction): $f = f_0 \sin\varphi_s / \sin\varphi_0$. Here, φ_s is the angle at which saltating particles impact the dune surface, φ_0 is the slope angle of the surface, and f_0 is the erosion rate or horizontal surface. However, if the dune surface $h(x)$ is sufficiently steep, some parts of it may be unreachable by saltating grains (shadowing), i.e. $f = 0$.

With an appropriate choice of rate functions Ξ, f, Q, and $\mathbf{J}$, eqns 8.36 can reproduce many observed features of dune formation, such as the initial instability of a flat state, asymmetry of the dune profiles, coarsening and interaction of dunes, etc., see Fig. 8.22.

A conceptually similar but more detailed models of non-stationary barchan dunes were developed in a series of papers (Kroy et al., 2002; Schwämmle and Herrmann, 2003; Hersen et al., 2004; Duran et al., 2005; Parteli et al., 2007). These models compute the evolution of the sand surface topography through the mass conservation equation (8.35), where the full flux of grains $\mathbf{J}$ is a sum of the saltation flux $\mathbf{J}_{\text{sal}}$,

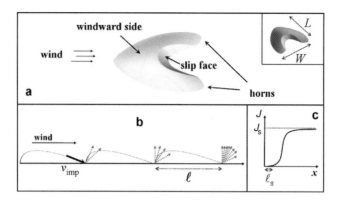

Fig. 8.23 (a) Illustration of a barchan dune showing the windward side, horns, and slip face. The inset shows the definitions of dune width W and length L. A slip surface develops when the the slope of the surface exceed static repose angle for sand, $\varphi_s \approx 34°$. (b) Schematics of the main elements of saltation process: the number of ejected grains (indicated by dashed arrows) is proportional to the velocity of the impacting grain, $v_{\rm imp}$, l is the mean saltation length. (c) Typical dependence of the sand flux J on the horizontal distance x from the grain impact point. The function J increases exponentially for small x and saturates at the characteristic saturation length l_s, from Parteli et al. (2007).

reptation flux $\mathbf{J}_{\rm rep}$, and the avalanching flux $\mathbf{J}_{\rm aval}$:

$$\partial_t h = \nabla \cdot [\mathbf{J}_{\rm sal} + \mathbf{J}_{\rm rep} + \mathbf{J}_{\rm aval}]. \tag{8.37}$$

The saltation flux $\mathbf{J}_{\rm sal}$ acts predominantly on the windward side of the barchan (see Fig. 8.23(a)) and is directed along the wind. Since saltating grains can eject several additional grains when they land on a dune surface, the saltation mechanism is self-amplified, see Fig. 8.23(b). However, since the wind spends its momentum on grains acceleration, the flux eventually saturates after a certain saturation length l_s, see Fig. 8.23(c). This saturation effect is usually described by the quasistationary *charge equation*, which in the simplest linear form (Hersen et al., 2004) reads

$$\partial_x J_{\rm sal} = l_s^{-1}(J_s - J_{\rm sal}), \tag{8.38}$$

where $J_{\rm sal}$ is the magnitude of the saltation flux, and J_s is the saturated flux magnitude that depends on the wind shear stress ς above the dune. The latter can be accurately computed assuming logarithmic velocity profile of the atmospheric turbulent boundary layer above a smooth hill. However, when dunes become sufficiently steep, the boundary layer separates on the lee side of the dune, and wind forms a big eddy (recirculation bubble). So the turbulent wind stress is computed not for the actual dune profile $h(x,y,t)$, but for the "dune envelope" $h_e(x,y,t)$ which coincides with h up to the point where the boundary layer separates, and then extrapolates it along the separation streamline. The corresponding expression for the saturated flux J_s has the form (Hersen et al., 2004)

$$J_s = J_0 \left[1 + A \int \frac{{\rm d}x'}{\pi x'} \partial_x h_e(x - x', y, t) + B \partial_x h_e(x, y, t)\right], \tag{8.39}$$

where J_0 is the saturated flux on a flat surface for a given wind speed, and A, B are tunable phenomenological constants. The integral term in the rhs of eqn 8.39 describes non-local effects of the surface curvature on the wind profile, and the last term describes the local slope effects in the first approximation: positive slope erodes faster than the negative one. Since reptating particles are generated by saltating ones, the reptation flow is proportional to the magnitude of the saltation flow, but it is directed toward the dune steepest descent:

$$\mathbf{J}_{\text{rep}} = -D J_{\text{sal}} \nabla h, \qquad (8.40)$$

where D is a dimensionless parameter characterizing the mobility of grains. According to this definition, reptation acts as a non-linear diffusive process.

The point where the boundary layer separates is hard to find theoretically. Empirically, it is chosen where the slope exceeds a certain critical value ($\sim 14°$). Inside the recirculation bubble, the wind stress and the saltation flux are assumed to be zero. But due to the deposition of grains on the windward side, the slope of the lee side continuously increases. Eventually, when the slope angle exceeds the static repose angle φ_s, the avalanching flux is introduced. This flux leads to relaxation of the slope back to the repose angle:

$$\mathbf{J}_{\text{aval}} = -C \left(|\nabla h|^2 - (\tan \varphi_s)^2 \right) \Theta \left(|\nabla h|^2 - (\tan \varphi_s)^2 \right) \nabla h \qquad (8.41)$$

Here, $\Theta(x)$ is the Heaviside function. Parameter C is usually chosen large, so the slope on that *slip face* (see Fig. 8.23(a)) is always close to φ_s.

With a proper choice of fitting parameters, the model captures many features of natural dunes. For example, starting from an arbitrary localized (e.g. conical) conical shape, an evolving dune quickly acquires a self-similar shape with horns on its lee side and all spatial dimensions proportional to each other. Since the flux at the crest of the dune is almost independent of its size, small dunes propagate faster than large ones. In fact, the velocity with which the barchan propagates, is inversely proportional to its size, in agreement with observations. Furthermore, similar but somewhat more elaborate models (Duran et al., 2005; Parteli et al., 2007) not only predict the shape of barchan dunes, but also the dynamics of barchan collisions and growth (breeding), see, e.g. Fig. 8.24. However, long-term evolution of multiple dunes within the model presented above shows collisional instability and coarsening: most dunes disappear and one large dune remain. This is in a direct contradiction with the observations showing that dunes form so-called *corridors* along the dominant wind direction, in which most of the dunes have roughly similar sizes, and no coarsening is observed. Furthermore, in adjacent corridors sizes of dunes can be markedly different (see (Hersen et al., 2004)). These puzzling features of collective long-term dune dynamics still await theoretical explanation.

8.4.3 Aquatic ripples and dunes

The phenomenon qualitatively similar to the dune formation occurs in an oscillatory fluid flow above a granular layer: sufficiently strong flow oscillations produce so-called vortex ripples on the surface of the granular layer. These ripples are familiar to any

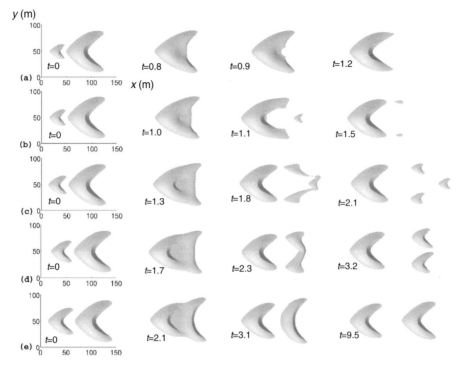

Fig. 8.24 Collision of two barchans with two different initial volumes V_H (large barchan) and V_h (small barchan) for $V_h/V_H = 0.06$ (a), 0.08 (b), 0.12 (c), 0.17 (d) and 0.3 (e) using open boundary condition: coalescence (a), breeding (b) and (c), budding (d), and solitary-wave behavior (e). The time (t) here is in months. The initial volume and height of the big barchan are 6×10^3 m^3 and 5 m, respectively, whereas the heights of the smaller barchan are 1.8 m (a), 1.9 m (b), 2.2 m (c), 2.6 m (d) and 3.1 m (e), from Duran et al. (2005).

beach-goer. Vortex ripple formation was first studied by Ayrton (1910) and Bagnold and Taylor (1956), and recently by Stegner and Wesfreid (1999), Scherer et al. (1999) and many others.

In the majority of laboratory experiments the ripples are formed by oscillatory fluid flow that mimics the periodic action of water waves, see Fig. 8.25. It was found that ripples emerge via a hysteretic transition, and are characterized by a near-triangular shape with slope angles close to the angle of repose. The characteristic size of the ripples l is directly proportional to the displacement amplitude of the fluid flow A (with a proportionality constant ≈ 1.3) and is roughly independent of the frequency ω of the driving.

The formation of underwater ripples is an intrinsically multiphase flow process. In additional to the fluid Reynolds number (which is typically of the order of 10^3, and therefore the flow is turbulent), the presence of sand introduces at least 3 new dimensionless quantities: the ratio of the density of sand ν_s the the density of water ν_w, $s = \nu_s/\nu_w \approx 2.65$, the ratio of the sand settling velocity w_s to the magnitude of

Patterns in solid–fluid mixtures 235

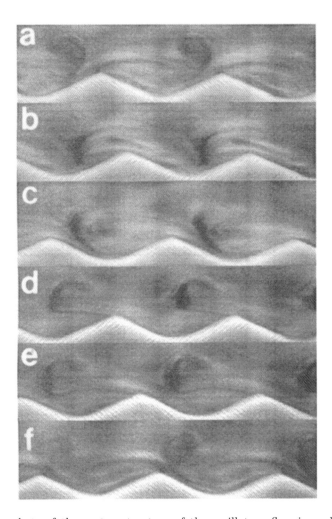

Fig. 8.25 Snapshots of the vortex structure of the oscillatory flow in underwater ripple formation. The experiments were carried out in an annual container, with the amplitude of the fluid displacements $A = 60$ mm, frequency $f = 0.05$ Hz: (a) $\omega t = 0$, (b) $\omega t = 0.16\pi$, (c) $\omega t = 0.4\pi$, (d) $\omega t = 0.77\pi$, (e) $\omega t = 0.90\pi$, (f) $\omega t = \pi$, $\omega = 2\pi f$. Hydrodynamic vortices above the ripples were visualized with the Kalliroscope solution injected into the fluid, from Scherer *et al.* (1999).

flow velocity $v_f = A\omega$, $W = w_s/A\omega$, and the Shields number Sh characterizing the ratio of the viscous drug to the gravitational force acting on a single grain

$$\text{Sh} = \frac{\varsigma}{(\nu_s - \nu_w)gd}, \qquad (8.42)$$

where d is the grain diameter and ς is the shear stress at the bed surface. For laminar flows the stress and the Shields number can be found from the solution of the Stokes problem. For a more typical case of turbulent flows, however, no analytic expression exists. The following empiric expression is often used for the maximum (over the oscillation period) shear stress $\varsigma_{max} = \max\{\varsigma\}$ (Andersen et al., 2001)

$$\varsigma_{max} \approx 0.02\nu_w \left(\frac{A}{d_N}\right)^{0.25} (A\omega)^2, \tag{8.43}$$

where d_N is the length scale of the order of grain size characterizing the roughness of the bed surface.

The sediment transport can be approximately described by the following empiric expression for the flux of grains

$$J = \alpha(\text{Sh} - \text{Sh}_c)^\beta \sqrt{g(s-1)d^3}, \tag{8.44}$$

where $\text{Sh}_c \approx 0.06$ is the critical Shields number for turbulent boundary layers, and model fitting parameters $\alpha \approx 8$ and $\beta \approx 1.5$ are obtained by comparison with experiment.

In order to model the ripple formation, the first step is to calculate the turbulent fluid flow over a ripple using some standard model of turbulence (Andersen et al., 2001). Then, the sediment transport can be calculated from eqn 8.44. Figure 8.26 illustrates the flow structure and the Shields number Sh for the ripple size $l = 1.15A$. There are two mechanisms generating the shear stress on the bed, namely, the converging flow on the "windward" side of the ripple (left in Fig. 8.26) and the *separation bubble* formed in the "lee" side (right in Fig. 8.26), where the flow near the bed surface is directed opposite to the mean flow direction. The bubble moves into the trough of the ripple at non-dimensional time (phase) $\omega t = 90°$, where it stays until at $\omega t = 150°$ it is thrown over the crest as the flow reverses. These stresses typically are several times stronger than for a flat bed.

The shear stresses transport sand from the trough towards the crest, steepening the ripple profile until the slopes of the ripple reach the repose angle, and avalanches occur. As a consequence, most slopes of the fully developed ripples are close to the angle of repose and have an almost triangular shape. Ripples interact by exchanging granular material between neighbors over the troughs. The amount of this granular mass flow is closely related to the size and the strength of the separation bubble.

While the interaction of ripples is a complicated and multiscale phenomenon, many features can be captured by a very simple one-dimensional discrete coarse-grained model (Andersen et al., 2001). In this model the ripples are triangular and symmetric and are characterized by their length l_i. The mass transfer between two ripples with length l_1 and l_2 is determined from the assumption that the mass transport during a half-period of driving depends only on the size of the ripple that creates the separation bubble. Under this approximation, the evolution of ripple i is described by the following simple model following from the sand mass-exchange rules:

$$\partial_t l_i = -f(l_{i-1}) - f(l_{i+1}) + 2f(l_i). \tag{8.45}$$

The non-linear function $f(l)$ characterizes the amount of sand transported over the trough during the first half-wave period (see eqn 8.44)

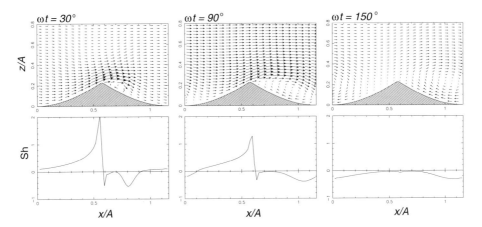

Fig. 8.26 Top row: the flow over a ripple at three instants in the first half of the wave period. The system consists of a single ripple in a system with periodic boundary conditions and length $l = 1.15$. Bottom row: the spatial profiles of the Shields number on the surface at the corresponding times and for $\text{Sh}_{\max} = 0.13$, from Andersen *et al.* (2001).

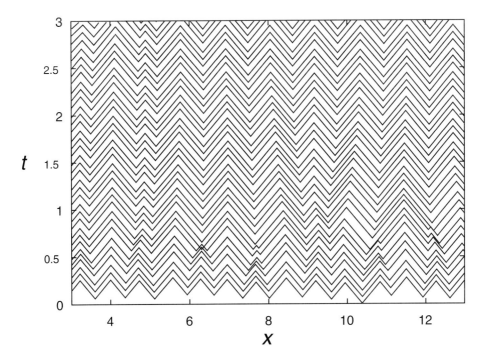

Fig. 8.27 Illustration of the dynamics of the model (eqn 8.45) using the piecewise-linear interaction function $f(\lambda)$ and $l_{\max} = 1.35$. The time scale is arbitrary. As time progresses, small ripples coarsen and large ripples grow, from Andersen *et al.* (2001).

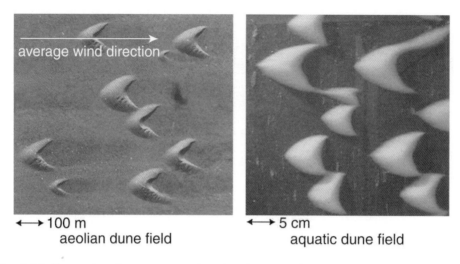

100 m
aeolian dune field

5 cm
aquatic dune field

Fig. 8.28 Comparison between natural aeolian barchans (southern Morocco, aerial photo) and laboratory scale aquatic barchans. While the spatial scale are vastly different, the crescent shapes are quite similar, from Hersen et al. (2002).

$$f(l) \sim - \int_0^{\pi/\omega} J(x_{\text{tr}}, t) \mathrm{d}t \qquad (8.46)$$

where x_{tr} is the position of the trough. Function $f(l)$ has the following properties: $f(0) = f(l_{\max}) = 0$ and reaches its maximum value at $l = l_{\min}$, $0 < l_{\min} < l_{\max}$. Andersen et al. (2001) showed that both smooth continuous and piecewise-linear approximations of the function $f(l)$ yield qualitatively similar results: the existence of a stable band of wave numbers limited by secondary instabilities and coarsening of small ripples, see Fig. 8.27, in agreement with experiments.

On a more detailed level the formation of ripples can be described by a continuum model that takes into account both the local shear stress (which is deduced from the the fluid flow profile over a two-dimensional deformed bed) and the local bed slope. Langlois and Valance (2005) suggested the following phenomenological expression for the horizontal mass flux of sand $\mathbf{J} = (J_x, J_y)$:

$$\mathbf{J} = J_{\text{b}} \left[\left| \boldsymbol{\vartheta}_{\text{h}} - \frac{\text{Sh}_{\text{c}}}{\mu_{\text{s}}} \nabla h \right| - \text{Sh}_{\text{c}} \right]^{3/2} \frac{\boldsymbol{\vartheta}_{\text{h}} - \frac{\text{Sh}_{\text{c}}}{\mu_{\text{s}}} \nabla h}{\left| \boldsymbol{\vartheta}_{\text{h}} - \frac{\text{Sh}_{\text{c}}}{\mu_{\text{s}}} \nabla h \right|}, \qquad (8.47)$$

where $\boldsymbol{\vartheta}_{\text{h}}$ is the horizontal projection of dimensionless surface shear stress, Sh_{c} the critical value of the Shields number to set grains in motion, μ_{s} is the internal friction coefficient of material, $J_{\text{b}} = c\sqrt{(s-1)gd^3}$, and c is the fitting parameter. The transport law above incorporates, on a more detailed level, the rheology of granular matter, e.g. nonzero repose angle given by the value of friction coefficient μ_{s}. The bed shear stress is obtained from the solution of the incompressible Navier–Stokes equations for viscous fluid. Again, accurate calculations are only possible for a laminar flow, which is not typical in the ripple-formation process. From the knowledge of the grain flux **J**, the ripple height can be obtained by solving the mass conservation equation (eqn 8.35).

Linear stability analysis of eqn 8.35 with transport law (eqn 8.47) coupled to the Navier–Stokes equations shows that the ripple formation is initiated by an exponentially growing longitudinal mode. The weakly non-linear analysis taking into account resonance interaction of only three unstable modes reveals a variety of steady two-dimensional ripple patterns drifting at a certain speed in the direction of the fluid flow.

Like ripples, dunes can also form underwater. Experiments in underwater dune formation have been performed by Betat *et al.* (1999) and Hersen *et al.* (2002). While water-driven and wind-driven dunes and ripples have similar shapes, the sizes of dunes differ by at least 3 orders of magnitude, see Fig. 8.28. The difference is explained by the entirely different balance between gravity and viscous drag in air and water. In particular, the saltation process critical for the formation of aeolian dunes is irrelevant for the aquatic dunes.

8.4.4 Erosion and sedimentation patterns

In this section we briefly discuss a variety of complex erosion and sedimentation patterns occurring in underwater granular flows. In natural conditions, these processes occur on very large scales (kilometers). However, recent laboratory experiments performed with very fine powders and under controlled conditions allowed scaling down of the patterns from kilometers to a few centimeters without changing the essential physics of the pattern-forming processes.

Spectacular erosion patterns in sediment granular layers were observed in laboratory experiments with underwater flows (Daerr *et al.*, 2003; Malloggi *et al.*, 2006). In particular, a fingering instability of flat avalanche fronts was observed for both dry and underwater avalanches, see Fig. 8.29 and Section 6.3.1. While these fingering patterns occur at very different length scales, the mechanism of the fingering is probably similar, and is related to the dependence of the avalanche front velocity on the granular mass carried by the avalanche.

Erosion and deposition processes are critical for the understanding of Earth surface morphology. For example, they are responsible for formation of spectacular drainage networks, such as shown in Fig. 2.14.

The erosion processes have been intensively studied in geophysics and planetary science (Smith and Bretherton, 1972; Dietrich *et al.*, 1992; Dunne, 1980). It is believed that many erosive features on Mars, such as those shown in Fig. 8.30 resulted from subsurface flows (Aharonson *et al.*, 2002). However, recent laboratory-scale experiments with hollow dry grains (ceramic beads) by Shinbrot *et al.* (2004) demonstrated remarkably similar flow and erosion patterns to those observed on Mars, see Fig. 8.31.

The slope instability and channelization driven by subsurface flows were studied in recent laboratory-scale experiments (Schorghofer *et al.*, 2004; Lobkovsky *et al.*, 2004). In these well-controlled experiments the slope of a sand pile and the pressure of the water entering the sand pile from below were varied. As a result, three modes of sediment mobilization were observed: surface erosion, fluidization, and slumping. Un-

240 *Granular materials with complex interactions*

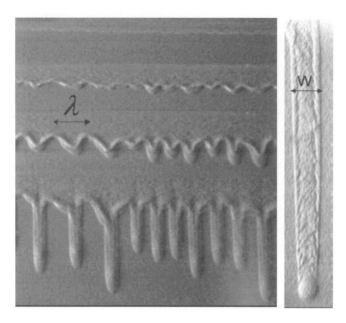

Fig. 8.29 Fingering instability of a planar avalanche in the underwater flow, the image on the left zooms on an individual finger, from Malloggi et al. (2006).

like the mechanisms of aquatic ripple formation, the erosion is controlled together by shear stresses of the above-surface ("overland") flow and the subsurface ("ground") flow. Whereas above-surface flow alone can mobilize surface grains and cause erosion instability only if the water flux exceeds a threshold, subsurface flows cause this threshold to disappear when the slope becomes steeper than a certain critical angle. Interestingly, this critical angle is much smaller than the maximum angle of stability.

Duong et al. (2004) studied experimentally and numerically the formation of periodic arrays of knolls in a slowly rotating horizontal cylinder filled with granular suspension; see Fig. 8.32. The solidified sediment knolls coexist with freely circulating fluid. For numerical modelling, assuming the low Reynolds number of the flow, the authors used the Stokes approximation for the suspension velocity (Landau and Lifshitz, 1987), mass conservation for both fluid and solid fractions, and constitutive relations for solid phase settling velocity and viscosity of the suspension. This settling velocity was taken to be the Stokes terminal velocity, and the effective viscosity of the suspension η_s was assumed to depend on the solid volume fraction ϕ and diverge at the random close-packing density ϕ_c,

$$\eta_s = \eta_0(1 - \phi/\phi_c)^{-b}. \tag{8.48}$$

Here, η_0 is the clear fluid viscosity and b is an empirical constant. This simple model reproduced the experiment surprisingly well. An interesting question in this context is whether there is a connection between this phenomenon and patterns in "dry" horizontally rotating cylinders studied in the experiment by Shen (2002), see Fig. 6.27.

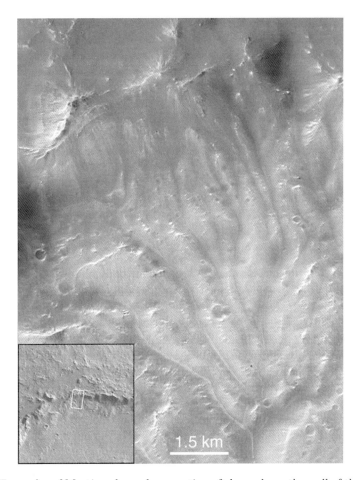

Fig. 8.30 Examples of Martian channels: a portion of channels on the wall of the Bakhuysen crater is shown, from Malin and Carr (1999).

8.5 Electrically driven granular media

Another broad class of long-range interactions in granular materials includes forces of an electromagnetic nature. Small dielectric or conducting particles easily acquire electric charge through rubbing and interact through competing long-range electromagnetic and short-range contact forces. Many industrial technologies face the challenge of assembling and separating single- or multicomponent micro- and nanosize ensembles. Traditional methods, such as mechanical vibration and shear, are ineffective for very fine powders due to their spontaneous charging and ensuing agglomeration. Electric forces often change the statistical properties of granular matter, such as energy dissipation rate (Scheffler and Wolf, 2002), velocity distributions in granular gases (Aranson and Olafsen, 2002; Kohlstedt et al., 2005), agglomeration rates in suspensions (Dammer and Wolf, 2004), etc.

On the other hand, electrically charged particles offer new possibilities for their

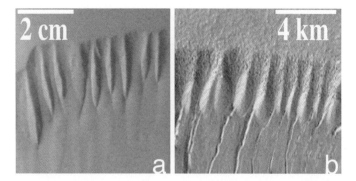

Fig. 8.31 Qualitatively similar erosion patterns observed in laboratory experiments with light hollow dry grains and on Mars. (a) Rounded hillocks separating flowing gullies after the collapse of a flat surface of a layer of hollow beads; (b) "Alcoves" in Martian polar pits that are often presented as evidence for surface water flow, from Shinbrot et al. (2004).

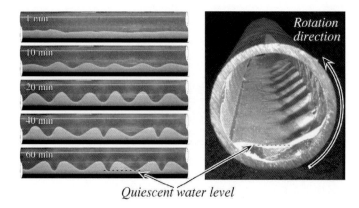

Fig. 8.32 Self-supporting knolls formed in water/glass beads suspension in a horizontally rotating cylinder (side and end views). from Duong et al. (2004).

external excitation compared to traditional techniques of mechanical excitation, such as vibration and shear. It enables one to deal with extremely fine non-magnetic and magnetic powders that cannot be effectively driven by other means. Electrically driven granular media were investigated in a series of experimental and theoretical studies (Aranson et al., 2000; Aranson et al., 2002; Sapozhnikov et al., 2003b; Sapozhnikov et al., 2004), where the method of electric excitation relied on the collective interactions between particles due to a competition between short-range collisions and long-range electromagnetic forces.

The experimental setup is illustrated by Fig. 8.33. In most experimental realizations, several grams of monodispersed conducting microparticles were placed in a 1.5-mm gap between two horizontal 30×30 cm^2 glass plates covered by transparent conducting layers of indium-doped tin dioxide. The experimental cell could be evacuated, filled with air, dry nitrogen, or a certain dielectric liquid such as toluene. Typically

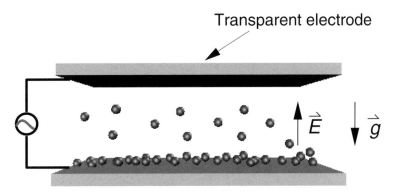

Fig. 8.33 Schematics of experimental setup for electric excitation of conducting granular matter.

45-μm copper or 120-μm bronze spheres were used, however experiments were also performed with much smaller, 1-μm, gold and silver particles, (Sapozhnikov et al., 2004). An electric field perpendicular to the plates was created by a high-voltage source (0–3 kV) connected to the conductive inner surface of each plate.

The basic principle of the cell operation in constant (dc) electric field is as follows. A particle acquires an electric charge when it is in contact with the bottom conducting plate. In an air-filled or evacuated cell, the particle remains immobile at the bottom plate if the electric field E is smaller than the first critical field E_1. For $E > E_1$ an isolated particle leaves the bottom plate and travels to the upper plate since the upward force induced by the electric field exceeds gravity g. Then the particle reverses charge upon contact, and is attracted back to the bottom plate. This process repeats in a cyclical fashion. However, by applying an alternating (ac) electric field $E = E_0 \sin(2\pi f t)$, and adjusting its frequency f, one can control the particle elevation by effectively turning them back before they collide with the upper plate.

Thus, individual particles begin to move for $E > E_1$. However, if several particles are in contact on the plate, screening of the electric field reduces the force on individual particles, and they remain immobile even if $E > E_1$. A simple calculation shows that for the same value of the applied voltage the force acting on an isolated particle exceeds by a factor of two the force acting on the particle inside a dense monolayer. If the field is larger than a second critical field value, $E_2 > E_1$, all particles leave the plate, and the system of many particles transforms into a uniform gas-like phase. When the field is decreased below E_2, but still above E_1, localized clusters of immobile particles (precipitate) spontaneously nucleate on the bottom plate, see Fig. 8.34 (Aranson et al., 2000; Sapozhnikov et al., 2005). Initially, a large number of small cluster is formed. Then, small clusters gradually shrink and large clusters grow, while the total area occupied by the clusters remains practically constant. This is a hallmark of the *interface-controlled* Ostwald-type ripening (Bray, 1994; Meerson, 1996), see also Section 4.4 where a similar phenomenon in monolayers of vertically vibrated spherical particles was described.

In experiments (Aranson et al., 2000; Sapozhnikov et al., 2005), the number of

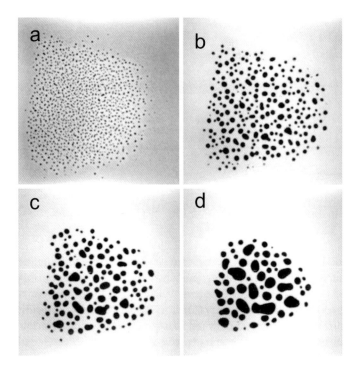

Fig. 8.34 Snapshots illustrating coarsening and coalescence of granular clusters in electric cell at times (in seconds) $t = 0$ (a), 1×10^4 (b), 2×10^4 (c) and 5×10^4 s (d). The applied dc electric field $E = 2.33$ kV/cm, in a 30×30 cm^2 experimental cell filled with dry nitrogen, with 45-μm copper spheres, from Sapozhnikov et al. (2005).

clusters N and the total area of the cluster phase A_{tot} were measured as functions of time. These studies were conducted in a rather large cell (30×30 cm^2) and with small (45 μm) conducting (copper or bronze) particles. Under these conditions up to several thousand small clusters are formed initially. Measurements yielded the following asymptotic scaling law for the number of clusters N versus time t (Fig. 8.35):

$$N \sim \frac{1}{t}. \tag{8.49}$$

Accordingly, the average cluster area $\langle A \rangle$ increases with time as $\langle A \rangle \sim t$, whereas the total area $A_{\text{tot}} = N \langle A \rangle$ remains essentially constant.

8.5.1 Phase-field theory of cluster formation

A theoretical description of coarsening in an electrostatically driven granular system can be formulated in terms of a continuum phase-field-type equation for the number densities of immobile particles (precipitate or solid) ν_s and bouncing particles (gas) ν_g (Aranson et al., 2000; Aranson et al., 2002):

$$\partial_t \nu_s = D_s \nabla^2 \nu_s + \Xi(\nu_s, \nu_g) \tag{8.50}$$

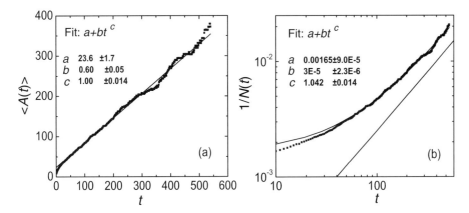

Fig. 8.35 Average cluster area $\langle A(t) \rangle$ (a) and inverse number of clusters $1/N(t)$ vs. time in an air-filled cell. The straight line in (b) shows the theoretical prediction $1/N \sim t$, from Sapozhnikov et al. (2005).

$$\partial_t \nu_g = D_g \nabla^2 \nu_g - \Xi(\nu_s, \nu_g) \tag{8.51}$$

where $\Xi(n_s, n_g)$ is a function characterizing the solid/gas conversion rate, and D_s, D_g are the diffusion coefficients of the dense (solid) and gas phases, respectively. The effectiveness of the solid/gas transitions is controlled by the local gas density ν_g. One of the main assumptions of the theory is that the mean free path of particles is large, and, therefore, the gas diffusion coefficient D_g is much larger than the solid state diffusion, $D_g \gg D_s$. Consequently, in this approximation the gas density ν_g becomes uniform over the whole system, and eqn 8.51 can be omitted. Then, the solid-phase diffusion D_s can be scaled away, the density ν_s can be normalized by close-packed density ν_c, and eqn 8.50 assumes the form

$$\partial_t \nu_s = \nabla^2 \nu_s + \Xi(\nu_s, \nu_g) \tag{8.52}$$

The gas density ν_g is coupled to the solid fraction density ν_s due to the total mass conservation constraint

$$S\nu_g + \int\int \nu_s(x,y) \mathrm{d}x \mathrm{d}y = M, \tag{8.53}$$

where S is the total area, and M is the total number of particles. Function $\Xi(\nu_s, \nu_g)$ is chosen in such a way as to provide bistable local dynamics of the concentration corresponding to the hysteresis of the gas/solid transition. More specifically, in some range of gas concentrations corresponding to phase coexistence the function $\Xi(\nu_s, \nu_g)$ should possess three zeros, two of which correspond to stable fixed points at $\nu_s = 0$ (pure gas) and $\nu_s = 1$ (close-packing density). These stable equilibria should be separated by some unstable fixed point ν_* such that $0 < \nu_* < 1$. The phase equilibrium then occurs when the Maxwell area rule is satisfied (compare with the van-der-Waals model of phase separation considered in Section 4.2.2),

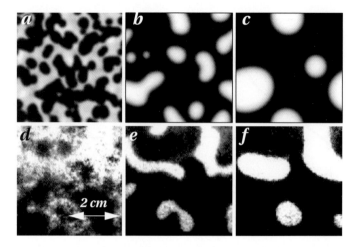

Fig. 8.36 Phase separation and coarsening dynamics in electrically vibrated grains: (a)–(c) Numerical solution of eqns 8.52 and 8.53, white corresponds to dense clusters, black to dilute gas; (d)–(f) Experimental results, from Aranson et al. (2002).

$$\int_0^{\nu_*} \Xi(\nu_s, \nu_g) d\nu_s = -\int_{\nu_*}^1 \Xi(\nu_s, \nu_g) d\nu_s. \tag{8.54}$$

Generally, the phase equilibrium occurs only for some specific value of the gas concentration ν_g. While on a qualitative level the description is rather insensitive to the specific choice of the function $\Xi(\nu_s, \nu_g)$, in order to obtain more quantitative agreement with the experiment this function needs to be derived from underlying physical mechanisms (such as the electrostatic screening of interaction in dense clusters, etc). Specifically, we use the following function (Aranson et al., 2002):

$$\Xi(\nu_s, \nu_g) = (\nu_s - \nu_*) \times \begin{cases} \nu_s, & \text{if } 0 \geq \nu_s < \nu_* \\ C\nu_g(1 - \nu_s), & \text{if } \nu_* \geq \nu_s \leq 1 \end{cases} \tag{8.55}$$

Here, constants $C = O(1)$ and ν_* are the only model parameters and can be expressed through the value of the applied electric field, the gap between the plates and the particle size, etc. The above description captures the essential physics of the system: phase separation instability in early stages and consequent formation and coarsening of clusters at later stages, see Fig. 8.36.

8.5.2 Sharp-interface limit

Further simplification of the theoretical description provided by eqns 8.52, 8.53 and (8.55) is possible in the so-called *sharp-interface limit* when the size of clusters is much larger than the width of interfaces between clusters and granular gas. In this rather generic situation eqn 8.52 can be reduced to equations for the cluster radii R_i (assuming that clusters have circular form):

$$\frac{dR_i}{dt} = \kappa \left(\frac{1}{R_c(t)} - \frac{1}{R_i} \right), \tag{8.56}$$

where R_c is a critical cluster size that is a certain function of the gas concentration n_g, κ is the effective surface tension. The surface tension κ of the granular solid/granular gas interface in this experiment was measured directly by Sapozhnikov et al. (2003a). The form of eqn 8.56 accounts for two competing effects: shrinkage of the cluster due to surface tension (term $\sim 1/R_i$) and expansion of the cluster due to absorption of particles from the gas phase (term $\sim 1/R_c(t)$). The clusters will shrink in finite time if $R_i < R_c$.

The critical radius R_c couples eqn 8.56 with the gas density ν_g through the the conservation law eqn 8.53 that in two dimensions reads

$$\nu_g S + \pi \sum_{i=1}^{N} R_i^2 = M. \tag{8.57}$$

The dynamics described by eqns 8.56 and 8.57 can be summarized as follows. Droplets (clusters) with $R > R_c(t)$ grow at the expense of droplets with $R < R_c(t)$ that shrink. At the late-time asymptotic stage the critical radius grows with time, and as a result a droplet that was growing at an early time begins to shrink at a later time. Eventually, only the biggest droplet will survive the coarsening process.

The statistical properties of Ostwald ripening can be understood in terms of the probability distribution function $f(R, t)$ of cluster sizes. Following (Lifshitz and Slyozov, 1958; Lifshitz and Slyozov, 1961; Wagner, 1961) and neglecting cluster merger, one obtains in the limit $N \to \infty$ that the probability distribution $f(R, t)$ satisfies the continuity equation

$$\partial_t f + \partial_R \left(\dot{R} f \right) = 0. \tag{8.58}$$

In the limit of small gas density the mass conservation (eqn 8.57) yields an additional constraint:

$$\pi \int_0^\infty R^2 f(R, t) \mathrm{d}R = M. \tag{8.59}$$

Equations 8.58 and 8.59 have a self-similar solution of the form

$$f(R, t) = \frac{1}{t^{3/2}} F\left(\frac{R}{\sqrt{t}}\right). \tag{8.60}$$

The total number of clusters $N = \int_0^\infty f \mathrm{d}R$ scales as $1/t$, which appears to be in good agreement with the experiment; see Fig. 8.35.

While the Lifshitz–Slyozov–Wagner theory yields the correct scaling for the number of clusters as a function of time, the experimental cluster-size distribution function appears to be in a strong disagreement with the theory, see Fig. 8.37. In the Lifshitz–Slyozov–Wagner theory the scaling function F assumes the universal Wagner form

$$F = C_0 \phi \frac{\xi}{(\xi - \xi_m)^4} \exp\left(-\frac{2}{\xi_m - \xi}\right) \tag{8.61}$$

where $\xi = R/\sqrt{t}$, $\xi_m = [1 + 2e^2 \mathrm{Ei}(-2)]^{-1/2} \approx 1.9$, ϕ is the area fraction, and $C_0 = \pi^{-1}[1/(2e^2) + \mathrm{Ei}(-2)]^{-1} \approx 17$ is the normalization constant, and $\mathrm{Ei}(z)$ is the

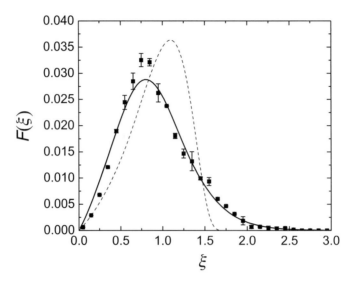

Fig. 8.37 Scaled cluster-size distribution function $F(\xi)$ with $\xi = R/\sqrt{t}$. The squares show experimental results, the dotted line shows analytic result from the Lifshitz–Slyozov–Wagner theory (eqn 8.61), and the solid line shows F obtained from the modified theory accounting for binary coalescence based on eqn 8.62, from Sapozhnikov et al. (2005).

exponential integral. For $\xi > \xi_m$ the scaling function $F(\xi) = 0$. Thus, the theory predicts the finite-support distribution with a cutoff at $\xi = \xi_m$ (dotted line) whereas the experiment yields the function with an exponential tail.

The main reason for the inaccuracy of the above model is its disregard of the coalescence of clusters seen in Fig. 8.34. A much better agreement with the experiment can be obtained when binary coalescence of clusters is incorporated in the Lifshitz–Slyozov–Wagner theory (Conti et al., 2002; Sapozhnikov et al., 2005). This modified theory accounts for the binary coalescence of clusters (growing clusters merge upon touching each other) within the mean-field framework of Ostwald ripening. The continuity equation (eqn 8.58) for the function f gives rise to the following kinetic equation:

$$\partial_t f + \partial_R (\dot R f) = -\frac{1}{2} \int_0^\infty \int_0^\infty 2M(R_1, R_2) \qquad (8.62)$$
$$\times \left[\delta(R - R_1) + \delta(R - R_2) - \delta\left(R - \sqrt{R_1^2 + R_2^2}\right) \right] \times f(R_1, t) f(R_2, t) \, dR_1 \, dR_2$$

where $M(R_1, R_2) = \pi (R_1 + R_2)(\dot R_1 + \dot R_2) \theta(\dot R_1 + \dot R_2)$, $\theta(\ldots)$ is the Heaviside step function, and $\dot R_1$ and $\dot R_2$ are expressed through eqn 8.56, the details of the derivation can be found in (Conti et al., 2002). The δ-function terms in the rhs of eqn 8.62 (collision integral) describe annihilation of two round clusters with radii $R_{1,2}$ and creation of one new cluster with radius $R = \sqrt{R_1^2 + R_2^2}$ as a result of coalescence.

Remarkably, self-similar solutions to eqn 8.62 admit the same scaling anzatz (eqn 8.60) as the original equation without coalescence (eqn 8.58). Correspondingly, the scaling solution is described by a non-linear integrodifferential equation that is obtained from eqn 8.62, subject to the cluster area conservation $\phi = M/S = $ const. This integrodifferential equation can be reduced to an integral equation and solved iteratively. Figure 8.37 presents the scaled cluster-size distribution function, obtained by the iteration procedure (Conti et al., 2002). This function corresponds to the same area fraction $\phi = 0.092$ as in the experiment. Obviously, the agreement is much better than for the Wagner distribution. Notice that the only parameter of the scaled cluster-size distribution that enters the theory is the area fraction ϕ. Therefore, the comparison of the theory and experiment does not involve *any* fitting parameters.

8.5.3 Liquid-filled cell: vortices and dynamic patterns

In all experiments with electrically driven granular media described above the viscous drag, and correspondingly hydrodynamic interactions between the grains were not critical. However, the hydrodynamic interactions come into play if the cell is filled with more viscous liquid.

In order to explore the effects of additional hydrodynamic interactions between the particles, Sapozhnikov et al. (2003b) performed experiments with electrically driven granular media immersed in a weakly conducting non-polar liquid. In the majority of experiments the cell was filled with non-polar organic dielectric liquid such as toluene (viscosity about one third of the viscosity of water, dielectric constant ≈ 2). To control the conductivity of the solution, a small amount of pure ethanol was mixed with the base liquid (toluene). Ethanol molecules dissociate in toluene and are therefore responsible for a very weak ionic conductivity (corresponding Debye radius is of the order of 1 micrometer, i.e. about 1000 larger than that of water).

The experiments with different applied electric fields and ethanol concentrations revealed a plethora of new static and dynamic patterns not present in the gas-filled or evacuated cell, ranging from static hexagonal lattices to dynamic toroidal vortices and pulsating rings, see Fig. 2.23. The phase diagram illustrating main dynamic and static regimes is shown in Fig. 8.38.

For relatively low concentrations of ethanol (below 3%), the qualitative behavior of the liquid-filled cell is not very different from that of the gas-filled cell described above: clustering of immobile particles and coarsening were observed between two critical field values $E_{1,2}$ with the clusters being qualitatively similar to those of the gas-filled cell. However, when the ethanol concentration is increased, the phase diagram becomes asymmetric with respect to the direction of the electric field. Critical field magnitudes, $E_{1,2}$, are larger when the electric field is directed downward ("+" on the upper plate) and smaller when the field is directed upward ("−" on the upper plate). This difference increases with ethanol concentration. The observed asymmetry of the critical fields is apparently due to an excess negative charge in the bulk of the liquid due to asymmetric dissociation of the ethanol and residual water molecules in the dielectric toluene solution. The dissociation produces highly mobile and reactive positive H^+ ions and slower and less reactive negative $C_2H_5O^-$ and OH^- ions. So, these negative ions are distributed throughout the cell by the current, while positive ions (H^+) are absorbed by the electrodes. It appears that the liquid contains almost exclusively ions of the same (negative) sign (albeit in a very small concentration).

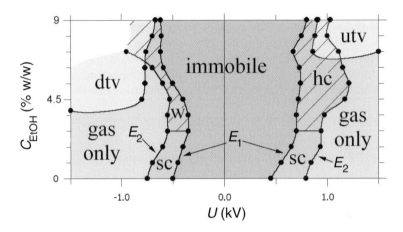

Fig. 8.38 Experimental phase diagram illustrating main dynamic patterns in the liquid-filled cell. Here, U is the applied voltage (positive values correspond to plus on the top plate), C is the concentration of ethanol, E_1 and E_2 denote values of critical electric fields. Domains sc denote "static clusters", w, Wigner crystal, hc, honeycombs, utv and dtv, up/down toroidal vortices, from Sapozhnikov et al. (2003b).

Correspondingly, this excess bulk charge in the liquid increases the effective electric field acting on the positively charged particles sitting on the bottom plate when a positive voltage is applied to that plate and correspondingly decreases the effective electric field in the case of an applied negative voltage.

The situation changes dramatically for higher ethanol concentrations: increasing applied voltage leads to the formation of two new immobile phases: honeycomb (Fig. 2.23(b)) for the downward direction of the applied electric field, and two-dimensional crystal-type states of individual particles for the upward direction (called by analogy with electron systems, a "Wigner" crystal).

A further increase of the ethanol concentration leads to the appearance of a novel dynamic phase – a condensate of collectively moving particles (Figs. 2.23(c) and (d)) where almost all particles are engaged in a circular vortex motion in the vertical plane, resembling Rayleigh–Bénard convection that occurs in a thin layer of ordinary fluid heated from below. The condensate coexists with the dilute gas of bouncing particles. The direction of rotation is determined by the polarity of the applied voltage: particles stream towards the center of the condensate near the top plate for the upward field direction and vice versa. The localized condensate droplets have the structure of toroidal vortices (called here correspondingly up (utv) and down (dtv) toroidal vortices depending on the direction of the applied voltage). The subsequent evolution of the condensate depends on the direction of the electric field. For the downward field, large structures become unstable due to the spontaneous formation of voids (Fig. 2.23(d)). These voids exhibit complex intermittent dynamics. In contrast, for the upward field, large vortices merge into one, forming an asymmetric object that

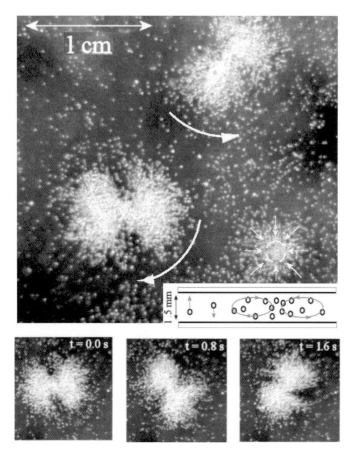

Fig. 8.39 Rotating "binary star" vortices. Upper panel: two horizontally rotating down toroidal vortices (dtv). Arrows indicate direction of horizontal rotation of the vortices. Small arrows show particle trajectories at the top plate for a single vortex. Inset: vertical cross-section of the vortex flow illustrating a toroidal structure of the object. Lower panel: time evolution of one rotating asymmetric vortex, from Sapozhnikov et al. (2003b).

rotates in the horizontal plane, as shown in Fig. 8.39.

Pattern formation in this system is caused by self-induced electro-hydrodynamic microvortices created by the particles in a weakly conducting liquid. These microvortices create large-scale hydrodynamic vortex flows that overwhelm electrostatic repulsion between like-charged particles through attractive dipole-like hydrodynamic interactions. Surprisingly, while the underlying physics is apparently very different, somewhat similar microvortices emerge in aqueous charge-stabilized colloidal suspensions of dielectric microparticles when electrohydrodynamic forces due to constant applied fields compete with gravity (Yeh et al., 1997; Grier, 2003).

8.5.4 Theory of pattern formation in the liquid-filled cell

Despite complicated physical mechanisms involved in the pattern formation in the liquid-filled cell, important insights can be obtained by the suitable generalization of the theory of phase separation developed for an air-filled cell in Section 8.5.1. Again, the theory can be formulated in terms of conservation laws for the number densities of immobile particles (solid or precipitate) ν_s and bouncing particles (gas) ν_g averaged over the thickness of the cell (compare with the corresponding equations for the air-filled cell eqns 8.50 and 8.52):

$$\partial_t \nu_s = \nabla \cdot \mathbf{J}_s + \Xi, \quad \partial_t \nu_g = \nabla \cdot \mathbf{J}_g - \Xi. \tag{8.63}$$

Here, $J_{s,g}$ are the mass fluxes of solid and gas fractions, respectively, and the function Ξ again describes gas/solid conversion, which depends on $\nu_{s,g}$, electric field E and local ionic concentration c. While for low value of ionic concentration c the function Ξ is given by eqn 8.55, for higher values of c, the exchange rate Ξ depends explicitly on c and, according to Aranson and Sapozhnikov (2004), can be expressed as $\Xi \sim c(\nu_g - \tilde{\mu}(E,q)\nu_s/(\nu_s + \nu_g))$, where $\tilde{\mu}(E,q)$ is a certain function of electric field E and charge density q.

The fluxes are written as:

$$\mathbf{J}_{s,g} = D_{s,g}\nabla \nu_{s,g} + \alpha_{s,g}(E)\mathbf{v}_\perp \nu_{s,g}(1 - \beta(E)\nu_{s,g}), \tag{8.64}$$

where $\mathbf{v}_\perp$ is the horizontal hydrodynamic velocity, $D_{s,g}$ are solid/gas fractions diffusivities. The last terms $\sim \alpha_{s,g}$ in these expressions describe particle advection by the liquid. The factor $(1 - \beta(E)\nu_{s,g})$ describes the saturation of flux at large particle densities $\nu \sim 1/\beta$ due to the decrease of void fraction. Without advection terms ($\alpha_{s,g} = 0$) the model reduces to eqns 8.50 and 8.52) used for the gas-filled cell (Aranson et al., 2002).

In order to close the description, eqns 8.63 need to be coupled to the Navier–Stokes equation for the vertical component v_z of hydrodynamic velocity $\mathbf{v}$. This equations can be written in the following form:

$$\nu_0(\partial_t v_z + \mathbf{v}\nabla v_z) = \eta \nabla^2 v_z - \partial_z p + E_z q, \tag{8.65}$$

where ν_0 is the density of liquid (we set $\nu_0 = 1$), η is the viscosity, p is the pressure, and q is the charge density. The last term describes the electric force acting on the charged liquid. The horizontal velocity $\mathbf{v}_\perp$ is obtained from v_z using the incompressibility condition $\nabla \mathbf{v} = \partial_z v_z + \nabla_\perp \mathbf{v}_\perp = 0$ in the approximation that the vertical vorticity $\Omega_z = \partial_x v_y - \partial_y v_x$ is small compared to the inplane vorticity. This assumption allows one to find the horizontal velocity as a gradient of the quasipotential ψ: $\mathbf{v}_\perp = -\nabla_\perp \psi$, see for detail Aranson and Sapozhnikov (2004).

For appropriate choices of the parameters, eqns 8.63–8.65 reproduce most of the patterns observed in the experiment, including pulsating rings, rotating multipetal vortices and honeycomb lattices, see Figs. 2.23 and 8.40. Furthermore, the theory also yields a qualitatively correct phase diagram shown in Fig. 8.41 on the plane of the principal control parameters (the amplitude of electric field and the concentration of ethanol).

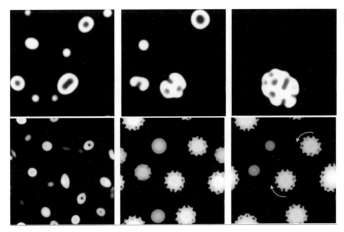

Fig. 8.40 Sequence of snapshots illustrating evolution of pulsating rings (top row) and rotating vortices (bottom row) obtained from numerical solution of eqns 8.63 and 8.65, from Aranson and Sapozhnikov (2004).

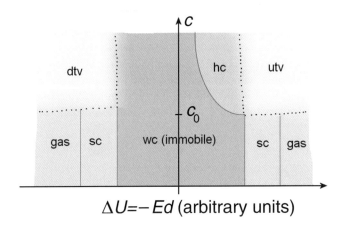

Fig. 8.41 Qualitative phase diagram, $\Delta U = -Ed$ is applied voltage (plus on top plate), c is concentration of an additive (ethanol). Domain **sc** denotes static clusters, **wc** denotes stable homogeneous precipitate (Wigner crystals), **hc** honeycombs, **utv** and **dtv** up/down toroidal vortices, compare with experimental phase diagram shown in Fig. 8.38, from Aranson and Sapozhnikov (2004).

8.6 Magnetic particles

Compared to predominantly radially symmetric monopole electrostatic forces due to particle charges that lead to agglomeration, interaction between magnetic particles is strongly anisotropic and therefore more complicated. Besides an immediate interest for the physics of granular media, studies of magnetic granular media may provide an additional insight into the behavior of dipolar hard-sphere fluids where the nature

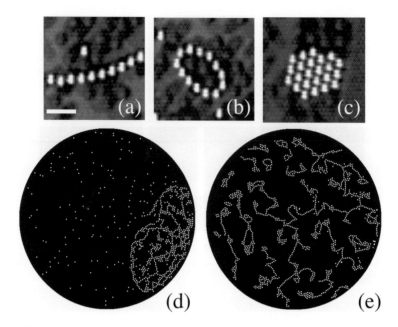

Fig. 8.42 Illustration of the typical structures in vibro-fluidized magnetic granular medium: chains (a), rings (b), and dense crystalline clusters (c). Rings appear to be the most stable configuration. The scale bar denotes 1 cm. (d) Snapshot of the system at filling fraction $\phi = 0.09$, where the granular temperature T is lowered to the transition temperature T_s from the gas state after 1092 s (slow quench). (e) The system at filling fraction $\phi = 0.15$ after a rapid quench from the gas state into the network state, from Blair and Kudrolli (2003).

of solid–liquid transitions is still debated (de Gennes and Pincus, 1970; Levin, 1999). Vibrated or electrically fluidized magnetic particles can also be viewed as a macroscopic model of a ferrofluid, where similar experiments are difficult to perform.

Magnetic forces favor tail-to-head alignment of the magnetic dipole moments of the particles. The idealized interaction between two dipolar hard spheres i and j separated by the distance r_{ij} is characterized by the potential energy U

$$U = U_{\mathrm{HC}} + \sum_{i \neq j} \frac{1}{r_{ij}^3}(\boldsymbol{\mu}_i \cdot \boldsymbol{\mu}_j) - \frac{3}{r_{ij}^5}(\boldsymbol{\mu}_i \cdot \mathbf{r}_{ij})(\boldsymbol{\mu}_j \cdot \mathbf{r}_{ij}), \tag{8.66}$$

where U_{HC} corresponds to the hard-core repulsion, $\boldsymbol{\mu}$ is the particle magnetic dipole moment (we assume for simplicity that particles are hard magnets, i.e. the dipole moments do not change due to interaction), and $\mathbf{r}_{ij}$ is the radius-vector connecting the centers of two dipoles. As a result of this anisotropic interaction a highly nontrivial phenomenology can be anticipated. A number of experimental and theoretical studies were performed recently with ensembles of magnetic grains energized either by mechanical vibration or by alternating electric or magnetic fields.

Blair et al. (2003), Blair and Kudrolli (2003), Stambaugh et al. (2003), Stambaugh et al. (2004) performed experimental studies with vibrofluidized magnetic particles.

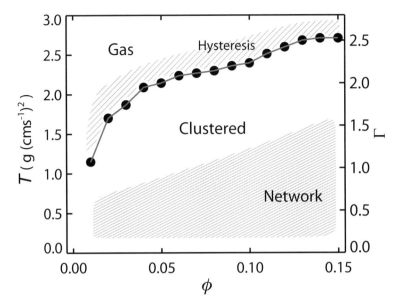

Fig. 8.43 Phase diagram illustrating various regimes in magnetic granular media, T is the granular temperature determined from the width of the velocity distribution, ϕ is the surface coverage fraction of glass particles, from Blair and Kudrolli (2003).

Several interesting phase transitions were reported, in particular, the formation of dense two-dimensional clusters and loose quasione-dimensional chains and rings, such as those shown in Fig. 8.42. In order to vary the corresponding granular temperature of the particle ensemble, Blair et al. (2003) used a mixture of magnetic and non-magnetic (glass beads) particles of equal mass. The glass particles played the role of the "thermal bath" or "phonons", their concentration allowed an adjustment of the typical fluctuation velocity of the magnetic subsystem. The phase diagram delineating various regimes in this system is shown in Fig. 8.43. While the phase diagram shows some similarity with equilibrium dipolar fluids (such as phase coexistence), there are some significant differences due to the non-equilibrium character of the granular magnetic system.

Experiments with vibrofluidized magnetic granular media were conducted with a small number (about 10^3) of large particles due to the limitations of the mechanical vibrofluidization. To overcome these limitations, Snezhko et al. (2005) used 90-µm conducting magnetic nickel microparticles subjected to an electric excitation; schematics of the setup and details of the electric driving technique are described in Section 8.5. In addition to vertical electric field used mostly to shake the conducting particles (in fact, controlling the "granular temperature"), uniform constant (dc) external vertical magnetic field was applied to tune the interaction between particles: by varying the magnetic field one can control the alignment of magnetic dipoles along the field direction, and, in turn, the length of chains formed by the magnetic particles.

The electromagnetic system permitted experiments with a very large number of

256 *Granular materials with complex interactions*

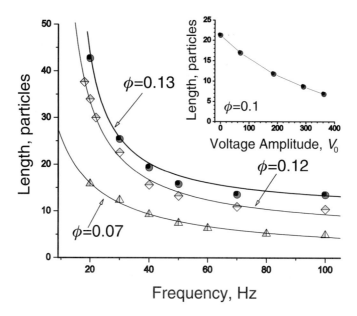

Fig. 8.44 Typical chain length vs. frequency of an applied 15-Oe ac magnetic field for different amounts of 90-μm nickel particles in the cell. Solid lines are fits to the expression $y = A + B/(x - f_0)$. *Inset:* Chain length vs. applied ac (100Hz) electric field amplitude for magnetically driven ($H_{\max} = 15$ Oe; $f = 25$ Hz) system (filling fraction $\phi \simeq 0.10$), from Snezhko et al. (2005).

particles (of the order of 10^6) and a large aspect ratio of the experimental cell. Thus, the transition between small chains and large networks (Fig. 2.21) could be addressed in detail. The typical length of the chains as a function of the driving frequency of the magnetic field is shown in Fig. 8.44. An abrupt divergence of the chain length was found when the frequency of the field oscillations was decreased, resulting in the formation of a giant interconnected network. As one can see from the inset in Fig. 8.44, such electric shaking controls the typical chain length, similar to the effect of "phonon" or non-magnetic beads in (Blair et al., 2003).

In Section 2.5 we briefly mentioned that alternating (ac) magnetic field can also be used to energize magnetic microparticles. In experiments by Snezhko et al. (2006) and Belkin et al. (2007), 90-μm nickel particles suspended on the liquid–air interface were subjected to an oscillating vertica magnetic field and in a certain range of parameters they self-assembled in remarkable snake-like objects, see Fig. 2.24. The formatio nof these patters can be qualitativbely understood through the following instability mechanism. A constant (dc) vertical magnetic field of sufficient magnitude generates a triangular lattice of particles due to the repulsive dipole-dipole interactions between magnetically aligned particles. However, in an alternating magnetic field, as the particles rotate to align their magnetic moment with the direction of the magnetic field, the particles drags the surrounding fluid and consequently affect neighboring particles. If particles are close enough, their head-to-tail dipole-dipole attraction overcomes the repulsion promoted by the external field, and a chain of particles is formed with resulting magnetic moment along the chain. Each chain further promotes the self-assembly process of another chain by producing local surface wave-like oscillations. From a general perspective, the excitation of surface waves by oscillating magnetic particles responding to an alternating magnetic field is analogous to Faraday ripples in vertically vibrated liquids and granular layers discussed in Chapter 5.

Quantitatively, Belkin et al. (2007) proposed the following phenomenological model of snake formation which couples the motion of particles with the high-frequency surface waves excited by oscillating particles and the self-generated large-scale hydrodynamic flow. Generation of parametric waves by magnetic particles on the surface of

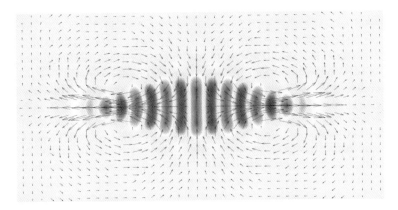

Fig. 8.45 Gray-scale image of the amplitude of surface waves ($|\psi|$) superimposed with hydrodynamic flows (arrows) obtained by numerical solution of eqns. 8.67–8.69. Parameters are: $q = 1, b = 2, \gamma = 2.5, D = 1, \beta = 5, \omega_* = 3, \eta_f = 200$ in domain of 200×200 dimensionless units, from Belkin et al. (2007).

fluid is described by the following set of equations for the complex amplitude of the surface waves ψ and the density of magnetic particles ν:

$$\partial_t \psi + (\mathbf{v}\nabla)\psi = -(1 - i\omega_*)\psi + (a + ib)\nabla^2\psi - |\psi|^2\psi + \gamma\psi^*\Phi(\nu), \quad (8.67)$$
$$\partial_t \nu + \nabla(\nu\,\mathbf{v}) = D\nabla^2\nu - \beta\nabla\left(\nu\nabla|\psi|^2\right). \quad (8.68)$$

Here ω_*, a, b are model constants, and parameter γ characterizes the magnitude driving (square of amplitude of external magnetic field). Equation 8.67 differs from the standard Ginzburg–Landau model for parametrically-excited surface waves (see, e.g. Coullet et al. (1990)), by the factor $\Phi(\nu)$ in the parametric driving term. This factor that characterizes the dependence of the forcing term on the density of particles is chosen to be proportional to ν for small particle densities and saturating for higher densities. Such a dependence accounts for the fact that the effect of forcing should vanish for low particle density ($\nu \to 0$) (since the fluid itself is non-magnetic) and should saturate for high particle density. Equation 8.68 is the mass conservation law for the magnetic particle density, where D is a phenomenological diffusion constant, and parameter β denotes the amplitude of the advection term due to transport (herding) of particles by waves. Equations 8.67 and 8.68 are rather similar to the equations studied in the context of oscillons in thin vibrated granular layers, see Section 5.4, eqns. 5.3 and 5.4. However, the main difference with the oscillon equations is that initially the amplitude of driving in the corresponding parametric Ginzburg–Landau equation *increases* with the increase in number density of magnetic particles ν (since the fluid supporting particles is non-magnetic, no driving occurs for particles-free regions). These two equations are coupled to the large-scale hydrodynamic mean flow velocity $\mathbf{v}$:

$$\partial_t \mathbf{v} + (\mathbf{v}\nabla)\mathbf{v} + \frac{\nabla p}{\nu_f} = \eta_f \nabla^2 \mathbf{v} + \mathbf{F}, \qquad (8.69)$$

where p is the pressure, ν_f and η_f denote the fluid density and viscosity, and $\mathbf{F}$ is averaged over a period of magnetic field force caused by the oscillations of particles on the fluid surface. For simplicity, it is assumed that the large-scale hydrodynamic velocity $\mathbf{v}$ is two-dimensional and incompressible. The force term $\mathbf{F}$ is expressed via complex amplitude of the surface waves ψ. The simplest slowly-varying term built from ψ, $\nabla\psi$, and their conjugates has the form $\mathbf{F} \sim (\psi^*\nabla\psi - \psi\nabla\psi^*)/2i$.[8]

This system of three equations successfully captured formation of localized snake-like objects as well as generation of 4 large vortices at the tails of the snake, compare Figs. 2.24 and 8.45.

---

[8] Note that terms like $\mathbf{F} \sim \nabla|\psi|^2$ are not relevant because they do not generate vortical flows in eqn. 8.69.

9
Granular physics of biological objects

Until this last chapter, we were concerned with patterns in inanimate or "dead" granular matter. However, the concepts of granular physics are beginning to take hold in biology. Indeed, the macroscopically discrete nature of biological organisms and populations makes them "granular" according to our loose definition, albeit usually with rather complex interaction rules. Biological patterns have always attracted much interest among physicists. Morphogenesis is one of the most important themes in biology, and it also happens to be one of the central issues in non-equilibrium physics. The fundamental issue here is to understand how local interaction of individual components leads to the observed collective behavior and the formation of highly organized systems. In Nature, this self-organization is found on many different scales, from biomolecules and single cells to schools of fish and herds of animals. Complex patterns can be caused by a variety of mechanisms, from "simple" physical effects, such as steric repulsion in dense populations, hydrodynamic entrainment, to highly complex biological mechanisms, like chemotaxis (drift of micro-organisms in the direction of chemical gradients), quorum sensing, or even "social" interactions due to direct visual, audio, or chemical communication between individual species. A comprehensive overview of patterns in biological systems is certainly beyond the scope of this book.

In this chapter we restrict ourselves to relatively simple situations where biological pattern formation is mostly determined by the same physical mechanisms that are at work in non-living granular systems, such as hydrodynamic entrainment, steric repulsion, inelastic collisions, etc. We will demonstrate how these mechanisms lead to nematic ordering of rod-like bacteria, self-organization of microtubules and molecular motors, collective locomotion in ensembles of swimming bacteria.

The structure of this chapter is the following. We begin with a purely biomechanical ordering of rod-like cells in confined environments, then we discuss ordering of microtubule filaments interacting via molecular motors. Then, we turn to the collective dynamics of *self-propelled* objects, starting from the abstract "microscopic" discrete model of flocking (Vicsek *et al.*, 1995) and its continuous counterpart (Toner and Tu, 1998). We conclude this chapter by discussing experiments and theoretical modelling of patterns of motile bacteria swimming in thin fluid films.

9.1 Nematic ordering of growing colonies of non-motile bacteria

Micro-organisms, despite their name, are small but macroscopic objects and may interact inelastically, which makes them somewhat similar to grains. Of course, their motion and interaction can be very complicated because they are self-propelled ob-

260 *Granular physics of biological objects*

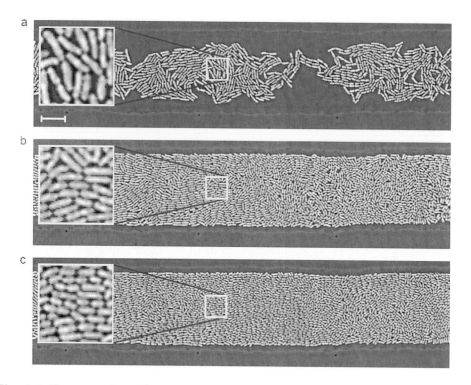

Fig. 9.1 Three snapshots illustrating *E. coli* monolayer growth and ordering in a quasi-two-dimensional open microfluidic cavity taken at 60, 90, 138 min from the beginning of the experiment. The length of a bar is 10 μm, from Volfson *et al.* (2008).

jects and can communicate via chemical signals that in turn can affect their motility, adhesion, and other properties. On the other hand, many bacteria are non-motile and therefore can only diffuse and interact via hydrodynamic fields or direct contacts. One important difference between living cells and ordinary grains, however, is that unlike grains, living cells grow and divide. The growth and division of cells can have a profound effect on the structure of bacterial populations, and in particular can lead to their long-range ordering.

9.1.1 Microfluidic experiments

Volfson *et al.* (2008) studied ordering of non-motile rod-like bacteria *Escherichia coli* (or *E. coli*) in controlled conditions of a thin microfluidic chamber, see Fig. 9.1. The thickness of the chamber was only 1 μm which is about one cell diameter, so bacteria could only form a monolayer. Friction with the upper and lower walls prevented bacteria from moving unless they were pushed by their neighbors. In the beginning, only a few randomly oriented cells were inserted in a chamber, and they remained motionless during the initial phase of growth. But after several cell divisions, cells filled up the chamber (which had the form of an open channel, so cells could be pushed away and washed out with fluid flow). As cells grew and proliferated, an expansion flow ensued

Nematic ordering of growing colonies of non-motile bacteria   261

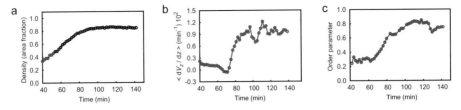

Fig. 9.2 Time series of the mean density, longitudinal velocity gradient, and the order parameter for the experiment illustrated by Fig. 9.1, from Volfson et al. (2008).

and oriented cells along itself. Eventually, a quasistationary ordered regime was established, with constant density and velocity distributions along the channel (Fig. 9.1(c)). Figure 9.2 shows time series of density, velocity shear and the order parameter Q that characterizes the ordering of cells in the chamber. The order parameter here is defined as $Q = \left[\langle \cos 2\varphi \rangle^2 + \langle \sin 2\varphi \rangle^2\right]^{1/2}$, where φ is the angle between the cell axis and the channel axis, and brackets denote averaging over the whole system (compare with the similar definition of the order parameter for apolar rod-like particles in Section 8.2.4). This order parameter ranges from 0 for a completed disordered colony to 1 for a colony that is perfectly aligned.

This experiment suggests that the nematic ordering of cells is driven by the self-generated growth-induced expansion flow. This is in marked contrast with thermal systems such as liquid crystals and polymers, where nematic ordering is driven by steric exclusion of rod-like molecules and a corresponding entropy maximization (Onsager, 1949).[1] Discrete-element simulations confirm this picture. The soft-particle molecular dynamics algorithm that Volfson et al. (2008) used in simulations is very similar to the one used for simulations of rod-like grains (see Section 8.1.2), with the only difference that the length of the spherocylinders grows exponentially with a certain rate α until it reaches a maximal length l_m, which is drawn from a Gaussian distribution near a fixed length l_0 with the coefficient of variation 0.3. During the division event each mother cell is replaced by two coaxial daughter cells of half-length $l_m/2$ that touch each other. These daughter cells then continue to grow and move independently; the process of growth and division repeats, resulting in an exponential growth of the population. These simulations were used to explore the effect of the channel length on the resulting orientation of the colony (Fig. 9.3). In short channels, complete order is established, with almost all cells oriented along the channel, whereas in long channels the population remains in a partially disordered state characterized by large "swirls." This can also be seen from the velocity field obtained by coarse graining the velocities of the individual cells (Figs. 9.3(b) and (d)). The arrows indicate the direction of the velocity field; the velocity is nearly unidirectional in short channels but exhibit significant swirling in long channels. The magnitude of the velocity grows linearly from the middle of the channel to its edges, which is consistent with mass conservation that stipulates that the local cell growth has to be balanced by the mass expansion toward

---

[1] The Onsager theory was discussed in Section 8.1.1.

262 *Granular physics of biological objects*

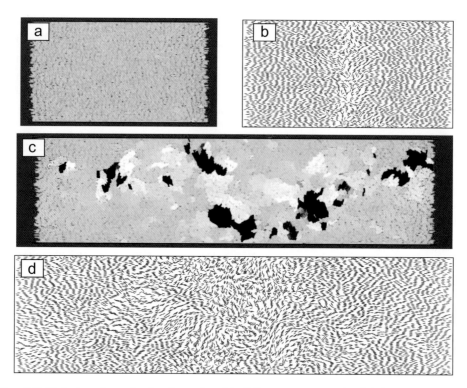

Fig. 9.3 Ordering dynamics in channels with different aspect ratio; (a) Orientation of individual cells (graycoded, different shades correspond to different orientations, with medium gray denoting the orientation along the channel) in the system with aspect ratio $A = 2.0$ and a constant growth rate; (b) Velocity field for the same case as in (a), where unit velocity vectors show the velocity direction for each cell; (c) The same as (a) for a system twice as long ($A = 4.0$). Defects of the orientation are constantly created in the middle of the channel and advected by the flow toward the open boundaries; (d) The same as (b) for the system with $A = 4.0$. The flow is no longer laminar, and there is no apparent correlation between orientation and velocity magnitude, from Volfson *et al.* (2008).

the open ends of the channel, and so the velocity gradient must be equal to the cell growth rate α.

9.1.2 Continuum model of nematic ordering

The mechanism of nematic ordering can be understood from the continuum equations of nematodynamics (Doi and Edwards, 2003) suitably generalized to include the effects of cell growth and division. Each cell is characterized by a unit vector $\mathbf{u}$ specifying the orientation of the cell ($\pm \mathbf{u}$ directions are equivalent since cells are assumed apolar). While the mean orientation $\langle \mathbf{u} \rangle$ (averaged over a mesoscopic volume containing many cells) is zero due to the reflection symmetry of cells, the distribution of cell directions can become anisotropic. The anisotropy of the distribution is characterized by the tensor order parameter $\mathbf{Q}$ with components

$$Q_{ij} = \langle u_i u_j - d_s^{-1}\delta_{ij}\rangle, \quad (9.1)$$

where $i,j \in x,y$, δ_{ij} is the Kroneker symbol, and brackets denote averaging over the d_s-dimensional mesoscopic volume. Tensor $\mathbf{Q}$ is symmetric and traceless, and in two dimensions ($d_s = 2$) it can be written in the form (see also Section 8.2.4):

$$Q_{ij} = Q(2n_i n_j - \delta_{ij}), \quad (9.2)$$

where n_i are components of the Frank director (the unit vector aligned with the mean orientation of the cells in a mesoscopic volume), and $Q \equiv \sqrt{Q_{xx}^2 + Q_{xy}^2}$ is the scalar order parameter defined in the previous section.

In the theory of liquid crystals, the continuous description of nematics is typically based on the equations for the director field and the velocity field (Doi and Edwards, 2003). Such a description works well below the critical point and when the scalar order parameter is everywhere close to unity. However, close to the isotropic–nematic transition, the magnitude of the order parameter changes significantly, and more general equations of nematohydrodynamics incorporating the orientational order parameter have to be used (Olmsted and Goldbart, 1992; de Gennes and Prost, 1993). We already made use of this description in Section 8.2.4 on swirling of granular rods. Here, we adopt this model with important modifications resulting from the differences between rod-like molecules and macroscopic rod-like cells. One such difference is that cells are much less influenced by thermal fluctuations, and accordingly, the free-energy minimization plays a minor role in the dynamics of the order parameter. Second, unlike granular rods, living cells grow and divide, and this process profoundly affects the collective dynamics.

The set of governing equations of "cellular nematodynamics" consists of the mass balance equation

$$\frac{\partial \nu}{\partial t} + \nabla(\nu \mathbf{v}) = \alpha \nu, \quad (9.3)$$

momentum balance equation

$$\frac{D\nu \mathbf{v}}{Dt} = -\nabla \cdot \boldsymbol{\sigma} - \mu \nu \mathbf{v}, \quad (9.4)$$

and the order parameter equation

$$\frac{DQ_{\alpha\beta}}{Dt} = \kappa_{\alpha\gamma}^{[a]} Q_{\gamma\beta} - Q_{\gamma\beta}\kappa_{\alpha\gamma}^{[a]} + B\kappa_{\alpha\beta}^{[s]} + \Gamma H_{\alpha\beta} \quad (9.5)$$

where $D/Dt = \partial_t + \mathbf{v}\nabla$ is the material derivative, ν is density, $\mathbf{v}$ is velocity, $\boldsymbol{\sigma}$ is the stress tensor, and Γ and B are kinetic coefficients. In eqn 9.3, the rhs describes exponential growth of cell mass with the rate α, and the last term in eqn 9.4 describes velocity-dependent friction of cells with the top and bottom walls of the chamber. In the order-parameter equation, $\kappa_{\alpha\beta}^{[a]} = (\partial_\alpha v_\beta - \partial_\beta v_\alpha)/2$ and $\kappa_{\alpha\beta}^{[s]} = (\partial_\alpha v_\beta + \partial_\beta v_\alpha)/2$ are symmetric-traceless and antisymmetric (vorticity) parts of the strain rate tensor $\kappa_{\alpha\beta} = \partial_\alpha v_\beta$, respectively.

The last term in eqn 9.5 describes the entropic relaxation of the non-conserved order parameter due to minimization of the free energy,

$$H_{\alpha\beta} = -\left[\frac{\delta \mathcal{F}}{\delta Q_{\alpha\beta}} - d_s^{-1}\text{Tr}\left(\frac{\delta \mathcal{F}}{\delta Q_{\alpha\beta}}\delta_{\alpha\beta}\right)\right], \qquad (9.6)$$

where Tr denotes the trace of the tensor. The free-energy functional, $\mathcal{F}$, has two parts,

$$\mathcal{F} = \mathcal{F}_{\text{LG}} + \mathcal{F}_{\text{F}}, \qquad (9.7)$$

where in two dimensions

$$\mathcal{F}_{\text{LG}} = \int d\mathbf{x}\nu(\mathbf{x},t)\left[A_2\text{Tr}(\mathbf{Q}\cdot\mathbf{Q}) + A_4\left(\text{Tr}(\mathbf{Q}\cdot\mathbf{Q}\cdot\mathbf{Q}\cdot\mathbf{Q})\right)^2 + ...\right] \qquad (9.8)$$

is the de Gennes–Landau–Ginzburg free energy, and

$$\mathcal{F}_{\text{F}} = \int d\mathbf{x}\nu(\mathbf{x},t)\left(K_1\left(\partial_\alpha Q_{\beta\gamma}\right)^2 + K_2\left(\partial_\alpha Q_{\alpha\beta}\right)^2\right) \qquad (9.9)$$

is the Frank elastic free energy and $K_{1,2}$ are the elastic constants. Here, A_2 and A_4 are parameters dependent upon the local cell density, $\nu(\mathbf{x},t)$.

Substituting eqn 9.2 into eqn 9.7, we derive

$$\mathcal{F} = \int d\mathbf{x}\nu(\mathbf{x},t)\left[A_2 Q^2 + A_4 Q^4 + K_1\left(\partial_\alpha Q_{\beta\gamma}\right)^2 + K_2\left(\partial_\alpha Q_{\alpha\beta}\right)^2\right]. \qquad (9.10)$$

To close this system of equations, a constitutive relation coupling the stress tensor to the strain and strain rate fields is needed. In the theory of liquid crystals (Olmsted and Goldbart, 1992), a linear constitutive relation is usually assumed,

$$\sigma_{\alpha\beta} = \beta_3\kappa_{\alpha\beta}^{[s]} - \beta_1 H_{\alpha\beta}^{[s]} + H_{\alpha\gamma}^{[s]}Q_{\gamma\beta} - Q_{\alpha\gamma}H_{\gamma\beta}^{[s]} - \frac{\delta\mathcal{F}}{\delta\partial_\alpha Q_{\lambda\nu}}\partial_\beta Q_{\lambda\nu} - p\delta_{\alpha\beta}. \qquad (9.11)$$

where the first four terms correspond to the irreversible (viscous) stress and the last two terms describe reversible stress due to distortion and isotropic pressure, $H_{\alpha\beta} = -\delta\mathcal{F}/\delta Q_{\alpha\beta}$ is the molecular field. The value of pressure, p, is, of course, density dependent and it diverges as the density approaches the close-packing limit. Note that in fact the value of the close-packing density itself depends on the order parameter: for $Q = 0$, the rods are completely disordered, and the close-packing density is low (it depends on the length of the rods l), whereas when $Q \to 1$, the close-packing density approaches 1. We model this behavior by choosing

$$p = P\exp[\xi(\nu - \nu_c)]. \qquad (9.12)$$

This relation implies that the pressure is exponentially small for $\nu < \nu_c$ and exponentially large at $\nu > \nu_c$ (P is $O(1)$ and ξ is large). The close-packing density ν_c is itself a function of the order parameter Q, the more ordered the population is, the higher is the close packing density. We approximate this dependency by the linear relation $\nu_c = \nu_c^d + (\nu_c^0 - \nu_c^d)Q^2$.

In low-Reynolds number regimes typical for cell motion, inertial effects can be neglected, and the momentum equation (eqn 9.4) yields the velocity of cells as a divergence of the stress tensor, $v_i = \mu^{-1}\partial_j \sigma_{ij}$. However, as we shall see, using the full momentum equation (eqn 9.4) yields a better agreement between the continuum theory and our numerical simulations.

The set of equations 9.3–9.5 forms the basis for a study of the cell growth and transport. However, to simplify the problem, in the following we take into account the fact that the cells are only weakly affected by thermal fluctuations and neglect thermodynamic effects entirely. Specifically, we neglect the free-energy contributions to the order-parameter dynamics and the stress tensor. Note that in this case eqn 9.5 does not guarantee that $Q < 1$ according to its definition. In order to uphold this condition heuristically, we multiply the rhs of eqn 9.5 by the normalization factor $1 - Q^2$. As a result, we arrive at the following set of equations:

$$\partial_t \nu + \nabla \cdot (\nu \mathbf{v}) = \alpha \nu \tag{9.13}$$

$$\frac{D\nu \mathbf{v}}{Dt} = -\nabla p - \mu \nu \mathbf{v} \tag{9.14}$$

$$\frac{DQ_{\alpha\beta}}{Dt} = (1 - Q^2)[\kappa^{[a]}_{\alpha\gamma} Q_{\gamma\beta} - Q_{\gamma\beta}\kappa^{[a]}_{\alpha\gamma} + B\kappa^{[s]}_{\alpha\beta}]. \tag{9.15}$$

According to this set of equations, for a constant α, the total mass of cells grows exponentially as $M \propto \exp(\alpha t)$. However, the outward flux of cells saturates the local growth of cell density at the expense of the exponential expansion of the area occupied by cells.

For the description of the expansion flow in a straight narrow channel of length L we can assume that all fields depend on time and coordinate x along the channel only, and reduce this full set to a one-dimensional model

$$\partial_t \nu + \partial_x (\nu v) = \alpha \nu \tag{9.16}$$
$$\partial_t q + v \partial_x q = B(1 - q^2)\partial_x v \tag{9.17}$$
$$\partial_t (\nu v) + v \partial_x (\nu v) = -\partial_x p - \mu \nu v, \tag{9.18}$$

where $q \equiv Q \equiv Q_{xx}, v \equiv v_x$, α is the constant cell growth rate, μ is the bottom friction coefficient (we assume that the friction force is proportional to the cell velocity independent of orientation), and the pressure p satisfies the constitutive relation (eqn 9.12) discussed above.

The solution in which the density $\nu(t)$ and the order parameter $q(t)$ are independent of x, and the velocity is a linear function of x, $v = v_0(t)x$ and the pressure is parabolic, $p = p_0(t)(1-x^2/L^2)$ satisfies eqns 9.16–9.18 at all times. The normalized "magnitudes" of this solution ν, v_0, q, p_0 satisfy the system of three ordinary differential equations

$$\dot{\nu} = \nu(1 - v_0) \tag{9.19}$$
$$\dot{q} = B(1 - q^2)v_0 \tag{9.20}$$
$$\dot{v}_0 = 2\nu^{-1}p_0 - (\mu + 1)v_0, \tag{9.21}$$

where we rescaled all variables by the growth rate α and the channel length L. Strictly speaking, the parabolic pressure profile and the uniform density distribution do not satisfy the constitutive relation. However, since the dependence of pressure on the density, is very sharp near the close-packing density of cells, the deviation of density from the uniform state, which accounts for a parabolic pressure distribution is very small and can be neglected. According to this set of equations, initially the density grows exponentially with rate α while the pressure remains low, and the velocity gradient is absent. Once the density approaches the random close-packing density ν_c, the pressure begins to rise rapidly, and it produces a rapidly growing velocity gradient (Fig. 9.2). This gradient orients the cells along the channel axis, which in turn affects the close-packing density. Eventually, the system reaches a stationary regime with a uniform density and a linear velocity profile, in which the cell growth is balanced by the outgoing flux of cells. In this regime, $v_0 = q = 1$ and $\nu \approx \nu_c^o$. In the original variables, the velocity gradient $v_0 = \alpha$ and the corresponding pressure $p_0 = (1+\mu)\nu_c^o \alpha^2 L^2/2$. Since the pressure scales as the square of the length of the channel, for long channels, it reaches very high values in the middle section of the channel. In reality, this high-pressure behavior is mitigated by the cell-growth saturation: highly pressurized cells stop growing (see (Shraiman, 2005)). To simulate this effect, we can replace the constant growth rate by a pressure-dependent one, $\alpha = \alpha_0[1 - (p/p_c)^2]$, so the cell growth terminates when the pressure reaches a critical value p_c. In fact, the pressure is not uniform across the channel, therefore the growth of cells will first saturate in the middle where the pressure is maximal, and then it will propagate to the periphery of the channel. However, in the simplified ordinary differential equation model we can ignore this subtlety. This model leads to similar dynamics, however, saturation occurs at much smaller values of the pressure and velocity gradient in the bulk of the colony (Fig. 9.3).

To test the validity of the continuum modelling, we can compare it with discrete-element (molecular dynamics) simulations. Figure 9.4 shows the results of this comparison for simulations in a short channel with pressure-independent growth and in a long channel with pressure-dependent growth. Except for replacing the constant growth rate α by $\alpha_0(1 - (p/p_c)^2)$, all parameters are kept the same. In both cases, the continuum model provides a good description of the colony dynamics, indicating that the cell density grows at a similar rate and saturates near the close-packed limit (Fig. 9.4(a)). However, both the pressure and the ensuing expansion flow in the pressure-dependent growth case are almost two times smaller (Figs. 9.4(b) and (d)), and the cellular ordering is proportionally slower (Fig. 9.4(c)). Interestingly, pressure-dependent growth leads to isotropization of local stresses, as evidenced by Fig. 9.4(d). This effect may be attributed to the ability of slower-growing cells to better adjust to local contact stresses.

In this section we have shown that purely biomechanical short-range interactions may lead to highly ordered structures in a growing bacterial colony. The nematic ordering of cells is mediated by the expansion flow generated by cellular growth. This phenomenon is fundamentally different from the nematic transition in liquid crystals and polymers (Doi and Edwards, 2003) and vibrated granular rods (Blair et al., 2003; Narayan et al., 2006; Aranson et al., 2007b), where ordering is driven by the combination of steric exclusion and fluctuations. The mechanism of the ordering transition here is related to cell growth and therefore is specific to living "granular matter."

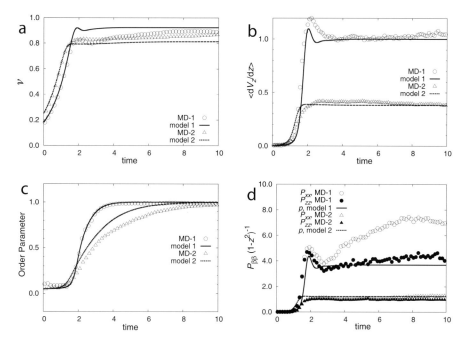

Fig. 9.4 Comparison of continuum and discrete-element simulations modelling of the biofilm dynamics: time traces of the amplitudes of (a) density, (b) velocity gradient, (c) order parameter, and (d) stress components; for constant growth-rate with aspect ratio $A = 2.0$ and pressure-dependent growth rate model with $A = 4.0$. Results of the discrete-element simulations are shown with symbols (circles correspond to $\alpha =$ const, triangles correspond to pressure-dependent α) and the results of the corresponding continuum modelling are shown with lines. From Volfson et al. (2008).

9.2 Self-organization of microtubules interacting via molecular motors

About 70% of the interior of a living cell is occupied by a complex, self-organizing viscoelastic fluid, the *cytosol*. Its main constituents (besides water) include cytoskeletal proteins that exist mainly in polymerized form as semiflexible actin filaments and stiff microtubules. Entangled networks of microtubules and actin filaments form the *cytoskeleton* that plays the role of a cell's "scaffold", stabilizing its morphology and determining its mechanical properties. The cell structure is organized and maintained by an efficient machinery that involves in addition to the cytoskeletal proteins also different constituents, such as active motor proteins and passive cross-linking proteins. Molecular motors are specialized proteins that can move on the cytoskeletal polymer scaffold and are used to perform directional intracellular transport and to reorganize the cytoskeleton itself. At first glance, all these processes are beyond the scope of the granular physics. However, from a physicist's perspective interaction of microtubules bears a strong similarity to the interaction of polar granular rods considered in the previous chapter. The significant linear size of the microtubules (up to tens of micrometers) makes them practically macroscopic objects little influenced by the thermal noise and thus obeying the laws of macroscopic classical mechanics rather than quantum physics. The specificity of this system lies in the mechanism of interaction of these rods that is mediated by molecular motors attaching to the tubules and "zipping" along them before they detach back into the surrounding fluid. In this section we discuss the consequences of this mechanism of tubule–motor interaction for self-organization of microtubules. We briefly touched upon this topic in Section 2.6.

9.2.1   *In-vitro* experiments

Microtubules are long, stiff and hollow polar filaments. Their length, self-regulated through asymmetric polymerization/depolymerization of tubulin dimers, can reach up to 30–50 µm. With the diameter only of the order of 30 nm, the thermal persistence length of a microtubule is of the order of few millimeters i.e. it significantly exceeds the filament length. Thus, microtubules are practically unbendable by thermal fluctuations. Molecular motors, however, can easily bend and even buckle microtubules. Linear molecular motors, such as kinesin, myosin, dynein and others, attach themselves to polar biofilaments such as microtubules or F-actin. Typical molecular motors (e.g. kinesin) march along the filament (microtubule) in discrete steps of about 8 nm with an average speed of up to 1 µm/s by sequentially binding and unbinding their two heads. This process is powered by hydrolyzing adenosine triphosphate (ATP), consuming one molecule per step. Between steps, the motor stalls until ATP binds to a kinesin head again. The opposite end of a molecular motor has a flexible tail that can attach to various structures (proteins, vesicles) and carry them along the filaments. A single motor can develop a significant force of the order 5–7 pN. Some motors can from multimotor clusters. These clusters have multiple binding sites and can bind to two filaments simultaneously and march along both of them. This motion leads to a large bending force acting on filaments (motor clusters are capable of developing even greater bending force than isolated motors). The interaction of microfilaments via molecular motors is the main underlying mechanism of the cytoskeleton dynamics.

Due to the complexity of the interaction of microtubules and molecular motors in a living cell, a number of *in-vitro* experiments were performed (Nédélec *et al.*, 1997; Surrey *et al.*, 2001) to study this process in isolation from other biophysical processes occurring *in vivo*. At small concentration, a solution of microtubules and molecular motors remains largely featureless, but with increase of concentration of both constitutive elements, an interesting transition occurs, and on the time scale of 20–60 min the filaments organize in ray-like *asters* or rotating *vortices* depending on the type and concentration of molecular motors, see Fig. 2.26. The typical spatial scale of the patterns is of the order of 100 µm, i.e. exceeding the size of individual microtubules by a factor of 5.

These *in-vitro* experiments spurred a large body of computational and theoretical work, mostly based on phenomenological models incorporating typical features of polar filament interaction and growth, see, e.g. (Lee and Kardar, 2001; Liverpool and Marchetti, 2003; Kruse *et al.*, 2004). In the remainder of this section, we discuss one such theoretical description that was inspired by the analogy between motor-mediated

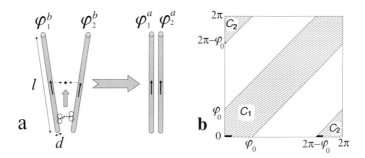

Fig. 9.5 (a) sketch of motor-mediated two-rod interaction for $\gamma = 1/2$, (b) integration regions $C_{1,2}$ for eqn 9.23.

interaction of filaments and inelastic collision of granular particles (Aranson and Tsimring, 2005).

9.2.2 Maxwell model for inelastic rods

Here we derive a model for the collective spatiotemporal dynamics of microtubules using the analogy between alignment of polar rods (microtubules) and inelastic collisions between grains. We show that the model exhibits an onset of orientational order for large enough density of microtubules and molecular motors, and, consequently, formation of vortices and then asters, in a qualitative agreement with experiment.

Molecular motors enter the model implicitly by specifying the interaction rules between two rods. Since the diffusion of molecular motors is hundreds of times higher than that of microtubules, as a first step we neglect spatial variations of the motor density. While variable motor concentration affects certain quantitative aspects of the pattern formation,[2] neglecting motor redistribution still captures the salient features of the phenomena. Each rod is assumed to be of length l and diameter $d \ll l$, and is characterized by the position of its center-of-mass $\mathbf{r}$ and orientation angle φ. Thus, we do not consider the individual dynamics of microtubules due to the asymmetric polymerization/depolymerization processes leading to the length oscillations (Flyvbjerg et al., 1996). This approximation is justified by the fact that in in-vitro experiments the polar ends of microtubules are often stabilized by adding taxol to the aqueous solution (Nédélec et al., 1997; Surrey et al., 2001).

Let us first consider the orientational dynamics only and ignore the spatial coordinates of the interacting rods (mean-field approximation). For simplicity we focus on binary collisions between the rods, which implies a dilute limit. Another assumption is that the probability of rod–rod interaction does not depend on their mutual orientation. These assumptions make the model of rod–rod interaction very similar to the Maxwell model of binary collisions in the kinetic theory of granular gases (Ben-Naim and Krapivsky, 2000). Since the motor residence time on a microtubule is only about a few seconds and is much smaller than the characteristic time of pattern formation

---

[2] Nédélec et al. (2001) provides evidence of motor accumulation in the centers of asters due to the advection of motors along microtubules

(about 1 h), a complicated interaction between two microtubules via a molecular motor can be modelled by an instantaneous "collision" in which two rods change their orientations. This event can be described by a map

$$\begin{pmatrix} \varphi_1^a \\ \varphi_2^a \end{pmatrix} = \begin{pmatrix} \gamma & 1-\gamma \\ 1-\gamma & \gamma \end{pmatrix} \begin{pmatrix} \varphi_1^b \\ \varphi_2^b \end{pmatrix} \qquad (9.22)$$

that relates orientations of rods before ($\varphi_{1,2}^b$) and after ($\varphi_{1,2}^a$) collision. The parameter γ characterizes inelasticity of collisions, and we introduce a certain maximum angle of interaction φ_0, so the "collision" may occur only if $|\varphi_2^b - \varphi_1^b| < \varphi_0 < \pi$. The angle between two rods is reduced after the collision by the factor $2\gamma - 1$. $\gamma = 0$ corresponds to an elastic collision (the rods exchange their angles, this case is apparently irrelevant for microtubule interaction), and $\gamma = 1/2$ corresponds to a totally inelastic collision: rods acquire identical orientation $\varphi_{1,2}^a = (\varphi_1^b + \varphi_2^b)/2$ (see Fig. 9.5(a)). Because of the 2π periodicity, we have to add the rule of collision between two rods with $2\pi - \varphi_0 < |\varphi_2^b - \varphi_1^b| < 2\pi$. In this case we have to replace $\varphi_1^{b,a} \to \varphi_1^{b,a} + \pi, \varphi_2^{b,a} \to \varphi_2^{b,a} - \pi$ in eqn 9.22. In the following we will mostly consider the case of totally inelastic collisions $\gamma = 1/2$ (so-called zipping of the filaments), and $\varphi_0 = \pi$. The generalization for arbitrary γ and φ_0 is straightforward.

The rationale behind selecting the fully inelastic case is the following. Simple micromechanical calculations of two stiff rods interacting with a single motor show that the relative decrease of the angle between the rods after the motor has passed along the tubules is only about 20–25% (Aranson and Tsimring, 2006a). This corresponds to the value of restitution coefficient $\gamma \approx 0.85$ that is rather far from the fully inelastic case $\gamma = 1/2$. However, if several motors bind to a pair of microtubules simultaneously, the "collision" becomes more inelastic. A similar effect is achieved by binding of a static cross-linker protein that works as a "hinge", or two motors moving in an opposite direction (in some experiments by Surrey et al. (2001) such a mixture was used). In all these cases the relative sliding of the microtubules is inhibited, and the motion is restricted to rotation only, leading ultimately to a complete alignment.

The probability $P(\varphi, t)$ to find a rod with the orientation φ at time t obeys the following master equation

$$\partial_t P(\varphi) = D_r \partial_\varphi^2 P(\varphi) + W_0 \int_{C_1} d\varphi_1 d\varphi_2 P(\varphi_1) P(\varphi_2) [\delta(\varphi - \varphi_1/2 - \varphi_2/2) - \delta(\varphi - \varphi_2)]$$

$$+ W_0 \int_{C_2} d\varphi_1 d\varphi_2 P(\varphi_1) P(\varphi_2) [\delta(\varphi - \varphi_1/2 - \varphi_2/2 - \pi) - \delta(\varphi - \varphi_2)], \qquad (9.23)$$

where W_0 is the "collision rate" proportional to the density of molecular motors, the diffusion term $\propto D_r$ describes thermal fluctuations of the rod orientation (rotational diffusion), and the integration domains C_1, C_2 are shown in Fig. 9.5(a). We dropped the argument t from $P(\varphi, t)$ in eqn 9.23 for brevity. In the following we rescale the variables

$$D_r t \to t_r, W_0 P/D_r \to P_r, \qquad (9.24)$$

and introduce a new variable $w = \varphi_2 - \varphi_1$. Note that in the new variables the renormalized probability P_r is proportional to the original probability P to find a rod with

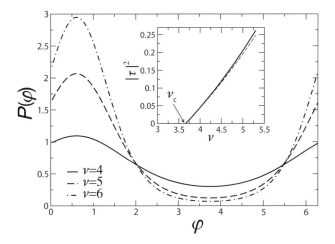

Fig. 9.6 Stationary solutions $P(\varphi)$ for different ν. Inset: the stationary value of $|\tau|$ vs. ν obtained from the Maxwell model (eqn 9.25), dashed line – truncated model (eqn 9.37), from Aranson and Tsimring (2005)

orientation φ multiplied by the collision rate W_0, so it is proportional to the product of the density of microtubules and motors.

From eqn 9.23 after substitution eqn 9.24 and integration of the δ-functions we arrive at

$$\partial_t P(\varphi) = \partial_\varphi^2 P(\varphi) + \int_{-\pi}^{\pi} dw \left[P(\varphi + w/2) P(\varphi - w/2) - P(\varphi) P(\varphi - w) \right] \quad (9.25)$$

(we dropped subscripts r from t and P for brevity).

We can introduce the rescaled number density

$$\nu = \int_0^{2\pi} P(\varphi, t) d\varphi, \quad (9.26)$$

which according to eqn 9.24 is in fact proportional to the density of rods multiplied by the density of motors.

Another quantity of interest is the coarse-grained orientation vector $\boldsymbol{\tau} = (\tau_x, \tau_y)$:

$$\tau_x = \langle \cos \varphi \rangle = \frac{1}{2\pi} \int_0^{2\pi} \cos \varphi P(\varphi, t) d\varphi,$$

$$\tau_y = \langle \sin \varphi \rangle = \frac{1}{2\pi} \int_0^{2\pi} \sin \varphi P(\varphi, t) d\varphi \quad (9.27)$$

In a two-dimensional situation it is convenient to introduce the "complex orientation" χ:

$$\chi = \tau_x + i\tau_y = \frac{1}{2\pi} \int_0^{2\pi} \exp(i\varphi) P(\varphi, t) d\varphi \quad (9.28)$$

As we will show below, this "complex orientation" χ plays the role of the order parameter.

9.2.3 Orientation transition

Let us consider the Fourier harmonics of the probability function:

$$P_k = \langle e^{-ik\varphi} \rangle = \frac{1}{2\pi} \int_0^{2\pi} d\varphi \, e^{-ik\varphi} P(\varphi, t) \qquad (9.29)$$

Obviously, the zeroth harmonic $P_0 = \nu/2\pi = \text{const}$ is the measure of the filament density ν, see eqn 9.26, and the real and imaginary parts of P_1 represent the components of the average orientation vector $\tau = (\tau_x, \tau_y)$: $P_1^* = \chi = \tau_x + i\tau_y$ (here, asterisk denotes complex conjugation), see eqn 9.27. Substituting eqn 9.29 into eqn 9.25 yields for each P_k

$$\dot{P}_k + k^2 P_k = 2\varphi_0 \sum_n \sum_m P_n P_m \left(S\left[\varphi_0(n\gamma + m(\gamma - 1))\right] - S[m\varphi_0] \right) \delta_{n+m,k}, \qquad (9.30)$$

where $\delta_{n+m,k}$ is the Kroneker symbol and $S[x] = \sin x / x$. Due to the angular diffusion term $\partial_\varphi^2 P$ in eqn 9.25 giving rise to the terms $-k^2 P_k$ in the equations for the Fourier components (eqn 9.30), the magnitudes of harmonics decay exponentially with $|k|$. Truncating all higher harmonics for $|k| > 2$ (i.e. setting $P_k = 0$ for $|k| > 2$), one obtains from eqn 9.30[3]

$$\dot{P}_0 = 0, \qquad (9.31)$$

$$\dot{P}_1 + P_1 = P_0 P_1 2\varphi_0 \left[S\left[\varphi_0(\gamma - 1)\right] + S[\varphi_0 \gamma] - S[\varphi_0] - 1 \right]$$
$$+ 2\varphi_0 P_2 P_1^* \left[S\left[\varphi_0(\gamma + 1)\right] + S\left[\varphi_0(\gamma - 2)\right] - S[2\varphi_0] - S[\varphi_0] \right], \qquad (9.32)$$

and

$$\dot{P}_2 + 4P_2 = P_0 P_2 2\varphi_0 \left[S\left[2\varphi_0(\gamma - 1)\right] + S[2\varphi_0 \gamma] - S[2\varphi_0] - 1 \right]$$
$$+ 2\varphi_0 P_1^2 \left[S\left[\varphi_0(2\gamma - 1)\right] - S[\varphi_0] \right]. \qquad (9.33)$$

Since the decay rate of the second harmonic P_2 is much larger than that of the first harmonic P_1, we can neglect the time derivative $\dot{P}_2$ in eqn 9.33 and obtain the algebraic relation $P_2 = AP_1^2$, with

$$A = \frac{S\left[\varphi_0(2\gamma - 1)\right] - S[\varphi_0]}{2/\varphi_0 - \left[S\left[2\varphi_0(\gamma - 1)\right] + S[2\varphi_0 \gamma] - S[2\varphi_0] - 1 \right] \nu/2\pi} \qquad (9.34)$$

that allows us to close the equation for P_1. Accordingly, the equation for the "complex orientation vector" χ orientation vector τ can be written as:

$$\dot{\chi} = b_0(\nu - \nu_c)\chi - A_0 |\chi|^2 \chi. \qquad (9.35)$$

with the coefficients A_0, b_0 in the form

$$A_0 = 2\varphi_0 A \left[S\left[\varphi_0(\gamma + 1)\right] + S\left[\varphi_0(\gamma - 2)\right] - S[2\varphi_0] - S[\varphi_0] \right]$$

---

[3]A more systematic approximation to the solution of eqn 9.30 was obtained by Ben-Naim and Krapivsky (2006) using partition of integers.

$$b_0 = \frac{\varphi_0}{\pi} \left[S\left[\varphi_0(\gamma-1)\right] + S[\varphi_0\gamma] - S[\varphi_0] - 1 \right], \quad \nu_c = b_0^{-1}. \tag{9.36}$$

It is easy to see that the equation for the *real* orientation vector τ has exactly the same form:

$$\dot{\tau} = b_0(\nu - \nu_c)\tau - A_0|\tau|^2\tau, \tag{9.37}$$

For fully inelastic collisions $\gamma = 1/2$ and for the interaction angle $\varphi_0 = \pi$ one obtains the following values: $b_0 = (4\pi^{-1} - 1) - 1 \approx 0.273$, critical density $\nu_c = \pi/(4-\pi) \approx 3.662$ and $A_0 = 8A/3 \approx 2.18$. For large enough densities $\nu > \nu_c$, a disordered state with $\tau = 0$ becomes unstable. The ordering instability leads to a spontaneous rod alignment. This instability saturates at the value of $|\tau|$ determined by ν. Figure 9.6 shows stationary solutions $P(\varphi)$ computed directly from eqn 9.25. Integrating these distributions we can obtain the moments including the components of τ. As seen from the inset, the corresponding values of $|\tau|$ are consistent with the truncated model (eqn 9.37) up to $\nu < 5.5$.

Thus, our analysis indicates that for large enough density of molecular motors and polar rods (microtubules), an initially isotropic mixture of rods will experience spontaneous ordering to the state with some preferred orientation determined by the initial conditions. The transition bears a strong resemblance to the second-order phase transition in equilibrium systems, e.g. the ferromagnetic transition. We also note that this transition is qualitatively different from the nematic transition in the system of hard rods with excluded volume interaction, see Section 8.1.1. The transition considered here is specific to the polar rods only and results in the onset of the *polar order*, whereas the thermodynamic Onsager transition yields only *nematic ordering*. As a result, the critical density ν_c in our situation is determined by both the density of motors and density of microtubules and does not have direct relation to the Onsager critical density determined by the density of rods only.

9.2.4 Spatial localization

To describe the spatial localization of interactions we introduce the probability distribution $P(\mathbf{r}, \varphi, t)$ to find a rod with orientation φ at location $\mathbf{r}$ at time t. The master equation for $P(\mathbf{r}, \varphi, t)$ can be written as

$$\partial_t P(\mathbf{r}, \varphi, t) = \partial_\varphi^2 P(\mathbf{r}, \varphi, t) + \partial_i D_{ij} \partial_j P(\mathbf{r}, \varphi, t) + \mathcal{I}(\mathbf{r}, \varphi, t), \tag{9.38}$$

where the first two terms describe angular and translational diffusion of tubules, and the last term (the "collision integral") describes the binary interaction among filaments via molecular motors.

The generally anisotropic tensor of translational diffusion coefficients can be written in the form

$$D_{ij} = \frac{1}{D_r} \left(D_\| n_i n_j + D_\perp (\delta_{ij} - n_i n_j) \right). \tag{9.39}$$

Here, $\mathbf{n} = (\cos\varphi, \sin\varphi)$ is the unit vector oriented along the tubule. The rotational D_r, parallel $D_\|$, and perpendicular $D_\perp$ diffusion coefficients of long stiff rods are known from the polymer physics:

$$D_{\|} = \frac{k_B T}{\xi_{\|}}, D_{\perp} = \frac{k_B T}{\xi_{\perp}}, D_r = \frac{4k_B T}{\xi_r}, \tag{9.40}$$

where $\xi_{\|} = 2\pi\eta_s l / \log(l/d)$, $\xi_{\perp} = 2\xi_{\|}$, $\xi_r \approx \pi\eta_s l^3 / 3\log(l/d)$ are corresponding viscous drag coefficients, and η_s is the shear viscosity of the solvent (Doi and Edwards, 2003).

Let us now turn to the collision integral $\mathcal{I}(\mathbf{r}, \varphi, t)$. In the following we focus entirely on the case of totally inelastic collision between the filaments, i.e. $\gamma = 1/2$. In this case the tubules acquire the same orientation along the bisector $\bar{\mathbf{n}} = (\cos(\bar{\varphi}), \sin(\bar{\varphi}))$. We also assume for the sake of simplicity that their center-of-mass positions, $\mathbf{r}_{1,2}$ align. Then, the interaction rules take the form:

$$\varphi_1^a = \varphi_2^a = \bar{\varphi} = \frac{\varphi_1^b + \varphi_2^b}{2},$$

$$\mathbf{r}_1^a = \mathbf{r}_2^a = \bar{\mathbf{r}} = \frac{\mathbf{r}_1^b + \mathbf{r}_2^b}{2}. \tag{9.41}$$

Hence, the collision integral in eqn 9.38 assumes the form

$$\mathcal{I}(\mathbf{r}, \varphi) = \int\int d\mathbf{r}_1 d\mathbf{r}_2 \int_{-\varphi_0}^{\varphi_0} d\varphi_1 d\varphi_2 W(\mathbf{r}_1 - \mathbf{r}_2, \mathbf{n}_1 - \mathbf{n}_2) P(\mathbf{r}_1, \varphi_1) P(\mathbf{r}_2, \varphi_2)$$
$$\times [\delta(\varphi - \bar{\varphi})\delta(\mathbf{r} - \bar{\mathbf{r}}) - \delta(\varphi - \varphi_1)\delta(\mathbf{r}_1 - \mathbf{r})], \tag{9.42}$$

where we performed the same rescaling as in eqn 9.25 and dropped argument t for brevity.

In this expression, the function $W(\mathbf{r}_1 - \mathbf{r}_2, \mathbf{n}_1 - \mathbf{n}_2)$ (or the *interaction kernel*) characterizes the probability of interaction between two filaments. It depends on the mutual positions and orientations of tubules, and can be estimated from the following considerations: (i) due to the translational and rotational invariance, the kernel should depend only on the differences $\mathbf{n}_1 - \mathbf{n}_2$ and $\mathbf{r}_1 - \mathbf{r}_2$; (ii) the kernel should be invariant with respect to permutations $\mathbf{n}_1 \to \mathbf{n}_2$, $\mathbf{r}_1 \to \mathbf{r}_2$; (iii) since the size of a motor is small compared to the length of a filament, it is reasonable to assume that two rods interact only if they intersect. While the precise form of the kernel is discontinuous and rather cumbersome (Aranson and Tsimring, 2006a), its approximation convenient for calculations is given by

$$W = \frac{W_0}{\pi b^2} \exp\left[-\frac{|\mathbf{r}_1 - \mathbf{r}_2|^2}{b^2}\right] \left(1 + \frac{\beta}{l}(\mathbf{r}_1 - \mathbf{r}_2) \cdot (\mathbf{n}_1 - \mathbf{n}_2)\right)$$
$$= W_I(|\mathbf{r}_1 - \mathbf{r}_2|)\left(1 + \frac{\beta}{l}(\mathbf{r}_1 - \mathbf{r}_2) \cdot (\mathbf{n}_1 - \mathbf{n}_2)\right), \tag{9.43}$$

with the effective interaction length b that is of the order of microtubule length l. The Gaussian form of the isotropic part of the kernel W_I implies that only nearby microtubules interact effectively due to the small size of molecular motors. The constant interaction rate W_0 as before can be scaled away by the renormalization of variables: $W_0 P \to P$. The anisotropic term $\sim \beta(\mathbf{r}_1 - \mathbf{r}_2) \cdot (\mathbf{n}_1 - \mathbf{n}_2)$ describes the dependence of the coupling strength on the mutual orientation of microtubules: "diverging" polar rods (such as shown in Fig. 9.5(a)) interact stronger than "converging" ones. This term, describing the anisotropic contribution to the kernel, is associated, for example, with the increase of motor density toward the polar end of the filament due to dwelling of motors (Aranson and Tsimring, 2006a). The constant β can be related to the dwell time of the motor at the polar end of the microtubule that is inversely proportional to the unbinding probability p_{end}. The experiments indicate that while the kinesin motors have very short dwell time, the NCD motor constructs used in the experiments of Surrey et al. (2001) possess considerable dwell time or, correspondingly, small value of p_{end}. Accordingly, the value of the anisotropy constant β is negligible for kinesin motors but is much bigger for NCD motors, $\beta \approx 0.001 - 0.1$ depending on the average motor density (Aranson and Tsimring, 2006a).

9.2.5 The generalized Ginzburg–Landau description

In the vicinity of the orientation transition $\nu \to \nu_c$ the master equation (eqn 9.38) can be systematically simplified by the means of the bifurcation analysis and reduced to much simpler equations for the coarse-grained density ν and orientation $\boldsymbol{\tau}$. As in the case of the spatially homogeneous orientation transition considered in Section 9.2.3, we perform Fourier expansion in φ and truncate the series at $|n| > 2$. Correspondingly, P_0 gives the local number density $\nu(\mathbf{r}, t)$, and $P_{\pm 1}$ the local orientation $\boldsymbol{\tau}(\mathbf{r}, t)$. Following the technique of Aranson and Tsimring (2006a), one can derive the system of the coupled equations for density and orientation (a rather tedious derivation of these equations can be found in Aranson and Tsimring (2006a))

$$\partial_t \nu = \nabla^2 \left[D_\nu \nu - \frac{B^2 \nu^2}{16} \right] + \frac{\pi B^2 H}{16} \left[3 \nabla \cdot \left(\boldsymbol{\tau} \nabla^2 \nu - \nu \nabla^2 \boldsymbol{\tau} \right) \right.$$
$$\left. + 2 \partial_i \left(\partial_j \nu \partial_j \tau_i - \partial_i \nu \partial_j \tau_j \right) \right] - \frac{7 \nu_0 B^4}{256} \nabla^4 \nu \qquad (9.44)$$

$$\partial_t \boldsymbol{\tau} = D_{\tau_1} \nabla^2 \boldsymbol{\tau} + D_{\tau_2} \nabla (\nabla \cdot \boldsymbol{\tau}) + b_0 (\nu - \nu_c) \boldsymbol{\tau} - A_0 |\boldsymbol{\tau}|^2 \boldsymbol{\tau}$$
$$+ H \left[\frac{\nabla \nu^2}{16 \pi} - \left(\pi - \frac{8}{3} \right) \boldsymbol{\tau} (\nabla \cdot \boldsymbol{\tau}) - \frac{8}{3} (\boldsymbol{\tau} \cdot \nabla) \boldsymbol{\tau} \right] + \frac{B^2 \nu_0}{4 \pi} \nabla^2 \boldsymbol{\tau}. \qquad (9.45)$$

Here, we rescaled the spatial variable $\mathbf{r}$ by l, introduced the normalized interaction length $B = b/l$ and the scaled kernel anisotropy parameter $H = \beta B^2$. The normalized diffusion coefficients D_ν, D_{τ_1} and D_{τ_2} are

$$D_\nu = \frac{D_\| + D_\perp}{2 D_r l^2} = \frac{1}{32}, \quad D_{\tau_1} = \frac{D_\| + 3 D_\perp}{4 D_r l^2} = \frac{5}{192}, \quad D_{\tau_2} = \frac{D_\| - D_\perp}{2 D_r l^2} = \frac{1}{96}.$$

The last two terms in eqns 9.44 and 9.45 are linearized near the mean density $\nu_0 = \langle \nu \rangle$. The last term in eqn 9.44 regularizes the short-wave instability when the diffusion term changes sign for $\nu_0 > \nu_{c1} = 1/4B^2$. As we show below, this instability leads to strong density variations associated with the formation of bundles of microtubules.

9.2.6 Aster and vortex solutions

For $B^2 H \ll 1$, i.e. small kernel anisotropy, the density modulations are small, and eqn 9.45 for the coarse-grained orientation of microtubules $\boldsymbol{\tau}$ decouples from the density equation (eqn 9.44). Assuming that the (uniform) density ν is close to critical ν_c, we again rewrite eqn 9.45 for the complex orientation variable $\chi = \tau_x + i\tau_y$:

$$\partial_t \chi = \epsilon \chi - A_0 |\chi|^2 \chi + D_1 \nabla^2 \chi + D_2 \bar{\nabla}^2 \chi^*$$
$$+ H \left[\left(\pi - \frac{8}{3} \right) \chi \mathrm{Re} \bar{\nabla} \chi^* + \frac{8}{3} (\mathrm{Re} \chi^* \bar{\nabla}) \chi \right], \tag{9.46}$$

where $\epsilon = b_0(\nu - \nu_c)$ is the small parameter, operator $\bar{\nabla} = \partial_x + i\partial_y$, and the diffusion coefficients $D_1 = 1/32 + \nu_0 B^2/4\pi$, $D_2 = D_{\tau_2}/2 = 1/192$. This equation generalizes the normal form equation (eqn 9.35) for the spatially nonuniform case. Equation 9.46 is similar to the generalized Ginzburg–Landau equation known in the context of superconductivity, superfluidity, non-linear optics, and general pattern formation, see for a review Aranson and Kramer (2002). Let us focus on the structure and dynamics of radially symmetric solutions of eqn 9.46 that can be sought in polar coordinates r, θ in the following generic form:

$$\chi = \sqrt{A_0/\epsilon} A(\tilde{r}, \tilde{t}) \exp(i\theta), \tag{9.47}$$

where we rescaled the radial coordinate $\tilde{r} = r/\sqrt{\epsilon}$ and time $\tilde{t} = t/\epsilon$ and introduced the new complex amplitude $A(\tilde{r}, \tilde{t})$. Substituting eqn 9.47 in eqn 9.46 we obtain

$$\partial_{\tilde{t}} A = D_1 \Delta_{\tilde{r}} A + D_2 \Delta_{\tilde{r}} A^* + \left(1 - |A|^2 \right) A$$
$$+ H \left(a_1 A \mathrm{Re} \nabla_{\tilde{r}} A + a_2 \partial_{\tilde{r}} A \mathrm{Re} A + \frac{i a_2 A \mathrm{Im} A}{\tilde{r}} \right), \tag{9.48}$$

with the following differential operators

$$\Delta_{\tilde{r}} = \partial_{\tilde{r}}^2 + \tilde{r}^{-1} \partial_{\tilde{r}} - \tilde{r}^{-2}; \nabla_{\tilde{r}} = \partial_{\tilde{r}} + \tilde{r}^{-1}, \tag{9.49}$$

and parameters

$$a_1 = (\pi - 8/3)/\sqrt{A_0} \approx 0.321, a_2 = 8/3\sqrt{A_0} \approx 1.81.$$

This system can be used to study transitions between asters and vortices. In the following we omit tildes above r and t.

In the standard Ginzburg–Landau model, corresponding to $D_2 = 0$ and $H = 0$ in eqn 9.46, there is a stationary solution $A = F(r) \exp[i\vartheta]$ with real $F(r)$ and an arbitrary phase constant ϑ. The value of ϑ determines the tilt angle of arrows to the azimuthal direction. It is easy to see that the values $\vartheta = 0, \pi$ correspond to aster solutions (sink or source), while any other value of ϑ would correspond to a vortex, see examples in Fig. 9.7. In the Ginzburg–Landau equation, ϑ is a free parameter and can be gauged away by a simple transformation. Thus, in this limit (zero diffusion anisotropy $D_2 \sim D_\| - D_\perp = 0$ and zero kernel anisotropy $H = 0$) there is no distinction between asters and vortices. However, non-zero diffusion and kernel anisotropy lift off this degeneracy and in fact make the phase ϑ a unique function of the radial coordinate. This in turn results in a non-trivial stability exchange between asters and vortices as the parameters of the system are varied.

Fig. 9.7 Schematic representation of orientation fields τ for three different values of ϑ: aster ($\vartheta = \pi$); generic vortex ($\pi/2 < \vartheta < \pi$) and ideal vortex ($\vartheta = \pm\pi/2$).

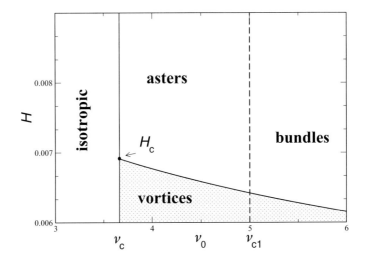

Fig. 9.8 The phase diagram of the rescaled density ν_0 (the product of motor and filament densities) versus kernel anisotropy parameter H for eqns 9.44 and 9.45. Above ν_c, the polar state is formed. Beyond ν_{c1}, an isotropic density instability in eqn 9.44 occurs. Depending on parameters, the density instability may also happen prior to the orientation instability, i.e. $\nu_{c1} < \nu_c$. Between ν_c and ν_{c1}, asters are stable above the critical (solid) line, while vortices are linearly stable below this line. The critical line terminates at the point $H = H_c$.

For a non-zero anisotropy parameter, $H > 0$, the aster solution still exists but it becomes unstable in a certain range of H. The corresponding bifurcation line can be found by linearizing eqn 9.48 near the aster solution, $A(r,t) = F(r) + i\varpi(r)\exp(\lambda t)$ where (real) function ϖ obeys $\hat{L}\varpi = \lambda\varpi$ with the operator

$$\hat{L} \equiv \bar{D}\Delta_r + \left(1 - F^2 - a_1 H \nabla_r F\right) - a_2 H F \nabla_r, \tag{9.50}$$

($\bar{D} = D_1 - D_2$) and should have zero derivative at $r = 0$ and decay at $r \to \infty$. This eigenvalue problem can be solved numerically by the matching-shooting method. The

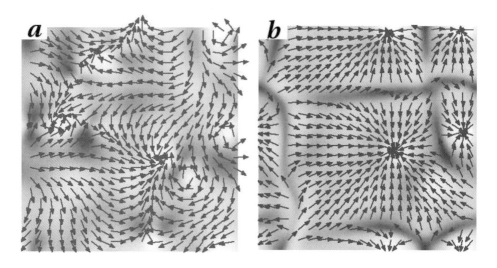

Fig. 9.9 Vectors of orientation τ for vortices ($H = 0.006$, left) and asters ($H = 0.125$, right) obtained from eqns 9.44 and 9.45. The graycode indicates the value of the density ν (white corresponds to maximum and black to zero), $B^2 = 0.05, \nu_0 = 4$, domain of integration 80×80, time of integration 1000. The density modulations are about 10% of the mean value ν_0.

result of this calculation (the critical value of the kernel anisotropy H_0 derived from the condition $\lambda = 0$ as a function of density ν_0) is shown in the phase diagram, Fig. 9.8. The vertical line at $\nu_0 = \nu_c$ corresponds to the transition to an oriented regime. The solid line $H_0(\nu_0)$ separates asters and vortices. The lines meet at the critical point $H_c = H_0(\nu_c)$. The phase diagram, shown in Fig. 9.8, is in qualitative agreement with the experiments (Surrey et al., 2001): for the low value of kernel anisotropy $H < H_c$ (corresponding to kinesin motors) increase of the density ν_0 first leads to the formation of vortices and then to asters. For $H > H_c$ (corresponding to NCD motors as discussed above) only asters are observed. It is noteworthy that the vortex solution has the non-zero phase ϑ at the core, but at large distances $r \to \infty$ the phase $\vartheta \to 0$, i.e. vortices and asters become indistinguishable far away from the core.

Beyond the linear stability analysis, the full system of eqns 9.44 and 9.45 can be studied numerically. Integration was performed in a two-dimensional square domain with periodic boundary conditions by the quasispectral method. For small H vortices are observed, and for larger H they are replaced by asters, in agreement with the above analysis. As seen in Fig. 9.9, asters have the unique orientation of the microtubules (here, towards the center). Asters with opposite orientation of τ are unstable. In large domains asters form a disordered network. Neighboring asters are separated by "shock lines" containing saddle-type defects. Starting from random initial conditions, simulations first exhibit a large number of small asters that gradually merge and annihilate. Eventually, annihilation slows down due to the exponential weakening of the interaction of asters.

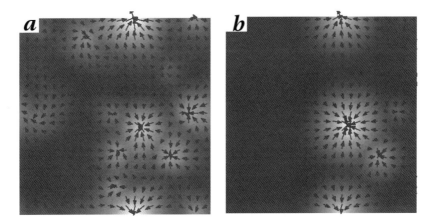

Fig. 9.10 Graycoded images of the motor density at $t = 200$ (left) and $t = 1000$ (right), white corresponds to the maximum density and black to the minimum density, domain size is 80×80, $\zeta_0 = 0.5$, $H = 0.125$, $D_0 = 5$, $m_0 = 1$. Arrows show the orientation of microtubules.

9.2.7 Inhomogeneous distribution of molecular motors

In the derivation of equations governing the organization of microtubules in Section 9.2.4 we assumed a homogeneous bulk distribution of molecular motors by fixing the rate $W_0 = \mathrm{const}$ in the interaction kernel W_I (eqn 9.43). This approximation was justified by the fact that the diffusion of motors is about two orders of magnitude larger than that of microtubules. However, experiments indicate that even despite fast motor diffusion, they tend to aggregate in the core regions of asters and vortices due to their directed transport along the microtubules (Surrey *et al.*, 2001; Nédélec *et al.*, 2001). The effects of the variable motor density were studied by Lee and Kardar (2001), Sankararaman *et al.* (2004), and the transport and accumulation of motors at the centers of asters and vortices was assumed to be a primary mechanism of self-organization. Here, we show that while taking into account the variable motor density indeed improves the agreement with experiment, in the framework of the above model it does not qualitatively change our previous conclusions obtained under the assumption of the uniform motor density. In particular, vortices are still obtained for low values of the kernel anisotropy H and small motor density, and asters for higher motor density.

Let us introduce two different concentrations of motors: concentration of free motors m_f diffusing freely in the fluid surrounding filaments and bound motors m_b attached to the filaments. The coarse-grained concentrations of bound and free motors obey the following advection–diffusion equations (Nédélec *et al.*, 2001; Lee and Kardar, 2001; Aranson and Tsimring, 2006*a*)

$$\partial_t m_f = D\nabla^2 m_f - \nu(P_{\mathrm{on}} m_f - P_{\mathrm{off}} m_b)$$
$$\partial_t m_b = -\zeta \nabla m_b \boldsymbol{\tau} + \nu(P_{\mathrm{on}} m_f - P_{\mathrm{off}} m_b), \qquad (9.51)$$

where $P_{\mathrm{on}}, P_{\mathrm{off}}$ are the rates of binding/unbinding of motor to the microtubules, D

is the bulk diffusion coefficient of molecular motors and ζ is the advection coefficient (note that advection velocity $\mathbf{V} = \zeta \boldsymbol{\tau}$). The term $-\nu(P_{\text{on}} m_{\text{f}} - P_{\text{off}} m_{\text{b}})$ describes the decrease of the concentration of free motors m_{f} with the total rate $\nu P_{\text{on}} m_{\text{f}}$ due to binding of free motors to microtubules and its increase with the rate $\nu P_{\text{off}} m_{\text{b}}$ due to unbinding of bound motors from microtubules.

We consider the case of smooth distributions of motor densities $m_{\text{f}}, m_{\text{b}}$ (due to the high value of the motor's diffusion) and large binding/unbinding rates. Then, the rhs of eqns 9.51 are dominated by the last term leading to the local balance relation between motor densities $m_{\text{f}}, m_{\text{b}}$

$$P_{\text{on}} m_{\text{f}} \approx P_{\text{off}} m_{\text{b}}. \tag{9.52}$$

Thus, eqns 9.51 can be reduced to a single equation for the total motor density $m = m_{\text{f}} + m_{\text{b}}$ that plays the role of a slow variable:

$$\partial_t m = D_0 \nabla^2 m - \zeta_0 \nabla m \boldsymbol{\tau}, \tag{9.53}$$

with renormalized diffusion $D_0 = D P_{\text{off}}/(P_{\text{off}}+P_{\text{on}})$ and advection rate $\zeta_0 = \zeta P_{\text{on}}/(P_{\text{off}}+P_{\text{on}})$.

In order to derive equations for the coarse-grained density of microtubules ν and orientation $\boldsymbol{\tau}$, we need to modify the expression for the interaction kernel (eqn 9.43) in order to include the effect of motor inhomogeneity. The simplest way to include the inhomogeneous motor density into the kernel is to replace the constant interaction rate W_0 assumed to be proportional to the motor density m in the symmetric part of the interaction kernel (eqn 9.43) by the expression $W_0 \sim m((\mathbf{r}_1 + \mathbf{r}_2)/2)$. Taking the motor concentration in the middle point $(\mathbf{r}_1 + \mathbf{r}_2)/2$ is necessary to preserve the mass conservation. Repeating the calculations outlined in the previous sections one can derive equations similar to eqns 9.44 and 9.45 but for the normalized tubule density $\nu = W_0 \int_0^{2\pi} P d\varphi$ that *does not* absorb the motor density m. However, the resulting equations are very cumbersome, especially for the transport term in eqn 9.44.

The calculations are simplified considerably in the limit of high motor diffusivity D, using the fact that in this limit the distribution of total motor density m becomes smooth. Then one can neglect the derivatives of m where appropriate, and the resulting equations assume the form:

$$\partial_t \nu = \nabla^2 \left[D_\nu \nu - \frac{m B^2 \nu^2}{16} \right] \tag{9.54}$$

$$- \frac{\pi B^2 H}{16} \left[3\nabla m \left(\boldsymbol{\tau} \nabla^2 \nu - \nu \nabla^2 \boldsymbol{\tau} \right) + 2 \partial_i m \left(\partial_j \nu \partial_j \tau_i - \partial_i \nu \partial_j \tau_j \right) \right] - \frac{7 \nu_0 m_0 B^4}{256} \nabla^4 \nu$$

$$\partial_t \boldsymbol{\tau} = D_{\tau 1} \nabla^2 \boldsymbol{\tau} + D_{\tau 2} \nabla (\nabla \cdot \boldsymbol{\tau}) + ((4/\pi - 1) m \nu - 1)) \boldsymbol{\tau} \tag{9.55}$$

$$- A_0 m^2 |\boldsymbol{\tau}|^2 \boldsymbol{\tau} - H m \left[\frac{\nabla \nu^2}{16\pi} - \left(\pi - \frac{8}{3} \right) \boldsymbol{\tau} (\nabla \cdot \boldsymbol{\tau}) - \frac{8}{3} (\boldsymbol{\tau} \nabla) \boldsymbol{\tau} \right] + \frac{B^2 \nu_0 m_0}{4\pi} \nabla^2 \boldsymbol{\tau}.$$

Here, motor density m is included in the lowest order in gradient expansion. Again for simplicity we replaced the motor density m by its mean value m_0 in the last terms in eqns 9.54 and 9.55.

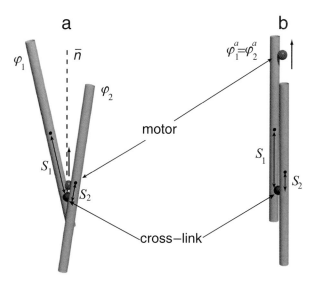

Fig. 9.11 Sketch of the interaction between two filaments, a cross-link and a molecular motor. After the interaction, the motor (shown as a sphere moving in the direction of the arrow) aligns the filaments along the bisector $\bar{n}$, but the midpoint positions do not coincide due to the cross-link.

Equations 9.53–9.55 were studied numerically. The values of the motor diffusion D_0 and motor advection ξ_0 can be estimated from the experiments, in our dimensionless units $D_0 \sim 1..5$ and $\xi_0 \sim 1$. Typical results are shown in Fig. 9.10. Not surprisingly, in agreement with experiment, motors tend to accumulate in the center of an aster or a vortex, as in Nédélec et al. (2001). Otherwise, the qualitative behavior of formation of asters and vortices remains the same. An initial multiaster state coarsens and leads to the formation of a network of large asters separated by domain walls.

9.2.8 Effect of cross-links

Intricate cytoskeletal networks are created and maintained by an efficient cell machinery involving not only various types of motor proteins but also passive cross-linking proteins. While the motors are mostly responsible for the assembly, reorganization of the cytoskeleton and transport of various molecular components, cross-links provide rigidity and elasticity of the cytoskeletal networks. They connect different filaments but do not move along them. It was recently understood that cross-links also play an important role in the self-organization of cytoskeletal networks (Smith et al., 2007). In particular, it was shown that cross-links facilitate the formation of bundles in the actin-myosin system: at high concentration of ATP, actin-myosin systems display an isotropic phase; in the course of depletion of ATP, however, myosin motors become static cross-links and initiate the formation of oriented bundles and cluster-like patterns. Reintroduction of ATP into the bundled state resulted in a dissolution of the structures and re-establishment of the isotropic state.

The effect of cross-links on the motor-induced interaction of filaments is two-fold. First, the simultaneous action of a static cross-link, serving as a hinge, and a motor moving along both filaments results in a fast and efficient alignment of filaments, as shown in Fig. 9.11. For microtubule–kinesin system the alignment time is of the order of 1 s. This justifies the assumption of fully inelastic and instant collisions assumptions for the rods interaction. Note that without cross-links the overall change in the relative orientation of the filaments is much smaller: the angle between filaments decreases only by 20–25 % (Aranson and Tsimring, 2006a). Complete alignment can also occur for the case of a simultaneous action of two motors moving in opposite directions, as in experiments on kinesin-NCD mixtures reported in Surrey et al. (2001), and even for several identical motors with a different speed due to the variability of the properties and the stochastic character of the motion (Visscher et al., 1999). Secondly, the cross-links inhibit relative sliding of rods in the course of alignment, restricting their motion to rotation only. Thus, in contrast to the situation considered in Section 9.2.4, and described by the interaction rules expressed by eqns 9.41, in the presence of a cross-link the midpoints of the rods would *not* coincide after the interaction. In fact, the distances $S_{1,2}$ from the midpoints to the cross-link point will not change, as is shown in Fig. 9.11.

To describe the interaction rules in the presence of a cross-link, we express the radius-vector of an arbitrary point on a filament $\mathbf{R}_i$ via the position of its midpoint $\mathbf{r}_i$, the filament orientation $\mathbf{n}_i$, and the distance from the center-of-mass S, $\mathbf{R}_i = \mathbf{n}_i S_i + \mathbf{r}_i$. The intersection point of two rods is given by the condition $\mathbf{R}^* = \mathbf{R}_1 = \mathbf{R}_2$ fixing the values of $S_{1,2}$ to

$$S_{1,2} = \frac{(\mathbf{r}_2 - \mathbf{r}_1) \times \mathbf{n}_{2,1}}{\mathbf{n}_1 \times \mathbf{n}_2} . \tag{9.56}$$

Due to the cross-link, the values of $S_{1,2}$ do not change during the interaction. Since the filaments become oriented along the bisector direction $\bar{\mathbf{n}}$, the distance of the two filament midpoints from the center of mass of the whole system will be $\Delta S = S_1 - S_2$. Therefore, instead of eqns 9.41), we obtain the following interaction rules

$$\varphi_1^a = \varphi_2^a = \bar{\varphi} = \frac{\varphi_1 + \varphi_2}{2} , \tag{9.57}$$

$$\mathbf{r}_{1,2}^a = \frac{\mathbf{r}_1 + \mathbf{r}_2}{2} \pm \frac{\bar{\mathbf{n}} \Delta S}{2} = \frac{\mathbf{r}_1 + \mathbf{r}_2}{2} \pm \frac{\bar{\mathbf{n}}((\mathbf{r}_1 - \mathbf{r}_2) \cdot \bar{\mathbf{n}})}{2 \cos \psi} , \tag{9.58}$$

with $\psi = (\varphi_1 - \varphi_2)/2$. Note that dropping the last term in eqn 9.58 the previous model, eqns 9.41, is recovered.

The interaction rules (eqns 9.57 and 9.58) can be used to evaluate the collision integral, eqn 9.42. Omitting lengthy calculations (see for details Ziebert et al. (2007)), after expanding the master equation (eqn 9.23) near the threshold of the orientation instability, we arrive at the following set of non-linear equations for the coarse-grained density ν and orientation τ:

$$\partial_t \nu = D_\nu \nabla^2 \nu - \zeta \nabla^4 \nu - \frac{\pi(1 - S(\varphi_0))\varphi_0 B^2}{16} \left[\nabla^2 \tau^2 - 2\partial_i \partial_j (\tau_i \tau_j)\right] , \tag{9.59}$$

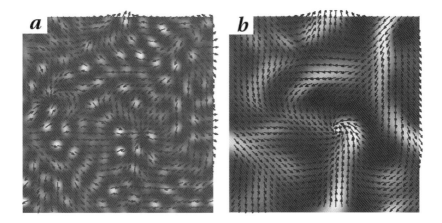

Fig. 9.12 Composite image of the density (black corresponds to low density ($\nu_{\min} \approx 3$), white to high density $\nu_{\max} \approx 13$) and the filament orientation field (arrows). (a) The model (eqns 9.44 and 9.45) by Aranson and Tsimring (2005) for $\nu_0 = 5$ in the region of the isotropic density instability. Here the filament orientation is uncorrelated with the density gradient. (b) The model with cross-links, eqns 9.59 and 9.60, for a density of $\nu_0 = 6$ displays pronounced bundles, and the local filament orientation is predominantly along the bundles. Other parameters values: $H = 0.005, B^2 = 0.6, \zeta = 0.04$.

$$\partial_t \boldsymbol{\tau} = b_0(\nu - \nu_c)\boldsymbol{\tau} - A_0|\boldsymbol{\tau}|^2\boldsymbol{\tau} + D_{\tau_1}\nabla^2\boldsymbol{\tau} + D_{\tau_2}\nabla\nabla\cdot\boldsymbol{\tau} + \frac{B^2\nu_0}{4\pi}\nabla^2\boldsymbol{\tau}$$
$$- H\left[\frac{1}{16\pi}\nabla\nu^2 - \left(\pi - \frac{8}{3}\right)\boldsymbol{\tau}(\nabla\cdot\boldsymbol{\tau}) - \frac{8}{3}(\boldsymbol{\tau}\cdot\nabla)\boldsymbol{\tau}\right]. \quad (9.60)$$

The constants A_0, b_0 and the critical density ν_c are functions of the maximum interaction angle φ_0 and the inelasticity coefficient γ given by eqns 9.36. In the following we again consider the case $\varphi_0 = \pi$ and $\gamma = 1/2$ as motivated above. Then $S[\varphi_0 = \pi] = 0$. The anisotropic contribution proportional to $H = \beta b^2/L^2 = \beta B^2$ is due to the polar distribution of motors along the interacting filaments, while the anisotropic contribution in the ν-equation is due to the cross-links. As in the case of eqn 9.44, the isotropic higher-order diffusion term $\zeta\nabla^4\nu$ is included for regularization purposes of the equation at very short wavelengths.

Let us briefly compare models without cross-links, eqns 9.44 and 9.45, and with cross-links, eqns 9.59 and 9.60. As was shown above, eqn 9.44 exhibits an isotropic density instability if $\nu > \nu_{c1}$ as calculated below. In the presence of cross-links the term in eqn 9.44 proportional to $\nabla^2\nu^2$ that is responsible for the density instability vanishes, and instead a new anisotropic term $\nabla^2\tau^2 - 2\partial_i\partial_j(\tau_i\tau_j)$ appears. This term couples density and orientation perturbations. As we will show below, this new cross-link-induced anisotropic coupling modifies the density instability so it becomes transverse to the direction of polar orientation (in both linear and non-linear regimes).

In order to investigate the linear stability of the homogeneous polar solution of eqns 9.59 and 9.60, describing a state with density ν_0 and polar orientation $\boldsymbol{\tau}_0$ given by $b_0(\nu - \nu_c) = A_0|\tau_0|^2$, without loss of generality we fix τ_0 along the x-direction: $\tau_0 = (|\tau_0|, 0)$. Linearizing the model equations around this state by making the anzatz $\{\nu, \tau_x, \tau_y\} = \{\nu_0, \tau_0, 0\} + \{\delta\nu, \delta\tau_x, \delta\tau_y\} \exp[\lambda(\mathbf{k})t + ik_x x + ik_y y]$, one can deduce the linear growth rate λ as a function of the modulation wave numbers k_x, k_y.

First, consider the case without cross-links which is described by eqns 9.44 and 9.45. There are three linear modes in the system. The two largest growth rates for long-wave perturbations are associated with the transverse orientational mode and to the mixed density–orientation mode. The third mode associated with the magnitude of the orientation vector, is always damped. To the leading order in k_x, k_y the growth rate for the transverse orientational mode reads

$$\lambda_\tau = -\left(D_{\tau_1} + \frac{B^2 \nu_0}{4\pi}\right) k_x^2 - \left(D_{\tau_1} + D_{\tau_2} + \frac{B^2 \nu_0}{4\pi}\right) k_y^2, \quad (9.61)$$

(for simplicity, we assume $H = 0$ here and below), and is always damped. For the mixed density mode one obtains

$$\lambda_\nu = -\left(D_\nu - \frac{B^2 \nu_0}{32}\right) (k_x^2 + k_y^2). \quad (9.62)$$

Thus, the density instability occurs at

$$\nu_0 > \nu_{c1} = \frac{32 D_\nu}{B^2}, \quad (9.63)$$

which as already described above to a leading order is isotropic, see phase diagram, Fig. 9.8. Note that depending on the model parameters the density instability may also occur prior to the orientation instability, i.e. ν_{c1} can be smaller than ν_c.

A similar analysis can be done in the presence of cross-links, eqns 9.59 and 9.60. While the orientational mode remains unchanged, for the mixed density mode one now obtains

$$\lambda_\nu = -\left(D_\nu + \frac{B^2 \pi^2 b_0}{16 A_0}\right) k_x^2 + \left(-D_\nu + \frac{B^2 \pi^2 b_0}{16 A_0}\right) k_y^2. \quad (9.64)$$

For perturbations in the x-direction, i.e. parallel to the polar orientation, the density mode is damped (the coefficient in front of k_x^2 is always negative). However, using the estimates from above, $b_0 \approx 0.273$, $A_0 = 2.18$, $D_\nu = 1/32$, and $B^2 \approx 2/3$, one obtains that the coefficient in front of k_y^2 is positive, signalling the instability with respect to transverse perturbations, i.e. with small k_x and finite k_y.

In order to study the system beyond the linear regime, numerical investigations of eqns 9.59 and 9.60 were performed in a $80L \times 80L$ square periodic domain. Small-amplitude noise was used as an initial condition for the orientation field τ, and also added to the uniform initial density $\nu = \nu_0$. The simulations were performed in the regime where the homogeneous oriented state is unstable with respect to density fluctuations. However, the manifestation of the instability is different with or without cross-links. Some representative numerical results are presented in Fig. 9.12. Without cross-links, the numerical solution shows that the filament orientation and density gradients are mostly uncorrelated (Fig. 9.12(a)). In contrast, with cross-links, the instability indeed resulted in the formation of anisotropic bundles with the filament orientation predominantly along the bundles (Fig. 9.12(b)). The bundles show a tendency to coarsen with time: small bundles coalesce into bigger bundles. The overall pattern is reminiscent of experimental observations of self-organization in both microtubules interacting with a mixture of motors of two different directions (kinesin and NCD) (Surrey et al., 2001) and experiments on actomyosin, where ATP-depleted myosin motors become cross-links (Smith et al., 2007).

The degree of alignment between the density gradient $\nabla \nu$ and the orientation $\boldsymbol{\tau}$ can be characterized by the alignment coefficient

$$C = 2\langle \sin^2 \varphi_{\nu\tau} \rangle - 1, \qquad (9.65)$$

where $\varphi_{\nu\tau}$ is the angle between orientation of $\nabla \nu$ and $\boldsymbol{\tau}$. The alignment coefficient $C = 1$ if the vectors $\nabla \nu$ and $\boldsymbol{\tau}$ are everywhere perpendicular, and $C = -1$ if they are parallel or antiparallel.

The value of the alignment coefficient is small, $C = -0.0045$, for the image shown in Fig. 9.12(a)a (no cross-links), confirming that the fields $\boldsymbol{\tau}$ and $\nabla \nu$ are practically uncorrelated. For the situation shown in Fig. 9.12(b) (cross-links), a much larger value of the alignment coefficient $C \approx 0.188$ was obtained, implying that the density gradient and the orientation are predominantly orthogonal, meaning that density modulations are transverse to the orientation within a bundle.

In conclusion of this section, let us offer a simple physical interpretation of the transverse instability of filament structure due to cross-links. In the absence of cross-links the motors tend to bring together the midpoint positions of microtubules, triggering an isotropic density instability. This instability is a direct analog of the aggregation or clustering in a gas of inelastic or sticky particles considered earlier in the context of inelastic collapse and clustering, see Chapter 4. With a cross-link holding two filaments together at the intersection point, however, the motion of the filaments along the bisector is suppressed, whereas the angular aggregation proceeds unopposed (furthermore, in fact it becomes much more effective). Thus, cross-links turn the isotropic instability into a transversal one.

9.3 Collective dynamics of self-propelled particles

Highly correlated collective motion of large colonies of organisms, such as flocks of birds, schools of fish, swarms of amoebae or motile bacteria, is a ubiquitous biological phenomenon. It occurs on vastly different scales, from kilometers (birds) to micrometers (bacteria). Despite clear differences between specific features of collective motion of various species, all these systems have one common feature: their collective motion is a result of local interactions of individual self-propelled objects. In this section we consider generic modelling of collective behavior of large populations of self-propelled particles.

9.3.1 Vicsek model

The recent progress in the understanding of collective behavior of self-propelled particles began with the work by Vicsek et al. (1995) who pointed out a deep analogy between flocking motion of self-propelled objects and ferromagentism: the velocity vector of an individual particle is similar to the magnetic spin of an iron atom in a ferromagnet. Then, the self-organized phase of a flock in which all objects, on average, are moving in the same direction, is analogous to the ferromagnetic phase of iron in which all spins, on average, point in the same direction (Toner et al., 2005). This realization allowed researchers to apply the powerful apparatus of the equilibrium condensed matter and statistical physics in this new context. Following the analogy with Ising model of ferromagnetism, Vicsek et al. (1995) suggested the first minimalistic model for flocking which operated with a discrete set of point particles ("boids"[4]) moving in a d-dimensional space according to certain short-range interaction rules. According to these rules each boid chooses the direction of motion by measuring mean direction of motion among its neighbors within a certain finite-size neighborhood. However, the boid's determination of the flight direction is affected by white Gaussian noise, so one can expect a transition from flocking to disoriented motion as the magnitude of noise is increased.

Specifically, Vicsek et al. (1995) proposed the following discrete-time algorithm: Each of the N particles situated at positions $\mathbf{r}_i$ in a two-dimensional plane, at an integer time t chooses the direction (angle $\theta_i(t+1)$ with respect to an arbitrarily chosen axis) to move during the next time step by averaging the directions of motion $\theta_j(t)$ of all particles located at time t within a circle of radius R_0 around $\mathbf{r}_i$. Update is assumed to be synchronous. The distance R_0 is assumed to be small compared to the characteristic size of the flock L. The direction in which the particle actually moves in the next time step differs from the average direction of its neighbors by a random angle $\xi_i(t)$, with zero mean and standard deviation ζ. The distributions of $\xi_i(t)$ is identical for all particles, time independent, and uncorrelated among different particles and different time steps. Each particle moves in the chosen direction with a constant speed v_0. The equations of motion can be cast in the form:

$$\theta_i(t+1) = \langle \theta_j(t) \rangle_n + \xi_i(t)$$
$$\mathbf{r}_i(t+1) = \mathbf{r}_i(t) + v_0(\cos(\theta_i(t+1)), \sin(\theta_i(t+1))) \qquad (9.66)$$
$$\langle \xi_i(t) \rangle = 0, \langle \xi_i(t)\xi_j(t') \rangle = \zeta^2 \delta_{ij}\delta_{tt'},$$

where $\langle ... \rangle_n$ is the average direction among "neighbors" within the interaction circle $|\mathbf{r}_i - \mathbf{r}_j| < R_0$.

Simulations of the model described by eqn 9.66 confirmed that for small noise magnitude particles form coherently moving "flocks" shown in Fig. 9.13. The form of the flocks depends on the interaction radius R_0, the number density ν, and the noise magnitude ζ. At small noise amplitude ζ and large densities the particles form coherently moving ordered flocks, see Fig. 9.13(a), whereas at lower densities and at higher noise the particles tend to form groups or clusters moving in random directions, such as shown in Fig. 9.13(b).

---

[4]The word "boid" was introduced into the scientific lexicon by Reynolds (1987) who developed the first artificial life code for simulations of flocks of birds, however, this word had long been a part of the Brooklyn dialect as a mispronunciation of the word "bird".

Collective dynamics of self-propelled particles 287

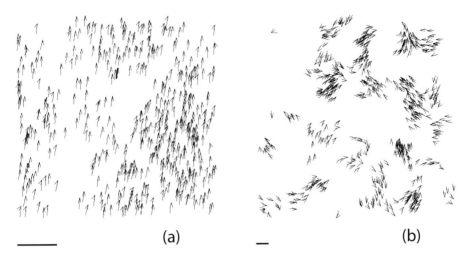

Fig. 9.13 Flocks of self-propelled particles in the Vicsek model for two values of the mean particle density. The vectors of particle velocities are indicated by small arrows, while their trajectories over the last 20 time steps are shown by short continuous curves. For comparison, the bar shows the range of the interaction. (a) At high density ($N = 300$, $L = 5$ and $\zeta = 0.1$) the motion is ordered; (b) For small density ($N = 300$, $L = 25$ and $\zeta = 0.1$) the particles form groups moving coherently in random directions, from Vicsek et al. (1995).

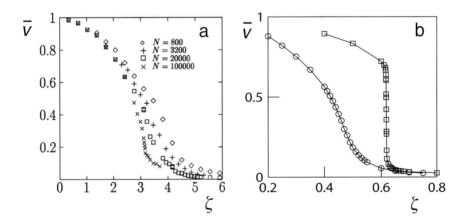

Fig. 9.14 (a) The average momentum (order parameter) of self-propelled particles in the two-dimensional Vicsek model in the steady state vs. the noise amplitude ζ for the number density $\nu = 2$ and four different system sizes [($\diamond$) $N = 800$, $L = 20$; (+) $N = 3200$, $L = 40$; ($\square$) $N = 20,000$, $L = 100$ and ($\times$) $N = 10^5$, $L = 223$], from Vicsek et al. (1995). (b) The order parameter $\bar{v}$ characterizing the onset of collective motion in the original Vicsek model (eqn 9.66) ($\bigcirc$) and modified model with vectorial noise (eqn 9.68), ($\square$) for parameter $v_0 = 0.5$, $L = 32$, $\nu = 2$, and equivalent statistics for both models, from Grégoire and Chaté (2004).

Polar, or "ferromagnetic" ordering of the flock can be characterized by the absolute value of average momentum $\bar{v}$ playing role of the order parameter (similar to the magnetization in a magnetic system):

$$\bar{v} = \left| \frac{1}{N} \sum_{i=1}^{N} \mathbf{v}_i \right|, \qquad (9.67)$$

where $\mathbf{v}_i = v_0(\cos\theta_i, \sin\theta_i)$ is the particle velocity vector. Calculations of $\bar{v}$ revealed that indeed in the range of noise amplitudes ζ there is global polar ordering of the flock characterized by non-zero value of the order parameter $\bar{v}$, see Fig. 9.14(a). On the basis of these simulations (Vicsek et al., 1995) suggested that the system undergoes the second-order phase transition with the increase of the noise amplitude ζ or decrease of the density ν. Accordingly, near the critical point the order parameter shows the following scaling

$$\bar{v} \sim (\zeta_c - \zeta)^\beta \text{ and } \bar{v} \sim (\nu - \nu_c)^\gamma,$$

with the scaling exponents $\beta = 0.45 \pm 0.07$ and $\gamma = 0.35 \pm 0.06$.

However, the nature of the continuous phase transition in the Vicsek-type model was challenged by Grégoire and Chaté (2004) and Chaté et al. (2007) who on the basis of large-scale simulations argued that in two-dimensional geometry any generic system of noisy locally interacting self-propelled particles with or without cohesion (i.e. attractive interaction between the particles) should exhibit a *discontinuous* phase transition.

To complement the original Vicsek model, Grégoire and Chaté (2004) proposed an alternative form of eqn 9.66 with a different form of noise (dubbed "vectorial noise" as opposed to the "angular noise" in the original model by Vicsek et al. (1995)) and the particle interaction rule in the equation for orientation angle θ_j:

$$\theta_j(t+1) = \arg\left[\alpha_0 \sum_{k \sim j} e^{i\theta_k(t)} + \alpha_1 \sum_{k \sim j} f_{jk}(t) e^{i\theta_{jk}(t)} + \zeta n_j(t) e^{i\xi_j(t)} \right]. \qquad (9.68)$$

Here, $n_j(t)$ is the number of neighbors of particle j at time t. The term $\sim \alpha_1$ in eqn 9.68 characterizes the additional "cohesive" force with amplitude $f_{jk}(t)$ that is repulsive up to an intermediate equilibrium distance R_e, short-range hard-core repulsive at the distance R_c and attractive up to the maximum interaction range R_0 (θ_{jk} is the direction of the vector connecting particles k and j):

$$f_{jk} = \begin{cases} -\infty & \text{if } r_{jk} < R_c \\ \frac{1}{4} \frac{r_{jk} - R_e}{R_a - R_e} & \text{if } r_c < r_{jk} < R_a \\ 1 & \text{if } R_a < r_{jk} < R_0 \end{cases}, \qquad (9.69)$$

where coefficients α_0, α_1 control the strength of alignment and cohesion, respectively.

According to Grégoire and Chaté (2004), Chaté et al. (2007), the modified model (eqn 9.68) and the original model (eqn 9.66) both exhibit a discontinuous transition to the collective motion, as illustrated by Fig. 9.14(b). The transition remains discontinuous even for $\alpha_1 = 0$ (zero cohesion) which coincides with the original Vicsek model.

Interesting behavior occurs for the non-zero value of the "cohesion" parameter α_1. For the intermediate α_1 ("liquid case"), the onset of motion is accompanied by the loss of cohesion: while small groups are moving smoothly without breaking up, larger groups gradually subdivide into several parts of roughly similar size linked by filamentary structures, as shown in Fig. 9.15. The filaments themselves are relatively static (see Fig. 9.15(d)), but they they can be displaced and broken by coherently moving subgroups.

Up to date there is no rigorous procedure that would allow one to identify the precise nature of the phase transition for the Vicsek-type models. However, one may speculate that the first-order phase transition in the Vicsek-type models is a manifestation of the ubiquitous clustering (or van der Waals) instability in the inelastic granular gas discussed in Chapter 4. As we will show below, the alignment of neighboring particles (or "following") is similar to inelastic collisions. Moreover, even in the absence of cohesion, inelastic collision often produce "apparent" attractive interaction between the particles. And, as we discussed in Chapter 4, the clustering instability, related to the bistable dependence of pressure or free energy on density, is generically subcritical, i.e. the transition is discontinuous.

9.3.2 Hydrodynamic description

As we mentioned above, it is difficult to rigorously deduce a continuum limit of the discrete-time Vicsek-type models. Here, we consider formulation and analysis of the hydrodynamic model of collective motion of self-propelled particles (Toner and Tu, 1998). Such a continuum model describing a large-scale behavior of the flock is derived on the basis of symmetry arguments. The relevant symmetries of the model are the invariance under rotation and translation. Translation invariance implies that displacing of all the particles by the same distance has no physical effect.[5] Rotation invariance means that there is no preferred direction of flock migration. However, systems of self-propelled particles should not obey Galilean invariance: indeed, most self-propelled objects in nature move in a medium on a substrate that provide a special Galilean reference frame. In the Vicsek model, since we assume that all particles maintain the same magnitude of velocity v_0 independent of their orientation, changing the velocities of all particles by some constant value does not leave the system invariant. The lack of the Galilean invariance gives rise to specific terms in the hydrodynamic equations for self-propelled particles that are absent in the Navier–Stokes equations for simple fluids.

Since we are not interested in the effects of smectic ordering in the course of motion, the hydrodynamic variables can be chosen the same as those of a simple fluid (Toner et al., 2005): the coarse-grained particle velocity field $\mathbf{v}$, and the coarse-grained density ν. Despite the fact that the hydrodynamic equations are derived from

---
[5] However, under certain conditions, translational invariance may be broken since smectic-like ordering can occur, similar to the case of vibrational transport of elongated particles with tapered ends considered in Section 8.2, see also Fig. 2.20(a).

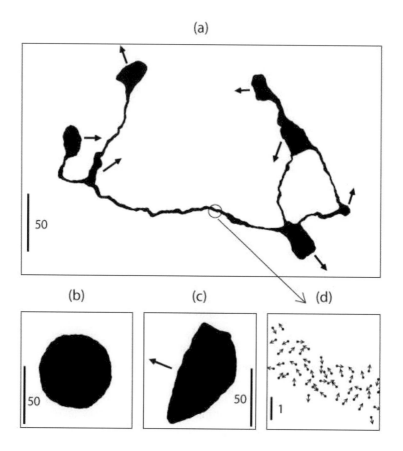

Fig. 9.15 Illustration of a typical cohesive group of 16384 particles, eqn 9.68, for $\nu = 1/16, \alpha_1 = 20, R_0 = 1, R_a = 0.5, R_e = 0.5, R_c = 0.2$ arrows indicate the direction of motion: (a) At the onset before loss of cohesion, $\alpha_0 = 1.78$; (b) Static phase, round shape, $\alpha_0 = 0.5$; (c) Moving phase, triangular shape, $\alpha_0 = 2.5$. (d) Closeup of a filament; no local ordering, from Grégoire and Chaté (2004).

symmetry arguments, one has to define "mesoscopic volume" in order to make the formal coarse-graining procedure self-consistent. This "mesoscopic" volume should be big enough to contain enough particles to make the averaging sensible, but should have a characteristic scale of no more than a few "microscopic" lengths (i.e. the interparticle distance, or the interaction range R_0).

Here, we need to comment that at least in the two-dimensional situations there is no clear separation of scales between microscopic and macroscopic length scales. In most of the examples considered in this chapter (self-organization of microtubules via molecular motors above, collective swimming of bacteria in thin films below), the scale of collective motion is of the order of 5–10 particle sizes only. Thus, the hydrodynamic description has serious limitations. This situation is very similar to the granular hydrodynamics when the restitution coefficient e is not too close to one: as we discussed earlier, in this situation formally there is no separation of scales, and granular hydrodynamics often breaks down. Additional conditions are required for the regularization of hydrodynamic equations, such as the order-parameter field considered in Section 6.2.4.

According to Toner et al. (2005), the hydrodynamic equations of motion can be written as follows (only lowest-order terms are kept):

$$\partial_t \mathbf{v} + \Lambda_1 (\mathbf{v} \cdot \nabla) \mathbf{v} + \Lambda_2 (\nabla \cdot \mathbf{v}) \mathbf{v} + \Lambda_3 \nabla |\mathbf{v}|^2$$
$$= \alpha \mathbf{v} - \beta |\mathbf{v}|^2 \mathbf{v} - \nabla p + D_b \nabla \nabla \cdot \mathbf{v} + D_T \nabla^2 \mathbf{v} + D_2 (\mathbf{v} \cdot \nabla)^2 \mathbf{v} + \mathbf{f} \quad (9.70)$$

$$p = p(\nu) = \sum_{n=1}^{\infty} \sigma_n (\nu - \nu_0)^n \quad (9.71)$$

$$\partial_t \nu + \nabla (\mathbf{v} \nu) = 0. \quad (9.72)$$

Here, ν_0 is the mean value of the local density, σ_n are expansion coefficients of the pressure p, and $D_{b,T,2}$ are diffusion coefficients. The value $\alpha > 0$ corresponds to the onset of the ordered, or in this case, oriented polar phase. The terms with $\Lambda_{1,2,3}$ are generalizations of the usual convective term for the coarse-grained field $\mathbf{v}$ in the Navier–Stokes equation. The absence of the Galilean invariance allows for all three combinations of one spatial gradient and two velocities that transform like vectors; in the Galilean-invariant case $\Lambda_2 = \Lambda_3 = 0$ and $\Lambda_1 = 1$. Parameters α and β enforce that in the ordered phase local velocity $\mathbf{v}$ approaches a non-zero asymptotic value $v_0 = \sqrt{\alpha/\beta}$. The diffusion terms describe spreading of localized perturbations due to the interaction between neighboring particles. The term $\mathbf{f}$ represents a fluctuating driving force that is assumed to be Gaussian and δ-correlated with standard deviation ζ,

$$\langle f_i(\mathbf{r}, t) f_j(\mathbf{r}', t') \rangle = \zeta^2 \delta_{ij} \delta^d(\mathbf{r} - \mathbf{r}') \delta(t - t'). \quad (9.73)$$

The pressure p tends to maintain the local density ν at its mean value ν_0. Equation 9.72 is the conservation law: it is assumed that self-propelled particles neither die nor reproduce. While this assumption is natural for inanimate particles such as vibrated rods, this is not always a good approximation for the biological objects. In the Section 9.1 we considered the situation when large-scale motion in a colony of non-motile *E. coli* bacteria arose solely due to the growth and division of cells.

Let us emphasize that despite generality, eqns 9.70–9.72 still do not capture all variety of collective motion occurring in systems of self-propelled particles. Depending on the physical situation, additional coupling to substrate or surrounding fluid may need to be taken into account, resulting in problem-specific instabilities, such as for example in the system of vibrated anisotropic grains (Section 8.2) or swimming bacteria that will be considered later in this chapter. One may also argue that velocity $\mathbf{v}$ is not always the most convenient hydrodynamic variable. In certain situations, the coarse-grained local orientation $\boldsymbol{\tau} = \langle \mathbf{n} \rangle$ can be more convenient for the theoretical analysis. As we will show in Section 9.4 in the case of collective motion in colonies of swimming bacteria, the relation between local orientation $\boldsymbol{\tau}$ and local velocity $\mathbf{v}$ can be rather non-trivial.

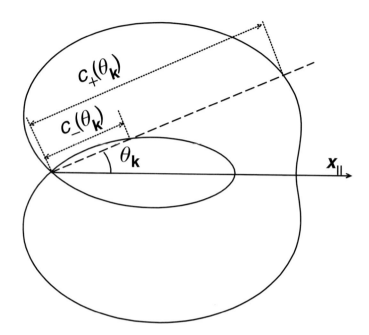

Fig. 9.16 Anisotropic sound propagation through an ordered system with non-zero mean orientation: a polar plot of the direction-dependent sound velocities $c_\pm(\theta_{\mathbf{k}})$, with the horizontal axis the along mean direction of motion $\mathbf{x}_\|$, from Toner and Tu (1998).

9.3.3 Linear stability of oriented state

An ordered oriented (or polar) state in eqns 9.70–9.72 exists for $\alpha > 0$. Following Toner et al. (2005), to examine the linear stability of the homogeneous (or uniform) oriented state, we can write the velocity field as $\mathbf{v} = v_0 \mathbf{x}_\| + \delta \mathbf{v}$, where $v_0 \mathbf{x}_\| = \langle \mathbf{v} \rangle$ and $v_0 = \sqrt{\alpha/\beta}$ is an average value of $\mathbf{v}$ in the ordered phase. We assume that in the oriented state the density variations near the uniform density ν_0 are small, $\delta\nu \equiv \nu - \nu_0 \ll \nu_0$. Then we can keep only one term in the pressure expansion (eqn 9.71): $p \approx \sigma_1(\nu - \nu_0)$, where σ_1 is the first pressure expansion coefficient.

The analysis (see Toner et al. (2005)) shows that there are two main modes of propagation of small periodic perturbations with wavevector $\mathbf{k}$ that are characterized by two different "sound" velocities $c_\pm$

$$c_\pm(\theta_{\mathbf{k}}) = \frac{\gamma + v_0}{2} \cos(\theta_{\mathbf{k}}) \pm c_2(\theta_{\mathbf{k}}), \tag{9.74}$$

where

$$c_2(\theta_{\mathbf{k}}) = \sqrt{\frac{(\gamma - v_0)^2 \cos^2(\theta_{\mathbf{k}})}{4} + \sigma_1 \nu_0 \sin^2(\theta_{\mathbf{k}})}, \tag{9.75}$$

and $\theta_{\mathbf{k}}$ is the angle between $\mathbf{k}$ and the direction of motion (i.e. the $\mathbf{x}_\|$-axis), $\gamma = \Lambda_1 v_0$. The sound propagation, described by eqn 9.74 is highly anisotropic, see for illustration Fig. 9.16. Both these modes are decaying, and, therefore, the solution to the linearized equations does not show any linear instability of the homogeneous oriented state. However, the long-range orientational order of the homogenous oriented state can be broken by fluctuations, at least in the case of the system dimension $d_s \leq 2$. To see that, one needs to calculate the mean-squared fluctuations of velocity $\langle \mathbf{v}_\perp^2 \rangle$ at a given point $\mathbf{r}$ and time t. The calculations yield (Toner et al., 2005)

$$\langle \mathbf{v}_\perp^2 \rangle = \int \frac{d^{d_s}\mathbf{k}}{(2\pi)^{d_s}} \langle v_i(\mathbf{k},t) v_i(-\mathbf{k},-t) \rangle$$

$$= \frac{\zeta^2}{2} \int \frac{d^{d_s}\mathbf{k}}{(2\pi)^{d_s}} \left(\frac{d_s - 2}{D_T \mathbf{k}_\perp^2 + D_\| \mathbf{k}_\|^2} + \frac{\tilde{\Phi}}{D_L \mathbf{k}_\perp^2 + D_\| \mathbf{k}_\|^2} \right). \quad (9.76)$$

Here, the factor $\tilde{\Phi}$ depends only on the direction $\mathbf{k}$, not its magnitude, and $D_\nu = \nu_0 \sigma_1/2\alpha$, D_b, D_T, $D_L = D_T + D_b$, and $D_\| = D_T + D_2 v_0^2$ are the corresponding diffusion constants. The complicated expression for $\tilde{\Phi}$ is given in Toner and Tu (1998); it suffices to note that $\tilde{\Phi}$ is a smooth, analytic function of $O(1)$ and it is non-vanishing for all $\mathbf{k}$. Note that the last integral in eqn 9.76 diverges at $|\mathbf{k}| \to \infty$ for $d_s \geq 2$ and at $|\mathbf{k}| \to 0$ for $d_s \leq 2$.

The divergence of the integral at $|\mathbf{k}| \to \infty$ ("ultraviolet" catastrophe) is artificial since the hydrodynamic theory apply for $|\mathbf{k}|$ larger than the inverse of the interaction length R_0 or the size of the particle. However, the "infra-red" divergence ($|\mathbf{k}| \to 0$) for $d_s \leq 2$ in eqn 9.76 is more serious since the hydrodynamic theory should be valid for small wave numbers $\mathbf{k}$, i.e. for slowly-varying modulation of the homogenous state. This divergence seemingly suggests that the long-range order for $d_s \leq 2$ is impossible since it is destroyed by arbitrarily small random fluctuations. More specifically, in two dimensions the integral in eqn 9.76 diverges logarithmically, and the original assumption that in infinite system the hydrodynamic velocity $\mathbf{v}$ can be written as a mean value $\langle \mathbf{v} \rangle$ plus a small fluctuation $\delta \mathbf{v}$ should be erroneous; the divergence of $\langle \mathbf{v}_\perp^2 \rangle$ implies that the velocity can sample through all possible directions, i.e. $\langle \mathbf{v} \rangle = 0$ for $d_s \leq 2$ if the system is sufficiently large. This conclusion based on the linear stability analysis agrees with the well-known Mermin–Wagner theorem (Mermin and Wagner, 1966) of the equilibrium statistical mechanics which states that in systems with $d_s \leq 2$ at thermal equilibrium and local interactions continuous symmetries cannot be spontaneously broken, and so a true long-range order is impossible. However, it apparently contradicts the ample numerical evidence that strongly suggests the existence of the true long-range order for sufficiently high density or small noise. The only rational solution of this apparent paradox is that the non-equilibrium nature of the system of self-propelled particles does not reveal itself in the linear-order and can only be captured in the framework of the full non-linear hydrodynamics. Indeed, a straightforward scaling analysis (Toner et al., 2005) shows that in a system driven by white noise, non-linear terms cannot be neglected for $d_s \leq 4$.

To obtain insights *beyond* the linear stability of the homogenous state, Toner and Tu (1998) assumed scale invariance and applied the renormalization group (RG) technique to full non-linear eqns 9.70–9.72. Based on this approach they showed that the diffusion constants themselves become wave number-dependent and in fact divergent at $|\mathbf{k}| \to 0$. This divergence negates the infra-red divergence of the velocity fluctuations and "stabilizes" the long-range order. Using RG-analysis Toner and Tu (1998) computed exactly all dynamical scaling exponents characterizing a transition to ordered state in $d_s = 2$. However, in light of the recent numerical work of Grégoire and Chaté (2004) the specific results of the RG analysis should be taken with a grain of salt: if the transition to long-range order is indeed discontinuous, one should not expect scale invariance to hold. Still, one thing is clear: it is the non-equilibrium nature of the system of self-propelled particles that works through non-linear terms and stabilizes the long-range order in $d_s = 2$ in a seeming violation of the Mermin–Wagner theorem. The theoretical question of the precise nature of the phase transition in the framework of full non-linear hydrodynamical description 9.70–9.72 is still open.

In fact, the nature of transition may depend crucially on details of inter-particle interactions. Dynamic instabilities arising in more specific models of self-propelled particles (e.g. due to long-range interactions, coupling to a substrate or other fields) may render this fluctuation-driven mechanism of the breakdown of long-range order irrelevant: dynamic instabilities can destroy the long-range order even in the absence of random fluctuations. This was, for example, the case in spontaneous swirl formation in the system of vibrated anisotropic grain considered in Section 8.2. Further insights into the collective dynamics of self-propelled particles are obtained from non-linear theories and simulations.

9.4 Collective swimming of motile bacteria in thin fluid films

Large assemblies of biological objects often develop a high degree of interaction that is expressed in a collective motion of the whole colony in a certain direction. This alignment occurs because each organism adjusts its motion according to the motion of its neighbors. In the previous section, we considered generic models describing flocking transition. However, specific systems have features that make their dynamics non-generic. Here we consider collective swimming of bacteria in a thin fluid film. Swimming bacteria are possibly the simplest example of self-propelled "bioparticles" driven by the input of mechanical energy through rotation of the helical flagella by motors in the cell membrane. In experiments on swimming bacteria *E. coli* in soap-like thin fluid films by Wu and Libchaber (2000) the increase of the number density of micro-organisms lead to small whorls and jets of co-operative swimming which in turn produced greatly enhanced diffusion and even superdiffusion of passive tracers, as is illustrated in Fig. 9.17.

More recent bacterial experiments in the restricted geometry of a thin film (Dombrowski *et al.*, 2004; Sokolov *et al.*, 2007) and simulations of "self-swimming rods" (Hernandez-Ortiz *et al.*, 2005) have demonstrated that the correlation length of collective motion can exceed the size of individual cells by more than an order of magnitude, and the collective flow speeds (of the order 50-100 µm/sec) similarly exceed the speed of individual bacteria (about 15-20 µm/sec). This collective swimming is found for sufficiently high number density (or volume fraction) of microorganisms; dilute suspensions show no collective flow and the correlation length is comparable to the bacterium size. One of the surprising observations is that the collective motion of bacteria is seemingly turbulent (Figure 9.18), despite the fact that at such low Reynolds numbers (for an individual microorganism of size $l = 5$ µm swimming with the speed

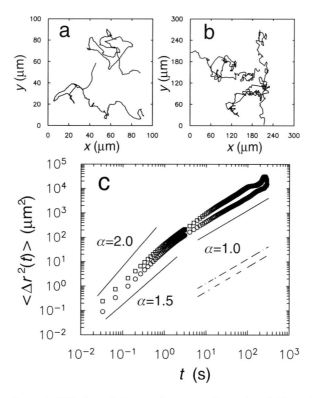

Fig. 9.17 The enhanced diffusion of tracers in a two-dimensional film with bacteria. (a) Typical 20-s trajectories of two 10-μm polystyrene beads; (b) Longer 3-min trajectories of same beads; (c) The mean square displacements measurements of polystyrene beads with diameters 4.5 (squares) and 10 μm (circles). The bead concentrations were $\approx 10^{-3}$ μm^{-2}. The solid lines with slopes $\alpha = 2.0, 1.5$, and $\alpha = 1.0$ are shown as guides to the eyes. The two dashed lines correspond to the expected thermal diffusion of 4.5 μm (upper line) and 10 μm (lower line) particles. Anomalous diffusion with the exponent $\alpha \approx 3/2$ occurs on early times, whereas long-time diffusion of beads appears normal, from Wu and Libchaber (2000).

$U = 20$ μm/sec in water the Reynolds number is Re $\approx 10^{-4}$) the conventional hydrodynamic turbulence can be completely ruled out. These studies clearly identified the hydrodynamic interactions between bacteria as the primary origin of collective flows.

9.4.1 Transition to a collective swimming

The first experiments by Wu and Libchaber (2000) and Dombrowski et al. (2004) demonstrated a transition to collective motion with the increase of the density of swimmers. This transition was signaled either by the increased tracer diffusion or by the direct observation of a large-scale vortex motion using particle-image velocimetry technique (PIV). However, beyond qualitative observations of the transition, the quantitative characterization of the spatiotemporal correlations and their dependence on the concentration of microorganisms was difficult because the number density of cells

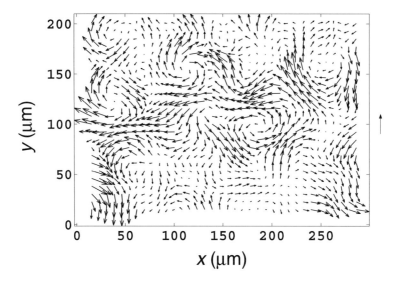

Fig. 9.18 Turbulent motion of dense bacterial suspensions of *Bacillus subtilis* in the two-dimensional geometry of a flat pendant drop. The image shows flows at the bottom of the pendant drop. Instantaneous bacterial swimming vector field is shown. The arrow at the right denotes a speed of 35 μm/s, the size of an individual bacterium is about 5 μm, from Dombrowski *et al.* (2004).

(the main control parameter responsible for the onset of collective motion) was hard to control, furthermore, it had a tendency to change in the course of the experiment. This difficulty was resolved by Sokolov *et al.* (2007) who studied the collective bacterial swimming in a thin-film geometry with a novel technique capable of adjusting and maintaining the number density of bacteria in a course of the experiment. This technique allowed the correlation length and the mean swimming speed to be measured over a range of densities within a single bacterial colony, greatly reducing statistical fluctuations due to inherent physiological differences among colonies. In contrast with the earlier theoretical and numerical work (Toner *et al.*, 2005; Grégoire and Chaté, 2004) discussed in Section 9.3, only a gradual increase of the correlation length with increasing density was found, and no sharp transition was observed. The absence of the sharp transition can be possibly explained as a noise-induced smearing of a dynamical phase transition, the main source of noise being strong fluctuations in the orientation of individual bacteria, as suggested earlier by Wu and Libchaber (2000). However, a more fundamental difference between the models and this experiment is that the first-order phase transition was predicted for systems of particles with short-range interactions (Vicsek-type models), whereas bacteria interact through long-range hydrodynamic forces.

Experiments by Sokolov *et al.* (2007) were conducted in suspensions of *Bacillus subtilis*, rod-shaped bacteria ∼4–5 μm long and with a diameter of ∼0.7 μm. The experimental setup was based on an open-air free-standing thin-film geometry similar to that used by Wu and Libchaber (2000), but with a number of important modi-

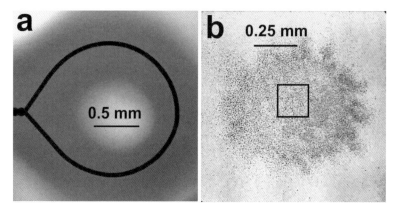

Fig. 9.19 Dynamic concentration of bacteria in free-standing thin-film sample. (a) Image illustrating decrease of pH near the electrodes as a result of transmission of electric current (darker regions correspond to lower pH value). (b) Concentrated bacteria (darker part of the image). Square indicates the field of view of the microscope, from Sokolov et al. (2007)

fications. A small drop of bacterial suspension was placed between four supporting, movable fibers: two platinum wires (to exclude contamination of the film) and two dielectric fibers. The drop was then stretched between the fibers up to the necessary thickness (about 1 μm) by pulling all the fibers apart with a control screw. In order to adjust the density of bacteria ν (or the "filling fraction") in the course of one experiment, a small platinum ring (diameter $\sim$ 1 mm) was gently lowered onto the stretched film containing the bacteria, and a small voltage ($\leq$ 2.3 V) was applied between the ring and the two platinum wires, thus creating electrolysis. The electrolysis caused a change of the pH level in the vicinity of the electrodes, which was monitored by the addition of the indicator fluid bromothymol blue to the solution. The distribution of the pH levels in the film is shown in Fig. 9.19(a), in which darker regions, corresponding to lower pH levels, expand on both sides of the ring after the application of electrical current. This expansion is due to the ionic diffusion. The change of pH in turn triggers a chemotactic response in bacteria: bacteria tend to swim away from the electrodes, towards the area of a more comfortable pH level ($\sim$ 7.2) in the middle of the ring, see Fig. 9.19(b). Conveniently, this technique stimulated a response only from motile bacteria; dead or otherwise non-motile cells remain behind. With this method it was possible to change dynamically the number density of bacteria by more than a factor of 5. To avoid the effects of the boundaries, images of the collective state were acquired from a small area 230 ×230 μm^2 in the middle of the ring restricted by the box in Fig. 9.19(b).

Experiments show that at low densities bacteria swim independently, however, after collisions bacteria tend to align their directions of swimming, see Fig. 9.20. This process is very similar to the "inelastic collision" considered before in the context of alignment of microtubules by molecular motors, compare with Fig. 9.5. The underlying reason for the inelastic collisions between the bacteria is the quadrupole nature of the bacterium velocity field (for the discussion of the flow field of individual swimmers see

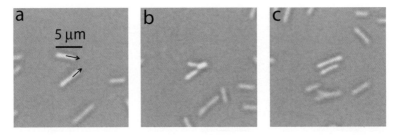

Fig. 9.20 Sequence of experimental images illustrating "inelastic collisions" between two swimming bacteria *Bacillus Subtilis* at low densities. Colliding bacteria (highlighted) swim from left to right: (a) before the collision; (b) collision; (c) after collision, from Aranson *et al.* (2007a)

Section 9.4.2). As we will show later, the inelastic collisions are a precursor for the onset of large-scale correlations and ultimately lead to the collective motion at higher densities.

Several representative images of bacterial patterns for different filling fractions (or number density) ν,[6] illustrating collective swimming are shown in Fig. 9.21. The corresponding velocity field **V** is extracted from consecutive images by particle imaging velocimetry with bacteria themselves serving the role of tracers. This technique has a resolution of about 2 μm/s, i.e. sufficient to resolve collective swimming patterns. As one sees in Fig. 9.21, large-scale whorls and jets of collective bacterial swimming are indeed observed with the increase of the number density ν.

In addition to the velocity field, the apolar bacterial director field **n** collinear with the vector of *orientation* τ was measured.[7] We can characterize the alignment between the bacterium orientation and alignment by the alignment coefficient C that is defined as follows. If we arbitrarily choose the unique orientation of the bacterium from the condition $(\tau \cdot \mathbf{V}) > 0$, the alignment coefficient can be defined as

$$C = \frac{\langle \cos \varphi \rangle - 2/\pi}{1 - 2/\pi}, \qquad (9.77)$$

where φ is the angle between **V** and τ (by definition, angle φ varies within $[0, \pi]$). For uncorrelated velocity and director fields, parameter C should be zero.

In the regime of well-developed large-scale flow in high-density suspensions, such as shown in Fig. 9.21(b), the alignment coefficient was rather small, $C < 8\%$. However, this seeming contradiction with the fact that bacteria swim in the direction of their orientation is easily resolved. Indeed, in a well-developed, chaotic flow the bacteria are advected by the large-scale fluid velocity field $\mathbf{v}_f$ created by *other* bacteria. Each bacterium swims in the direction of its orientation only in the *local frame* moving with coarse-grained fluid velocity $\mathbf{v}_f$, hence one should not expect a strong global correlation between **V** and τ as long as the fluid velocity field is chaotic.

---

[6]The number density $0 < \nu < 1$ is measured as a fraction of the full surface covered by bacteria.

[7]The "orientation" of the bacterium in principle can be determined by the location of the attachment point of its helical flagellum. However, observation of flagella is far beyond the resolution of optical microscopy.

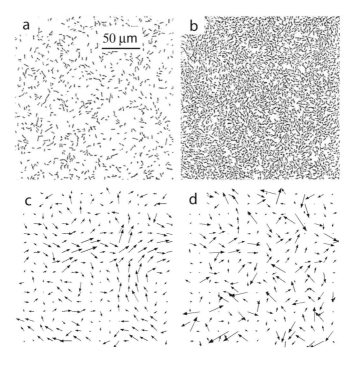

Fig. 9.21 Representative patterns of collective swimming for various densities: (a) Low density $\nu = 0.14$, no collective swimming; (b) High density $\nu = 0.47$, well-developed chaotic large-scale flows. Vector fields for velocity **V** (c) and orientation τ (d) for the high-density regime in image (b), from Sokolov *et al.* (2007)

The experimental technique described above allows one to systematically vary the bacterial density ν and measure various physical quantities as functions of ν. In particular, the typical (root mean squared) fluid velocity $\bar{V} = \sqrt{\langle \mathbf{V}^2 \rangle - \langle \mathbf{V} \rangle^2}$, and the radial velocity correlation functions $K(r)$, defined as

$$K(r) = \int d\mathbf{r}' \int_0^{2\pi} d\theta \left[\langle \mathbf{V}(\mathbf{r}') \cdot \mathbf{V}(\mathbf{r} + \mathbf{r}') \rangle - \langle \mathbf{V}(\mathbf{r}') \rangle^2 \right], \tag{9.78}$$

(θ is a polar angle) were measured during a single experiment in which the density was gradually increased. The corresponding measurements of the correlation length L and typical velocity $\bar{V}$ are shown in Fig. 9.22. The shape of the correlation functions $K(r)$ (not shown) and the value of the correlation length L are in agreement with previous measurements by Dombrowski *et al.* (2004).

The dependencies of correlation length L and velocity $\bar{V}$ shown in Fig. 9.22, feature no sharp transition with increasing density; only a smooth (although steep) increase of the velocity $\bar{V}$ and the correlation length L is observed. The overall changes in these quantities were about a factor of five. For even higher density ($\nu \to 1$, i.e. close to 90% surface coverage) jamming and complete termination of motion was observed.

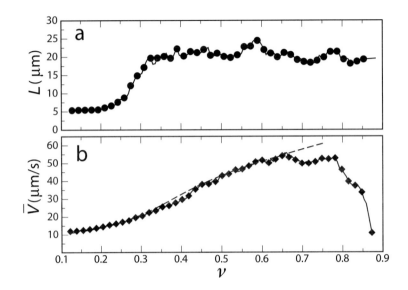

Fig. 9.22 Velocity correlation length L (a) and typical velocity $\bar{V}$ (b) and the functions of bacterial number density ν. Dashed line shows fit of $\bar{V}$ to solution to eqn 9.79 with $\nu_c \approx 0.28$.

These results differ from the predictions of simplified theories of collective motion in systems of self-propelled particles discussed in Section 9.3. The most important difference is of course the presence of the interstitial fluid that provides long-range interaction among particles and can fundamentally change the nature of the phase transition. In Section 9.4.3 we will discuss a mathematical model of collective swimming. In the absence of noise the model indeed exhibits a second-order phase transition, with a divergence of the correlation length L at the critical point. However, for strong enough noise, the transition smears, and only a smooth increase of L with density ν is observed, in qualitative agreement with these experiments. To support this conclusion, the experimental data for $\bar{V}$ vs. ν can be compared with those obtained from the normal form equation for a generic noisy second-order phase transition

$$\partial_t V = (\nu - \nu_{\text{cr}})V - V^3 + \xi(t), \qquad (9.79)$$

where ξ is Gaussian white noise with the intensity ζ, and ν_{cr} is a critical density. In the absence of noise one obtains the mean-field result $\bar{V} \sim \sqrt{\nu - \nu_{\text{cr}}}$. With noise, one obtains smearing of the transition near ν_{cr}. Figure 9.22 shows that the fit of a solution to eqn 9.79 is consistent with the experimental data shown in Fig. 9.22. In the experiments with swimming bacteria, the noise can arise from a number of sources, such as spontaneous orientation fluctuations of individual bacteria due to tumbling, small-scale hydrodynamic fluctuations due to flagellum rotation, size distribution of bacteria, etc.

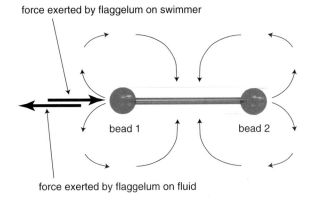

Fig. 9.23 Schematic representation of the dumbbell model of a swimmer. The flagellum is represented by a force exerted on one of the beads of the dumbbell, and a force in the opposite direction exerted by the dumbbell on the fluid. Thin arrows show the hydrodynamic velocity field generated by the moving dumbbell. The case $p = 1$ (pusher) is shown. For the case of puller ($p = -1$) the direction of all arrows is reversed.

9.4.2 Simulations of collective swimming

Detailed simulations of large populations of confined hydrodynamically interacting swimming particles at low Reynolds number were performed by Hernandez-Ortiz et al. (2005). These simulations used a "minimal swimmer model" to capture in the leading order far-field effects of hydrodynamic interactions among individual swimmers without specifying the detailed structure of the swimmer and its associated hydrodynamic flow.

An individual swimmer is modelled by a rigid, neutrally buoyant dumbbell comprised of two beads of radius a connected by a rigid rod of length l with unit orientation vector $\mathbf{n}$, see Fig. 9.23. The viscous drag on the swimmer is concentrated on the two beads. The propulsion is modelled by a "flagellum", which exerts a constant force of magnitude f in the $\mathbf{n}$ direction on one of the beads; an equal and opposite force is applied to the fluid. Depending on the sign of force f, the flagellum can either push or pull on the dumbbell; the "pushing" force corresponds to most spermatozoa and swimming bacteria such as *E. coli* and *Bacillus subtilis*, whereas some types of algae microorganisms (e.g. *Chlamydomonas*) can be modelled by the "pulling" force. The distinction between pulling and pushing cases is characterized by the polarity parameter p of the flagellar force: if $p = 1$, the flagellum pushes the swimmer; if $p = -1$, it pulls. Many micro-organisms, including *E. coli*, execute a complex "run-and-tumble" behavior, during which they change directions at random intervals corresponding to random switching of the polarity parameter p, whereas some other swimming bacteria, as *Bacillus subtilis*, practically don't tumble and are well described by the constant polarity case $p = 1$. To avoid these organism-specific effects, simulations of Hernandez-Ortiz et al. (2005) were performed for the simple pusher case $p = 1$ and under the assumption that in an unbounded domain each swimmer moves in a straight line with constant speed $v_{\text{sw}} = \frac{f}{2\eta_s} + O(a/l)$, where η_s is the Stokes viscous drag coefficient.

Far away from the domain boundaries swimming motion exerts no net force on the fluid, so in the far field the swimmer appears to be a moving symmetric force dipole (so-called stresslet) (Pedley and Kessler, 1992). The flow velocity field near the pusher has a characteristic quadrupole form as shown in Fig. 9.23. This particular structure of the pusher velocity field (the flow is directed towards the middle point of the rod connecting two beads) possibly explains the tendency of two close pushers to approach each other and align, as observed experimentally, see Fig. 9.19. In contrast, for the case of pullers, two nearby swimming objects tend to form "tail-to-head" state.

Near the boundaries the torque balance on the swimmer also needs to be considered. The torques may result in rather non-trivial effects, such as clockwise or anticlockwise circulation of swimmers near a rigid wall, as was observed in experiments with swimming sperm cells by Riedel et al. (2005), see Fig. 2.28, and with E. coli by DiLuzio et al. (2005).

Multiswimmer simulations (Hernandez-Ortiz et al., 2005) were performed in a three-dimensional rectangular slit $L_x \times L_y \times 2H$ box, periodic in x- and y-directions and confined by rigid walls in the z-direction. The interaction between swimmers was provided purely by the low Reynolds number hydrodynamics of the solvent, direct contacts between particles were ignored. The simulations confirmed that for large enough concentration of swimmers, the latter organize in collective swimming patterns of whorls and jets closely resembling experimental ones, see Fig. 9.24(a). Subsequently, Underhill et al. (2008) modified the computational model by adding steric repulsion among the particles to model the excluded volume interactions. While on the qualitative level the conclusions of the original work did not change, some significant differences were observed in the flow patterns, as exemplified in Fig. 9.24(b). The flow structure in the middle of the slab becomes more "three-dimensional", with well-pronounced sinks and courses. This is possibly explained by the fact that with modified interaction swimmers tend to concentrate near the walls of the slit for almost all concentrations, see Fig. 9.25(a). At least in small samples, the mean root square velocity of both tracers and swimmers increases monotonously with the number of particles, and no sharp transition is observed, see Fig. 9.25(b).

Thus, simulations of self-propelled swimmers clearly demonstrated that the underlying mechanism of large-scale motion is hydrodynamic long-range interactions rather than chemotaxis.[8] However, details of the flow pattern, the values of transport coefficients, and the nature of the transition to collective swimming appear to depend sensitively on the type of assumptions and model approximations.

9.4.3 Continuum description of collective motion

Experiments with swimming bacteria discussed above demonstrated the significance of self-induced hydrodynamic flows produced by flagella of swimming bacteria. In most of the experiments the characteristic scale of collective flows typically exceeded the size of individual organisms by at least an order of magnitude. However, despite significant information gained in experiments and numerical simulations, the underlying

---

[8]Chemotactic interaction resulting in direct motion of micro-organisms in response to concentration gradients of certain chemicals in their environment, is relevant for slow "crawling" bacteria on substrate (Keller and Segel, 1971).

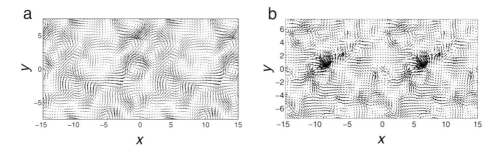

Fig. 9.24 Representative snapshots of the velocity field at the midplane of the slit for high swimmer density and $2H = 5$. Two periods in the x-direction and one period in the y-direction are shown. Swimmer length is $l = 1$, and polarity $p = 1$ (pushers). (a) Pure hydrodynamic interaction between the swimmers, from Hernandez-Ortiz et al. (2005); (b) Additional steric repulsion is added to the hydrodynamic forces, from Underhill et al. (2008)

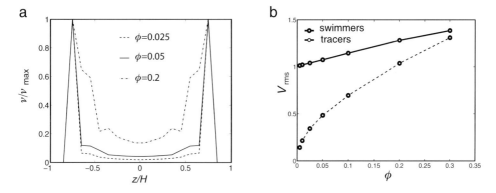

Fig. 9.25 (a) Concentration profiles ν vs. scaled wall-normal position z for different effective volume fractions ϕ. The volume fraction is defined as $\phi = U_m \nu$, ν is the concentration of swimmers, and $U_m = \pi l^3/6$ is the volume of a ball with the diameter l, from Underhill et al. (2008); (b) Mean square root velocity $V_{\rm rms}$ of swimmers and tracers vs. effective volume fraction ϕ, courtesy of Michael Graham.

mechanism of pattern formation in this system still requires theoretical understanding. In particular, the origin of a non-trivial characteristic scale of the bacterial flows and its dependence on system parameters remains unclear. The phenomenological models of interacting self-propelled particles discussed in Section 9.3 do not account for the long-range hydrodynamic interactions and therefore have only limited applicability to swimming bacteria.

In this section we consider the continuum model governing self-organization of ensembles of hydrodynamic self-propelled particles. The model is formulated in terms of a two-dimensional stochastic master equation for the probability density $P(\mathbf{r}, \varphi, t)$ of finding a bacterium at a certain orientation angle φ at the position $\mathbf{r}$ at time t derived from microscopic interaction rules. This model is analogous to that used for the description of self-organization of microtubules and molecular motors in Section 9.2, however, the underlying microscopic dynamics are different because of the self-propelled nature of bacteria. The master equation provides a link between the microscopic scale (individual bacteria) and the macroscopic scale (collective motion), which is described by the set of equations for local bacteria density and orientation. The macroscopic equations are obtained by coarse graining of the master equation. Since the bacteria swim in a viscous liquid, the system is coupled to the two-dimensional Navier–Stokes equation for the fluid flows with an additional forcing term due to bacteria swimming.

As in the case of microtubules, we approximate bacteria by identical polar rods of length l and diameter d_0. The model is based on the following elementary interaction rules:

- individual bacteria swim with velocity v_0 with respect to the ambient fluid in the direction given by its unit orientation vector $\mathbf{n} = (\cos\varphi, \sin\varphi)$;
- in the case of a "collision", two bacteria with the angles $\varphi_{1,2}$ continue to swim in the direction of the average orientation $\bar\varphi = (\varphi_1 + \varphi_2)/2$, i.e. we assume a fully inelastic collision. This assumption is justified by experimental observations, see Fig. 9.20. As we mentioned earlier, the underlying reason for the inelastic collisions between the bacteria is the quadrupole nature of the velocity field, see Fig. 9.23.
- the bacteria experience rotational and translational diffusion, mostly due to tumbling and small-scale hydrodynamic flows;
- the bacteria are advected and reoriented by the fluid flow with velocity $\mathbf{v}$.

The master equation describing the evolution of the probability $P(\mathbf{r}, \varphi, t)$ can be written in the form

$$\partial_t P + \nabla \cdot [(v_0 \mathbf{n} + \mathbf{v}) P] + \frac{\Omega}{2} \partial_\varphi P = D_r \partial_\varphi^2 P + \partial_i D_{ij} \partial_j P$$
$$+ \int\int d\mathbf{r}_1 d\mathbf{r}_2 \int_{-\pi}^{\pi} d\varphi_2 W(\mathbf{r}_1, \mathbf{r}_2) P(\mathbf{r}_1, \varphi_1) P(\mathbf{r}_2, \varphi_2) [\delta(\bar{\mathbf{r}} - \mathbf{r}, \bar\varphi - \varphi)$$
$$- \delta(\mathbf{r}_2 - \mathbf{r}, \varphi_2 - \varphi)] - \kappa_\gamma \left(\dot\gamma \cdot \mathbf{n} \cdot \frac{\partial P}{\partial \mathbf{n}} \right), \qquad (9.80)$$

(compare with eqn 9.38, the main difference here is self-transport and coupling to the self-generated hydrodynamic flow). The second and third terms in the lhs of eqn 9.80 account for the hydrodynamic advection of bacteria and their rotation by the flow with velocity $\mathbf{v}$ and vorticity $\Omega = (\partial_y v_x - \partial_x v_y)\hat{\mathbf{z}}$. The first two terms in the rhs of eqn 9.80 describe angular and translational diffusion of rods with the diffusion tensor $D_{ij} = D_\| n_i n_j + D_\perp(\delta_{ij} - n_i n_j)$. The expressions for $D_\|, D_\perp$ are given by eqn 9.40 where the thermodynamic temperature T is replaced by the effective temperature T_e. The effective temperature T_e arises from small-scale hydrodynamic flows and bacterial tumbling, and can considerably exceed the thermodynamic temperature. The last term in eqn 9.80 describes the coupling to the strain rate tensor $\dot\gamma$ ($\dot\gamma_{xy} = (\partial_x v_y + \partial_y v_x)/2$) (Kruse et al., 2004; de Gennes and Prost, 1993). While this term has some quantitative effect, e.g. on the instability threshold, it does not change the qualitative conclusions.

Thus, for simplicity one can set the coupling constant $\kappa_\gamma = 0$ and neglect the contribution of $\dot\gamma$. The integral term in the rhs of eqn 9.80 describes short-range binary interactions of rods, compare with the collision kernel (eqn 9.43) discussed in the context of cytoskeleton formation. The interaction kernel W is localized in space, and for the sake of simplicity we neglect the anisotropy of the kernel, which was essential for self-organization of microtubules, and use the following expression

$$W = \frac{W_0}{b^2\pi} \exp\left[-\frac{(\mathbf{r}_1 - \mathbf{r}_2)^2}{b^2}\right], \qquad (9.81)$$

where b and W_0 are the interaction cross-section and strength, respectively. This form implies that only nearby bacteria interact effectively. While hydrodynamic interactions usually decay algebraically (e.g. inversely proportional to the separation distance in three dimensions), exponential screening for shear perturbations can occur, for example, in the thin-film geometry when the film is in contact with a frictional substrate. For the free-standing thin-film geometry of, one could expect the free-slip boundary condition for the velocity on the top and the bottom surface of the film. However, a layer of surfactant (produced by bacteria themselves) quickly accumulates on both top and bottom surfaces of the film, and it plays the role of a frictional substrate, effectively implying the *non-slip* condition for the hydrodynamic velocity.

Following the analysis developed in Section 9.2.2 we introduce the coarse-grained density $\nu(\mathbf{r},t) = \int_{-\pi}^{\pi} d\varphi P$ and orientation $\boldsymbol{\tau}(\mathbf{r},t) = (1/2\pi)\int_{-\pi}^{\pi} d\varphi\, \mathbf{n} P$. The *spatially homogeneous* regime exhibits the onset of an oriented state above the critical density $\nu_c = (D_r/W_0)/(4/\pi - 1)$. Near this threshold, eqn 9.80 can be simplified by means of a standard bifurcation analysis, yielding a pair of coupled equations for ν and $\boldsymbol{\tau}$. Also, near the threshold, P depends slowly on the variable $\mathbf{r}$, so we keep only leading terms in the expansion in spatial gradients of the Fourier expansion of eqn 9.80 in φ, truncated at second order. Since the derivation is mostly similar to that considered in Section 9.2.4 in the context of microtubules, we skip the details. With rescaling $\mathbf{r} \to \mathbf{r}/l$, $t \to D_r t$, and $\nu \to W_0 \nu/D_r$, one obtains (compare eqns 9.44 and 9.45)

$$\partial_t \nu + \boldsymbol{\nabla}\cdot(\nu\mathbf{v}) = D_0\nabla^2\nu - v_0\pi\boldsymbol{\nabla}\cdot\boldsymbol{\tau} \qquad (9.82)$$

$$\partial_t\boldsymbol{\tau} + \mathbf{v}\cdot\boldsymbol{\nabla}\boldsymbol{\tau} + \frac{1}{2}\boldsymbol{\Omega}\times\boldsymbol{\tau} = (\epsilon\nu - 1)\boldsymbol{\tau}$$
$$-A_0|\boldsymbol{\tau}|^2\boldsymbol{\tau} + D_1\nabla^2\boldsymbol{\tau} + D_2\boldsymbol{\nabla}\boldsymbol{\nabla}\cdot\boldsymbol{\tau} - \frac{v_0}{4\pi}\boldsymbol{\nabla}\nu. \qquad (9.83)$$

Equation 9.82 describes advection of bacteria by the hydrodynamic velocity $\mathbf{v}$ and diffusive spreading with the diffusion coefficient D_0. Here $D_1 = (D_\| + D_\perp/2)/2D_r l^2$, $D_2 = (D_\| - D_\perp)/2D_r l^2$. In the rigid rods limit $D_1 = 5/192, D_2 = 1/96$. For small density ν and for the case of pure *thermal* diffusion of particles, the diffusion coefficients obey $D_0 = (D_\| + D_\perp)/2D_r l^2$. In the present context, this connection is not clear, especially for larger densities due to diffusive-type contributions from the collision integral in eqn 9.80. In experiments, there are no significant density fluctuations observed, so we treat $D_0 \gg D_{1,2}$ as an independent parameter in order to suppress

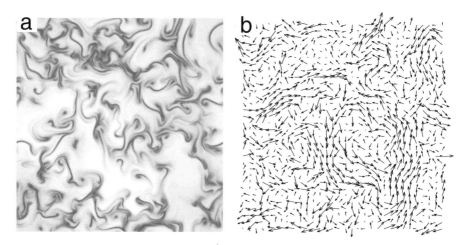

Fig. 9.26 Representative flow patterns obtained by solution of eqns 9.82, 9.83 and 9.85 for $\nu_0 = 3.8, D_0 = 50, \nu = 3, v_0 = 0.2, \alpha = 3, \beta = 0.5$, no noise, in the periodic domain of 200×200. (a) Grayscale image illustrates magnitude of the orientation field $|\tau|$ (white corresponds to the maximum of $|\tau|$, and black to $|\tau| = 0$); (b) Arrows depict the flow velocity $\mathbf{v}$ field, from Aranson et al. (2007a)

density variations. In eqn 9.83 the first term on the rhs describes the orientation instability, with $\epsilon = 0.276$, $A_0 = 2.81$ for fully inelastic particles, see eqns 9.36. Terms proportional to v_0 arise from bacterial swimming with respect to the fluid.

The inplane fluid velocity $\mathbf{v}$ obeys the Navier–Stokes equation with forcing due to bacterial swimming

$$\partial_t \mathbf{v} + \mathbf{v} \cdot \nabla \mathbf{v} = \eta \nabla^2 \mathbf{v} - \nabla p - \beta \mathbf{v} + \alpha \boldsymbol{\tau} \; , \tag{9.84}$$

with $\nabla \cdot \mathbf{v} = 0$ by incompressibility. In eqn 9.84, $\eta = \eta_0/D_r l^2$, where η_0 is the kinematic fluid viscosity, p is the pressure, and $\alpha \boldsymbol{\tau}$, with $\alpha \sim v_0$, models the forcing due to bacterial swimming. The origin of the terms $\sim \alpha, \beta$ deserves special discussion. The experiments by Wu and Libchaber (2000) and Sokolov et al. (2007) were performed in a free-standing thin-film geometry apparently suggesting slip-free boundary conditions for the hydrodynamic velocity on top and bottom surfaces of the film. However, as we discussed above, the surfactant produced due to bacterial metabolism accumulates on both surfaces of the fluid film and plays the role of semiflexible frictional walls.[9] This semiflexible layer results in a non-trivial velocity profile across the film and non-slip conditions at the top and the bottom surfaces giving rise to the damping term $\beta \mathbf{v}$. Similarly, the forcing term in eqn 9.84 is formally different from that for self-propelled particles immersed in bulk fluid proposed by Simha and Ramaswamy (2002), where the force is represented by the divergence of a certain three-dimensional stress tensor $\sigma_{ij} \sim \tau_i \tau_j$. Consequently, integration of three-dimensional divergence of the stress

---

[9] Indeed, in the course of experiment (3–5 min) the film often solidifies due to the accumulation of a surfactant layer rather than evaporates.

tensor $\partial_j \tau_i \tau_j$ over the film cross-section would produce a non-zero contribution in the form $\sim \tau$.

Remarkably, despite a significant difference in the underlying mechanisms of collective motion, eqns 9.83 and 9.84 for the orientation vector τ and velocity $\mathbf{v}$ are very similar to the model for shaken elongated particles, considered in Section 8.2.4, eqns 8.25 and 8.24. Note, however, that in eqn 8.25 the variable τ is not a true orientation vector but a combination of two components of the nematic alignment tensor.

To eliminate the pressure p and satisfy the incompressibility condition $\nabla \cdot \mathbf{v} = 0$ we introduce the stream function $\tilde{\varphi}$, with $v_x = \partial_y \tilde{\varphi}$, $v_y = -\partial_x \tilde{\varphi}$, and the vorticity $\Omega = \nabla^2 \tilde{\varphi}$. Then eqn 9.84 yields

$$\partial_t \Omega + \mathbf{v} \cdot \nabla \Omega = \eta \nabla^2 \Omega - \beta \Omega + \alpha \left(\partial_y \tau_x - \partial_x \tau_y \right). \tag{9.85}$$

Equations 9.82, 9.83, and 9.85 form a closed system. For flows with small Reynolds numbers the advection term $\mathbf{v} \cdot \nabla \Omega$ is much smaller than the viscous dissipation $\eta \nabla^2 \Omega$, but we keep it since a similar term is included in eqn 9.83. While the Reynolds number for individual swimming bacteria is exceedingly small (Re $\approx 10^{-4}$), for collective flows Re can be significantly larger. To simplify the analysis we consider the constant-density approximation $\nu = \nu_0$ valid for large D_0. Equations 9.83 and 9.85 have a steady-state solution corresponding to a homogeneous stream of bacteria in a certain direction (e.g. along x): $\tau_x = \tau_0 = ((\epsilon \nu - 1)/A_0)^{1/2}$, $\tau_y = v_y = 0$, $v_x = V = \alpha \tau_0/\beta$. The most unstable modes in the problem are longitudinal, and we examine the stability of this state with perturbations of the form $(\tau, \Omega) \sim \exp[\lambda t + ikx]$. Linearization of eqns 9.83 and 9.85 for $\nu = \nu_0$ shows that the equation for τ_x splits off, yielding the growth rate with a strictly negative real part, $\lambda_0 = -ikV - 2\tau_0^2 - (D_1 + D_2)k^2$, while equations for τ_y, Ω are coupled:

$$\lambda \tau_y = -ikV\tau_y - \frac{1}{2}\Omega \tau_0 - D_1 k^2 \tau_y \tag{9.86}$$

$$\lambda \Omega = -ikV\Omega - \eta k^2 \Omega - \beta \Omega - ik\alpha \tau_y. \tag{9.87}$$

These equations yield the other two growth rates λ:

$$\lambda_{1,2} = \frac{1}{2}\left(-(D_1 + \eta)k^2 - \beta - 2ikV \pm \sqrt{[(D_1 - \eta)k^2 - \beta]^2 - 2ik\tau_0 \alpha} \right). \tag{9.88}$$

The onset of the instability can be deduced by examining the limit $k \to 0$. An expansion of the real part of the growth rate $\operatorname{Re}\lambda$ in powers of k yields

$$\operatorname{Re}\lambda \simeq \left(\frac{\alpha^2 \tau_0^2}{\beta^3} - D_1 \right) k^2 + O(k^4), \tag{9.89}$$

so the long-wave instability occurs if the product $\alpha \tau_0$ exceeds a critical value $[D_1 \beta^3]^{1/2}$. Since $V = \alpha \tau_0/\beta$ is the collective steady-state swimming velocity, we can re-express the instability criterion simply as $V > V_d$, where $V_d = \sqrt{\beta D_1}$. Moreover, since $\beta \sim \eta/d^2$, where d is the film thickness, we find $V_d = \sqrt{\eta D_1}/d$. The selected wave number k_m

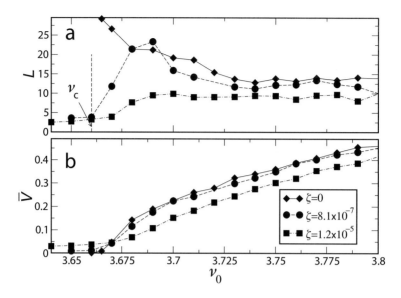

Fig. 9.27 Velocity correlation length L (a) and typical hydrodynamic velocity $\bar{V}$ (b) vs density ν for three different levels of noise ζ and for parameters of Fig. 9.26.

can be obtained in the limit of large collective swimming speed V. Expanding (9.88) for $\alpha \tau_0 \gg 1$ one obtains

$$\text{Re}\lambda \approx \frac{1}{2}\left(-(D_1 + \eta)k^2 - \beta + \sqrt{|k|\tau_0 \alpha}\right) + \dots \quad (9.90)$$

Then, from eqn 9.90 one finds for $(D_1 + \eta)k^2 \gg \beta$

$$k_m^{3/2} = \sqrt{\tau_0 \alpha}/4(D_1 + \eta) \sim (\nu - \nu_c)^{1/4} . \quad (9.91)$$

Thus, we see that the optimal wave number k_m increases with the increase of the density ν as $k_m \sim (\nu - \nu_c)^{1/6}$. Correspondingly, the instability length scale

$$L_m \sim 1/k_m \sim (\nu - \nu_c)^{-1/6}. \quad (9.92)$$

Since the instability length L_m is roughly related to the typical scale of the coherent structures, such as whorls and vortices, and, therefore, to the correlation length L, we see from eqn 9.92 that the correlation length L_m indeed changes very slowly with the density ν, consistent with experimental results.

To explore the evolution of the system beyond the linear stability analysis, numerical studies of the full system (eqns 9.82, 9.83, and 9.85) with periodic boundary conditions over a range of densities ν were conducted by Aranson et al. (2007a). A typical flow pattern and the distribution of orientational "order parameter" $|\tau|$ are shown in Fig. 9.26(a). The analysis shows that the correlation between the fields τ and $\mathbf{v}$ calculated over the entire domain of integration is close to zero (about 8–10%), in agreement with the experimental observations of Sokolov et al. (2007). However, there is always *local* dependence between τ and $\mathbf{v}$ through eqn 9.85.

In the course of numerical studies the following quantities were evaluated: the typical hydrodynamic velocity $\bar{V} = \sqrt{\langle \mathbf{v}^2 \rangle - \langle \mathbf{v} \rangle^2}$, and the velocity correlation function $K(r)$, averaged over the polar angle θ, see eqn 9.78. Figure 9.27 shows the typical hydrodynamic velocity $\bar{V}$, and the correlation length L computed from the velocity correlation function $K(r)$. The emerging picture of the transition again bears a strong resemblance to a second-order phase transition: the typical velocity $\bar{V} \sim \sqrt{\nu - \nu_c}$, and the correlation length diverges at $\nu \to \nu_c$, consistent with the prediction of eqn 9.91.

In order to include the effect of fluctuations, we added to the orientation equation (eqn 9.83) a random force $\xi(x, y, t)$, with correlation $\langle \xi(x, y, t) \xi^*(x', y', t') \rangle = 2\zeta\delta(x - x')\delta(y - y')\delta(t - t')$, where ζ is the noise amplitude. Results for various noise strengths are summarized in Fig. 9.27. Even a relatively small noise ($\zeta = 1.2 \times 10^{-7}$) smears the transition and removes the divergence of the correlation length. For strong enough noise ($\zeta \sim 10^{-5}$), one observes only a gradual increase of the correlation length with the density, in agreement with experiments in thin films, compare Figs. 9.22 and 9.27.

In conclusion of this section, both discrete-element simulations with simplified swimmers and the continuum model for large-scale flows generated by ensembles of swimming in thin films demonstrate the onset of coherence. These results provide strong evidence for a physical (hydrodynamic) origin of collective swimming, rather than chemotactic mechanisms of pattern formation (chemotactic interaction does not play any significant role in the experiments with swimming bacteria described above due to very fast mixing rates in the collective flow state). Swimming of bacteria in the direction of their orientation combined with inelastic collisions appears to be the underlying reason for the onset of large-scale chaotic flows.

References

Aegerter, C. M., Günther, R., and Wijngaarden, R. J. (2003). Avalanche dynamics, surface roughening, and self-organized criticality: Experiments on a three-dimensional pile of rice. *Phys. Rev. E*, **67**(5), 051306.

Aegerter, C. M., Lorincz, K. A., Welling, M. S., and Wijngaarden, R. J. (2004). Extremal dynamics and the approach to the critical state: Experiments on a three dimensional pile of rice. *Phys. Rev. Lett.*, **92**(5), 058702.

Aharonson, O., Zuber, M. T., Rothman, D. H., Schorghofer, N., and Whipple, K. X. (2002). Drainage basins and channel incision on mars. *Proc. Nat. Acad. Sci. U. S. A.*, **99**, 1780–1783.

Alexander, A., Muzzio, F. J., and Shinbrot, T. (2004). Effects of scale and inertia on granular banding segregation. *Granular Matter*, **5**(4), 171–175.

Ambroso, M. A., Kamien, R. D., and Durian, D. J. (2005). Dynamics of shallow impact cratering. *Phys. Rev. E*, **72**(4), 041305.

Andersen, K. H., Chabanol, M.-L., and van Hecke, M. (2001). Dynamical models for sand ripples beneath surface waves. *Phys. Rev. E*, **63**(6), 066308.

Andreotti, B., Claudin, P., and Douady, S. (2002). Selection of dune shapes and velocities Part 1: Dynamics of sand, wind and barchans. *Eur. Phys. J. B – Condens. Matter*, **28**(3), 321–339.

Aoki, K. M., Akiyama, T., Maki, Y., and Watanabe, T. (1996). Convective roll patterns in vertically vibrated beds of granules. *Phys. Rev. E*, **54**(1), 874–883.

Aradian, A., Raphael, E., and de Gennes, P.G. (2002). Surface flows of granular materials: A short introduction to some recent models. *C.R. Acad. Sci. Paris, Series IV*, **3**, 187.

Aranson, I.S. and Kramer, L. (2002). The world of the complex Ginzburg-Landau equation. *Rev. Mod. Phys.*, **74**(1), 99–143.

Aranson, I.S., Sokolov, A., Kessler, J.O., and Goldstein, R.E. (2007a). Model for dynamical coherence in thin films of self-propelled microorganisms. *Phys. Rev. E*, **75**(4), 040901.

Aranson, I. and Tsimring, L. (1999). Dynamics of axial segregation in a long rotating drum. *Phys. Rev. Lett.*, **82**, 4643–4646.

Aranson, I.S. and Tsimring, L.S. (2001). Continuum description of avalanches in granular media. *Phys. Rev. E*, **64**(2), 020301.

Aranson, I.S. and Tsimring, L.S. (2002). Continuum theory of partially fluidized granular flows. *Phys. Rev. E*, **65**(6), 061303.

Aranson, I.S. and Tsimring, L.S. (2005). Pattern formation of microtubules and motors: Inelastic interaction of polar rods. *Phys. Rev. E*, **71**(5), 050901.

Aranson, I.S. and Tsimring, L.S. (2006a). Theory of self-assembly of microtubules and motors. *Phys. Rev. E*, **74**(3), 031915.

Aranson, I.S., Volfson, D., and Tsimring, L.S. (2007b). Swirling motion in a system of vibrated elongated particles. *Phys. Rev. E*, **75**, 051301.

Aranson, I. S., Blair, D., Kalatsky, V. A., Crabtree, G. W., Kwok, W.-K., Vinokur, V. M., and Welp, U. (2000). Electrostatically driven granular media: Phase transitions and coarsening. *Phys. Rev. Lett.*, **84**(15), 3306–3309.

Aranson, I. S., Blair, D., Kwok, W., Karapetrov, G., Welp, U., Crabtree, G. W., Vinokur, V.M., and Tsimring, L. S. (1999a). Controlled dynamics of interfaces in a vibrated granular layer. *Phys. Rev. Lett.*, **82**, 731–734.

Aranson, I. S., Malloggi, F., and Clement, E. (2006). Transverse instability of avalanches in granular flows down an incline. *Phys. Rev. E*, **73**(5), 050302.

Aranson, I. S., Meerson, B, Sasorov, P. V., and Vinokur, V. M. (2002). Phase separation and coarsening in electrostatically driven granular media. *Phys. Rev. Lett.*, **88**(20), 204301.

Aranson, I. S. and Olafsen, J. S. (2002). Velocity fluctuations in electrostatically driven granular media. *Phys. Rev. E*, **66**(6), 061302.

Aranson, I. S. and Sapozhnikov, M. V. (2004). Theory of pattern formation of metallic microparticles in poorly conducting liquids. *Phys. Rev. Lett.*, **92**(23), 234301.

Aranson, I. S. and Tsimring, L. S. (1998). Formation of periodic and localized patterns in an oscillating granular layer. *Physica A*, **249**, 103–110.

Aranson, I. S. and Tsimring, L. S. (2003). Model of coarsening and vortex formation in vibrated granular rods. *Phys. Rev. E*, **67**(2), 021305.

Aranson, I. S. and Tsimring, L. S. (2006b). Patterns and collective behavior in granular media: Theoretical concepts. *Rev. Mod. Phys.*, **78**, 641–692.

Aranson, I. S., Tsimring, L. S., and Vinokur, V. M. (1999b). Hexagons and interfaces in a vibrated granular layer. *Phys. Rev. E*, **59**, 1327–1330.

Aranson, I. S., Tsimring, L. S., and V.M.Vinokur (1999c). Continuum theory of axial segregation in a long rotating drum. *Phys. Rev. E*, **60**, 1975–1987.

Argentina, M., Clerc, M.G., and Soto, R. (2002). van der Waals–Like Transition in Fluidized Granular Matter. *Phys. Rev. Lett.*, **89**(4), 44301.

Arndt, T., Siegmann-Hegerfeld, T., Fiedor, S. J., Ottino, J. M., and Lueptow, R. M. (2005). Dynamics of granular band formation: Long-term behavior in slurries, parameter space, and tilted cylinders. *Phys. Rev. E*, **71**, 011306.

Asakura, S. and Oosawa, F. (1958). Interaction between particles suspended in solutions of macromolecules. *J. Polymer Sci.*, **33**, 183–192.

Attard, P. and Parker, J. L. (1992). Deformation and adhesion of elastic bodies in contact. *Phys. Rev. A*, **46**(12), 7959–7971.

Aumaître, S., Kruelle, C. A., and Rehberg, I. (2001). Segregation in granular matter under horizontal swirling excitation. *Phys. Rev. E*, **64**, 041305.

Awazu, A. (2000). Size segregation and convection of granular mixtures almost completely packed in a thin rotating box. *Phys. Rev. Lett.*, **84**(20), 4585–4588.

Ayrton, H. (1910). The origin and growth of ripple-mark. *Proc. R. Soc. Lond., Ser. A*, **84**, 285.

Bagnold, R. A. (1954). Experiments on a gravity-free dispersion of large solid spheres in a Newtonian fluid under shear. *Proc. R. Soc. Lond., Ser. A*, **225**, 49–63.

Bagnold, R. A. (1956). *The Physics of Blown Sand and Desert Dunes*. Chapman and Hall, London.

Bagnold, R. A. and Taylor, G. (1956). Motion of waves in shallow water. interaction between waves and sand bottoms. *Proc. R. Soc. Lond., Ser. A*, **187**, 1–18.

Bak, P., Tang, C., and Wiesenfeld, K. (1987). Self-organized criticality: An explanation of $1/f$ noise. *Phys. Rev. Lett.*, **59**(4), 381–384.

Ball, R. C. and Blumenfeld, R. (2002). Stress field in granular systems: Loop forces and potential formulation. *Phys. Rev. Lett.*, **88**, 115505.

Batchelor, G.K. (1967). *An Introduction to Fluid Dynamics*. Cambridge University Press, Cambridge.

Belkin, M., Snezhko, A., Aranson, I.S., and Kwok, W.-K. (2007). Driven magnetic particles on a fluid surface: Pattern assisted surface flows. *Phys. Rev. Lett.*, **99**(15), 158301.

Ben-Jacob, E., Cohen, I., and Levine, H. (2000). Cooperative self-organization of microorganisms. *Adv. Phys.*, **49**(4), 395–554.

Ben-Naim, E., Chen, S. Y., Doolen, G. D., and Redner, S. (1999). Shocklike dynamics of inelastic gases. *Phys. Rev. Lett.*, **83**(20), 4069–4072.

Ben-Naim, E. and Krapivsky, P.L. (2006). Alignment of rods and partition of integers. *Phys. Rev. E*, **73**(3), 031109.

Ben-Naim, E. and Krapivsky, P. L. (2000). Multiscaling in inelastic collisions. *Phys. Rev. E*, **61**(1), R5–R8.

Benza, V. G., Nori, F., and Pla, O. (1993). Mean-field theory of sandpile avalanches: From the intermittent- to the continuous-flow regime. *Phys. Rev. E*, **48**, 4095–4098.

Betat, A., Frette, V., and Rehberg, I. (1999). Sand ripples induced by water shear flow in an annular channel. *Phys. Rev. Lett.*, **83**(1), 88–91.

Bizon, C., Shattuck, M. D., de Bruyn, John R., Swift, J. B., McCormick, W. D., and Swinney, Harry L. (1998a). Convection and diffusion in patterns in oscillated granular media. *J. Stat. Phys.*, **93**(3/4), 449–465.

Bizon, C., Shattuck, M. D., Newman, J. T., Umbanhowar, P.B., Swift, J. B., McCormick, W. D., and Swinney, H.L. (1997). Comment on "spontaneous wave pattern formation in vibrated granular materials". *Phys. Rev. Lett.*, **79**, 4713.

Bizon, C., Shattuck, M. D., and Swift, J. B. (1999). Linear stability analysis of a vertically oscillated granular layer. *Phys. Rev. E*, **60**(6), 7210–7216.

Bizon, C., Shattuck, M. D., Swift, J. B., McCormick, W. D., and Swinney, H. L. (1998b). Patterns in 3d vertically oscillated granular layers: Simulation and experiment. *Phys. Rev. Lett.*, **80**(1), 57–60.

Blair, D., Aranson, I. S., Crabtree, G. W., Vinokur, V., Tsimring, L. S., and Josserand, C. (2000). Patterns in thin vibrated granular layers: Interfaces, hexagons, and superoscillons. *Phys. Rev. E*, **61**(5), 5600–5610.

Blair, D.L. and Kudrolli, A. (2003). Clustering transitions in vibrofluidized magnetized granular materials. *Phys. Rev. E*, **67**(2), 21302.

Blair, D.L., Neicu, T., and Kudrolli, A. (2003). Vortices in vibrated granular rods. *Phys. Rev. E*, **67**(3), 31303.

Blair, D. L. and Kudrolli, A. (2001). Velocity correlations in dense granular gases. *Phys. Rev. E*, **64**(5), 050301.

Blair, D. L., Mueggenburg, N. W., Marshall, A. H., Jaeger, H. M., and Nagel, S.R.

(2001). Force distributions in 3d granular assemblies: Effects of packing order and inter-particle friction. *Phys. Rev. E*, **63**, 041304–1.

Blumenfeld, R. (2004a). Stress in planar cellular solids and isostatic granular assemblies: coarse-graining the constitutive equation. *Physica A*, **336**(3-4), 361–368.

Blumenfeld, R. (2004b). Stresses in Isostatic Granular Systems and Emergence of Force Chains. *Phys. Rev. Lett.*, **93**(10), 108301.

Boateng, A.A. and Barr, P.V. (1996). Modelling of particle mixing and segregation in the transverse plane of a rotary kiln. *Chem. Eng. Sci.*, **51**(17), 4167–4181.

Bocquet, L., Losert, W., Schalk, D., Lubensky, T. C., and Gollub, J. P. (2002). Granular shear flow dynamics and forces: Experiments and continuum theory. *Phys. Rev. E*, **65**, 011307.

Born, M. and Wolf, E. (1959). *Principles of optics: electromagnetic theory of propagation, interference, and diffraction of light*. Pergamon Press, London, N.Y.

Borzsonyi, T. and Ecke, R.E. (2006). Rapid granular flows on a rough incline: Phase diagram, gas transition, and effects of air drag. *Phys. Rev. E*, **74**(6), 061301.

Borzsonyi, T. and Ecke, R.E. (2007). Flow rule of dense granular flows down a rough incline. *Phys. Rev. E*, **76**(3), 031301.

Borzsonyi, T., Halsey, T. C., and Ecke, R. E. (2005). Two scenarios for avalanche dynamics in inclined granular layers. *Phys. Rev. Lett.*, **94**(20), 208001.

Bouchaud, J. P. (1994). Statics and dynamics of sandpiles: some phenomenological ideas. In *Lecture Notes in Physics* (ed. K. K. Bardhan, B. K. Chakrabarti, and A. Hansen), Volume Non-Linearity and Breakdown in Soft Condensed Matter, pp. 47–53. Springer, Berlin.

Bouchaud, J. P., Cates, M. E, Prakash, J. R., and Edwards, S. F. (1994). A model for the dynamics of sandpile surfaces. *J. Phys. I France*, **4**, 1383.

Bouchaud, J.-P., Cates, M. E., Prakash, J. R., and Edwards, S. F. (1995). Hysteresis and metastability in a continuum sandpile model. *Phys. Rev. Lett.*, **74**(11), 1982–1985.

Bougie, J., Kreft, J., Swift, J. B., and Swinney, H.L. (2005). Onset of patterns in an oscillated granular layer: Continuum and molecular dynamics simulations. *Phys. Rev. E*, **71**, 021301.

Bougie, J., Moon, S. J., Swift, J. B., and Swinney, H. L. (2002). Shocks in vertically oscillated granular layers. *Phys. Rev. E*, **66**(5), 051301.

Boutreux, T. and de Gennes, P. G. (1996). Surface flows of granular mixtures: I. General principles and minimal model. *J. Phys. I*, **6**, 1295.

Boutreux, T., Makse, H.A., and de Gennes, P. G. (1999). Surface flows of granular mixtures. *Eur. Phys. J. B - Condens. Matter*, **9**(1), 105–115.

Boutreux, T., Raphael, E., and de Gennes, P. G. (1998). Surface flows of granular materials: a modified picture for thick avalanches. *Phys. Rev. E*, **58**, 4692–4700.

Bray, A.J. (1994). Theory of phase-ordering kinetics. *Adv. Phys.*, **43**(3), 357–459.

Brazovskii, SA (1975). Phase transition of an isotropic system to an inhomogenous state. *Zh. Eksp. Teor. Fiz*, **68**(1), 175–85.

Brendel, L., Unger, T., and Wolf, D. E. (2004). Contact dynamics for beginners. In *Physics of Granular Media* (ed. H. H. and D. E. Wolf), pp. 325–344. Wiley-VCH, Weinheim.

Breu, A. P. J., Ensner, H.-M., Kruelle, C. A., and Rehberg, I. (2003). Reversing the Brazil-nut effect: competition between percolation and condensation. *Phys. Rev. Lett.*, **90**(1), 014302.

Brey, J. J., Dufty, J. W., Kim, C. S., and Santos, A. (1998). Hydrodynamics for granular flow at low density. *Phys. Rev. E*, **58**, 4638–4653.

Bridgwater, J., Cooke, M.H., and Scott, A.M. (1978). Inter-particle percolation: equipment development and mean percolation velocities. *Chem. Eng. Res. and Design*, **56**(a), 157–167.

Brilliantov, N.V. and Pöschel, T. (2004). Collision of Adhesive Viscoelastic Particles. In *The Physics of Granular Media* (ed. H. Hinrichsen and D. E. Wolf), Weinheim, Germany, pp. 189–209. Wiley VCH.

Brilliantov, N. V. and Pöschel, Th. (2004). *Kinetic Theory of Granular Gases*. Oxford University Press.

Brilliantov, N. V., Spahn, F., Hertzsch, J. M., and Pöschel, T. (1996). Model for collisions in granular gases. *Phys. Rev. E*, **53**(5), 5382.

Brito, R. and Ernst, M. H. (1998). Extension of Haff's cooling law in granular flows. *Europhys. Lett.*, **43**(15), 497–502.

Bromberg, Y., Livne, E., and Meerson, B. (2003). Development of a density inversion in driven granular gases. In *Granular Gas Dynamics* (ed. T. Pöschel and N. Brilliantov), Lecture Notes in Physics, Berlin, p. 251. Springer.

Budrene, E.O. and Berg, H.C. (1991). Complex patterns formed by motile cells of Escherichia coli. *Nature*, **349**(6310), 630–633.

Buehler, R. J., R. H. Wentorf, Jr., Hirschfelder, J. O., and Curtiss, C. F. (1951). The free volume for rigid sphere molecules. *J. Chem. Phys.*, **19**(1), 61–71.

Buhl, J., Sumpter, D. J. T., Couzin, I. D., Hale, J. J., Despland, E., Miller, E. R., and Simpson, S. J. (2006). From Disorder to Order in Marching Locusts. *Science*, **312**(5778), 1402–1406.

Burtally, N., King, P. J., and Swift, M.R. (2002). Spontaneous Air-Driven Separation in Vertically Vibrated Fine Granular Mixtures. *Science*, **295**(5561), 1877–1879.

Caps, H. and Vandewalle, N. (2001). Ripple and kink dynamics. *Phys. Rev. E*, **64**(4), 041302.

Carnahan, W. F. and Starling, K. E. (1969). Equation of state for nonattracting rigid spheres. *J. Chem. Phys.*, **51**, 635.

Cates, M. E., Wittmer, J. P., Bouchaud, J.-P., and Claudin, P. (1998). Jamming, force chains, and fragile matter. *Phys. Rev. Lett.*, **81**(9), 1841–1844.

Cattuto, C. and Marconi, U.M.B. (2004). Ordering Phenomena in Cooling Granular Mixtures. *Phys. Rev. Lett.*, **92**(17), 174502.

Cerda, E., Melo, F., and Rica, S. (1997). Model for subharmonic waves in granular materials. *Phys. Rev. Lett.*, **79**(23), 4570–4573.

Charles, C. R. J., Khan, Z. S., and Morris, S. W. (2006). Pattern scaling in axial segregation. *Granular Matter*, **8**(1), 1–3.

Chaté, H., Ginelli, F., and Grégoire, G. (2007). Comment on "phase transitions in systems of self-propelled agents and related network models". *Phys. Rev. Lett.*, **99**(22), 229601.

Chicarro, R., Peralta-Fabi, R., and Velasco, R. M. (1997). Segregation in dry granular systems. In *Powders & Grains 97* (ed. R. P. Behringer and J. T. Jenkins), pp. 479–481. Balkema, Rotterdam.

Chladni, E.F.F. (1787). *Entdeckungen über die Theorie des Klanges*. Weidmanns, Erben und Reich.

Choo, K., Baker, M. W., Molteno, T. C. A., and Morris, S. W. (1998). The dynamics of granular segregation patterns in a long drum mixer. *Phys. Rev. E*, **58**, 6115.

Choo, K., Molteno, T. C. A., and Morris, S. W. (1997). Traveling granular segregation patterns in a long drum mixer. *Phys. Rev. Lett.*, **79**(16), 2975–2978.

Ciamarra, M. Pica, Coniglio, A., and Nicodemi, M. (2005). Shear instabilities in granular mixtures. *Phys. Rev. Lett.*, **94**, 188001.

Claudin, P. and Bouchaud, J.-P. (1998). Stick-slip transition in the scalar arching model. In *Physics of Dry Granular Media* (ed. H. J. Herrmann, J.-P. Hovi, and S. Luding), p. 129. Kluwer Academic Publishers, Dordrecht.

Clément, E., Duran, J., and Rajchenbach, J. (1992). Experimental study of heaping in a two-dimensional "sandpile". *Phys. Rev. Lett.*, **69**(8), 1189.

Clément, E., Rajchenbach, J., and Duran, J. (1995). Mixing of a granular material in a bidimensional rotating drum. *Europhys.Lett.*, **30**(1), 7–12.

Clément, E., Vanel, L., Rajchenbach, J., and Duran, J. (1996). Pattern formation in a vibrated granular layer. *Phys Rev. E*, **53**(3), 2972–2975.

Conti, M., Meerson, B., Peleg, A., and Sasorov, P. V. (2002). Phase ordering with a global conservation law: Ostwald ripening and coalescence. *Phys. Rev. E*, **65**(4), 046117.

Conway, S.L., Shinbrot, T., and Glasser, B.J. (2004). A Taylor vortex analogy in granular flows. *Nature*, **431**, 433–437.

Cooke, M.H. and Bridgwater, J. (1979). Interparticle Percolation: a Statistical Mechanical Interpretation. *Industr. & Eng. Chem. Fund.*, **18**(1), 25–27.

Cooke, M. H., Stephens, D. J., and Bridgwater, J. (1976). Powder mixing – a literature survey. *Powder Technol.*, **15**, 1.

Cooke, W., Warr, S., Huntley, J. M., and Ball, R. C. (1996). Particle size segregation in a two-dimensional bed undergoing vertical vibration. *Phys. Rev. E*, **53**, 2812.

Coppersmith, S. N., Liu, C.-H., Majumdar, S., Narayan, O., and Witten, T. A. (1996). Model for force fluctuations in bead packs. *Phys. Rev. E*, **53**(5), 4673–4685.

Corwin, E. I., Jaeger, H. M., and Nagel, S. R. (2005). Structural signature of jamming in granular media. *Nature*, **435**, 1075–1078.

Costello, R. M., Cruz, K. L., Egnatuk, C., Jacobs, D. T., Krivos, M. C., Louis, T. S., Urban, R. J., and Wagner, H. (2003). Self-organized criticality in a bead pile. *Phys. Rev. E*, **67**(4), 041304.

Coullet, P., Lega, J., Houchmanzaden, B., and Lajzerowicz, J. (1990). Breaking chirality in nonequilibrium systems. *Phys. Rev. Lett.*, **65**, 1352–1355.

Crawford, C. and Riecke, H. (1999). Oscillon-type structures and their interaction in a Swift–Hohenberg model. *Physica D*, **129**, 83–92.

Cross, M. C. and Hohenberg, P. C. (1993). Pattern formation outside of equilibrium. *Rev. Mod. Phys.*, **65**, 851.

Cugliandolo, L. F., Kurchan, J., and Peliti, L. (1997). Energy flow, partial equilibration, and effective temperatures in systems with slow dynamics. *Phys. Rev.*

E, **55**(4), 3898–3914.

Cundall, P. A. and Strack, O. D. L. (1979). A discrete numerical model for granular assemblies. *Géotechnique*, **29**(1), 47–65.

Czirók, A. and Vicsek, T. (2000). Collective behavior of interacting self-propelled particles. *Physica A*, **281**(1–4), 17–29.

Daerr, A. (2001). Dynamical equilibrium of avalanches on a rough plane. *Phys. Fluids*, **13**(7), 2115–2124.

Daerr, A. and Douady, S. (1999). Two types of avalanche behaviour in granular media. *Nature*, **399**, 241–243.

Daerr, A., Lee, P., Lanuza, J., and Clément, E. (2003). Erosion patterns in a sediment layer. *Phys. Rev. E*, **67**(6), 65201.

Dammer, S. M. and Wolf, D. E. (2004). Self-focusing dynamics in monopolarly charged suspensions. *Phys. Rev. Lett.*, **93**(15), 150602.

Davidson, J. F., Scott, D. M., Bird, P. A., Herbert, O., Powell, A. A., and Ramsey, H. V. M. (2000). Granular motion in a rotary kiln: the transition from avalanching to rolling. *KONA Powder and Particle*, **18**, 149–155.

Davies, C. E. and Foye, J. (1991). Flow of granular material through vertical slots. *Trans. Ind. Chem. Eng. A*, **69**, 369–373.

de Bruyn, J.R., Bizon, C., Shattuck, M. D., Goldman, D., Swift, J. B., and Swinney, H.L. (1998). Continuum-type stability balloon in oscillated granular layers. *Phys. Rev. Lett.*, **81**, 1421–1424.

de Gennes, P.-G. and Pincus, P.A. (1970). Pair correlations in a ferromagnetic colloid. *Physik Kondens. Mat.*, **11**(3), 189–198.

de Gennes, P.-G. and Prost, J. (1993). *The Physics of Liquid Crystals*. Clarendon Press, Oxford.

Dietrich, W. E., Wilson, C. J., Montgomery, D. R., McKean, J., and Bauer, R. (1992). Erosion thresholds and land surface morphology. *Geology*, **20**, 675–679.

DiLuzio, W.R., Turner, L., Mayer, M., Garstecki, P., Weibel, D.B., Berg, H.C., and Whitesides, G.M. (2005). *Escherichia coli* swim on the right-hand side. *Nature*, **435**(6), 1271–1274.

Dinkelacker, F., Hübler, A., and Lüscher, E. (1987). Pattern formation of powder on a vibrating disc. *Biol. Cybern.*, **56**, 51–56.

Dippel, S. and Luding, S. (1995). Simulations on size segregation: Geometrical effects in the absence of convection. *J. Phys. I France*, **5**, 1527–1537.

Doi, M. and Edwards, S. E. (2003). *The Theory of Polymer Dynamics*. Oxford University Press.

Dombrowski, C., Cisneros, L., Chatkaew, S., Goldstein, R. E., and Kessler, J. O. (2004). Self-Concentration and Large-Scale Coherence in Bacterial Dynamics. *Phys. Rev. Lett.*, **93**(9), 98103.

Dorbolo, S., Volfson, D., Tsimring, L., and Kudrolli, A. (2005). Dynamics of a bouncing dimer. *Phys. Rev. Lett.*, **95**, 044101.

Douady, S., Andreotti, B., and Daerr, A. (1999). On granular surface flow equations. *Eur. Phys. J. B – Condens. Matter*, **11**, 131–142.

Douady, S., Andreotti, B., Daerr, A., and Clade, P. (2002). From a grain to avalanches: on the physics of granular surface flows. *Comptes Rendus Physique*, **3**(2),

177–186.

Douady, S., Fauve, S., and Laroche, C. (1989). Subharmonic instabilities and defects in a granular layer under vertical vibrations. *Europhys. Lett.*, **8**(7), 621.

Dunne, T. (1980). Formation and controls of channel networks. *Prog. Phys. Geogr.*, **4**, 211–239.

Duong, N.H.P., Hosoi, A.E, and Shinbrot, T. (2004). Periodic Knolls and Valleys: Coexistence of Solid and Liquid States in Granular Suspensions. *Phys. Rev. Lett.*, **92**(22), 224502.

Duran, J. (2000). Ripples in tapped or blown powder. *Phys. Rev. Lett.*, **84**(22), 5126–5129.

Duran, J. (2001). Rayleigh-Taylor instabilities in thin films of tapped powder. *Phys. Rev. Lett.*, **87**(25), 254301.

Duran, J., Rajchenbach, J., and Clément, E. (1993). Arching effect model for particle size segregation. *Phys. Rev. Lett.*, **70**(16), 2431–2434.

Duran, O., Schwämmle, V., and Herrmann, H. (2005). Breeding and solitary wave behavior of dunes. *Phys. Rev. E*, **72**(2), 021308.

Duru, P., Nicolas, M., Hinch, J., and Guazzelli, É. (2002). Constitutive laws in liquid-fluidized beds. *J. Fluid Mech.*, **452**, 371–404.

Edwards, S. F. and Grinev, D. V. (1999). Statistical mechanics of stress transmission in disordered granular arrays. *Phys. Rev. Lett.*, **82**, 5397–5400.

Edwards, S. F. and Oakeshott, R. B. S. (1989). The transmission of stress in an aggregate. *Physica D*, **38**, 88–92.

Efrati, E., Livne, E., and Meerson, B. (2005). Hydrodynamic Singularities and Clustering in a Freely Cooling Inelastic Gas. *Phys. Rev. Lett.*, **94**(8), 88001.

Eggers, J. and Riecke, H. (1999). Continuum description of vibrated sand. *Phys. Rev. E*, **59**(4), 4476–4483.

Ehrhardt, G. C. M. A., Stephenson, A., and Reis, P. M. (2005). Segregation mechanisms in a numerical model of a binary granular mixture. *Phys. Rev. E*, **71**(4), 41301.

Ehrichs, E. E., Jaeger, H. M., Karczmar, G. S., Knight, J. B., Kuperman, V. Yu., and Nagel, S. R. (1995). Granular convection observed by magnetic resonance imaging. *Science*, **267**, 1632–1634.

Elperin, T. and Vikhansky, A. (1998). Granular flow in a rotating cylindrical drum. *Europhys. Lett.*, **42**(6), 619–624.

Erikson, J. M., Mueggenburg, N. W., Jaeger, H. M., and Nagel, S. R. (2002). Force distribution in three-dimensional compressible granular packs. *Phys. Rev. E*, **66**, 040301(R).

Eshuis, P., van der Weele, K., van der Meer, D., Bos, R., and Lohse, D. (2007). Phase diagram of vertically shaken granular matter. *Phys. Fluids*, **19**(12), 123301.

Eshuis, P., van der Weele, K., van der Meer, D., and Lohse, D. (2005). Granular leidenfrost effect: Experiment and theory of floating particle clusters. *Phys. Rev. Lett.*, **95**(25), 258001.

Evesque, P. and Rajchenbach, J. (1989). Instability in a sand heap. *Phys. Rev. Lett.*, **62**(1), 44–46.

Evesque, P., Szmatula, E., and Denis, J. P. (1990). Surface fluidization of a sand pile.

Europhys. Lett., **12**, 623–627.

Falcon, E., Kumar, K., Bajaj, K. M. S., and Bhattacharjee, J. K. (1999*a*). Heap corrugation and hexagon formation of powder under vertical vibrations. *Phys. Rev. E*, **59**, 5716–5720.

Falcon, E., Wunenburger, R., Evesque, P., Fauve, S., Chabot, C., Garrabos, Y., and Beysens, D. (1999*b*). Cluster formation in a granular medium fluidized by vibrations in low gravity. *Phys. Rev. Lett.*, **83**, 440–443.

Falk, M. L. and Langer, J. S. (1998). Dynamics of viscoplastic deformation in amorphous solids. *Phys. Rev. E*, **57**(6), 7192–7205.

Faraday, M. (1831). On a peculiar class of acoustical figures; and on certain forms assumed by groups of particles upon vibrating elastic surfaces. *Phil. Trans. R. Soc. Lond.*, **52**, 299.

Fauve, S., Douady, S., and Laroche, C. (1989). Collective behaviours of granular masses under vertical vibrations. *J. Phys. (Paris)*, **50**, 187–189.

Feitosa, K. and Menon, N. (2002). Breakdown of energy equipartition in a 2D binary vibrated granular gas. *Phys. Rev. Lett.*, **88**(19), 198301.

Ferguson, A., Fisher, B., and Chakraborty, B. (2004). Impulse distributions in dense granular flows: Signatures of large-scale spatial structures. *Europhys. Lett.*, **66**(2), 277–283.

Fiedor, S.J. and Ottino, J.M. (2005). Mixing and segregation of granular matter: multi-lobe formation in time-periodic flows. *J. Fluid Mech.*, **533**, 223–236.

Fiedor, S. J. and Ottino, J. M. (2003). Dynamics of axial segregation and coarsening of dry granular materials and slurries in circular and square tubes. *Phys. Rev. Lett.*, **91**, 244301.

Finger, T., Voigt, A., Stadler, J., Niessen, H. G., L., Naji, and Stannarius, R. (2006). Coarsening of axial segregation patterns of slurries in a horizontally rotating drum. *Phys. Rev. E*, **74**(3), 031312.

Flyvbjerg, H., Holy, T. E., and Leibler, S. (1996). Microtubule dynamics: Caps, catastrophes, and coupled hydrolysis. *Phys. Rev. E*, **54**(5), 5538–5560.

Forterre, Y. (2006). Kapiza waves as a test for three-dimensional granular flow rheology. *J. Fluid Mech.*, **563**, 123–132.

Forterre, Y. and Pouliquen, O. (2001). Longitudinal vortices in granular flows. *Phys. Rev. Lett.*, **86**(26), 5886–5889.

Forterre, Y. and Pouliquen, O. (2002). Stability analysis of rapid granular chute flows: formation of longitudinal vortices. *J. Fluid Mech.*, **467**, 361–387.

Forterre, Y. and Pouliquen, O. (2003). Long-surface-wave instability in dense granular flows. *J. Fluid Mech.*, **486**, 21–50.

Frachebourg, L. and Martin, P. A. (2000). Exact statistical properties of the Burgers equation. *J. Fluid Mech.*, **417**, 323–349.

Fraerman, A. A., Melnikov, A. S., Nefedov, I. M., Shereshevskii, I. A., and Shpiro, A. V. (1997). Nonlinear relaxation dynamics in decomposing alloys: One-dimensional Cahn-Hilliard model . *Phys. Rev. B*, **55**, 6316–6323.

Frenkel, D. and Eppenga, R. (1985). Evidence for algebraic orientational order in a two-dimensional hard-core nematic. *Phys. Rev. A*, **31**(3), 1776–1787.

Frenkel, D., Mulder, B. M., and McTague, J. P. (1984). Phase diagram of a system of hard ellipsoids. *Phys. Rev. Lett.*, **52**(4), 287–300.

Frette, V. and Stavans, J. (1997). Avalanche-mediated transport in a rotated granular mixture. *Phys. Rev. E*, **56**, 6981–6990.

Galanis, J., Harries, D., Sackett, D.L., Losert, W., and Nossal, R. (2006). Spontaneous Patterning of Confined Granular Rods. *Phys. Rev. Lett.*, **96**(2), 28002.

Gallas, J. A. C., Herrmann, H. J., and Sokołowski, S. (1992a). Convection cells in vibrating granular media. *Phys. Rev. Lett.*, **69**(9), 1371.

Gallas, J. A. C., Herrmann, H. J., and Sokołowski, S. (1992b). Molecular dynamics simulation of powder fluidization in two dimensions. *Physica A*, **189**, 437–446.

Gallas, J. A. C., Herrmann, H. J., and Sokołowski, S. (1992c). Two-dimensional powder transport on a vibrating belt. *J. Phys. II*, **2**, 1389–1400.

Gallas, J. A. C. and Sokołowski, S. (1993). Grain non-sphericity effects on the angle of repose of granular material. *Int. J. of Mod. Phys. B*, **7**(9 & 10), 2037–2046.

Gao, D., Subramaniam, S., Fox, R. O., and Hoffman, D. K. (2005). Objective decomposition of the stress tensor in granular flows. *Phys. Rev. E*, **71**(2), 021302.

Garcimartin, A., Maza, D., Ilquimiche, J. L., and Zuriguel, I. (2002). Convective motion in a vibrated granular layer. *Phys. Rev. E*, **65**(3), 031303.

Garzó, V. and Dufty, J. W. (1999). Dense fluid transport for inelastic hard spheres. *Phys. Rev. E*, **59**, 5895–5911.

Geldart, D. (1973). Types of gas fluidization. *Powder Technol.*, **7**(5), 285–292.

Geminard, J.-C. and Laroche, C. (2003). Energy of a single bead bouncing on a vibrating plate: Experiments and numerical simulations. *Phys. Rev. E*, **68**(3), 031305.

Geng, J. and Behringer, R. P. (2005). Slow drag in two-dimensional granular media. *Phys. Rev. E*, **71**(1), 011302.

Gidaspow, D. (1994). *Multiphase Flow and Fluidization: Continuum and Kinetic Theory Descriptions*. Academic Press, San Diego.

Goddard, J. D. and Alam, M. (1999). Shear-flow and material instabilities in particulate suspensions and granular media. *Partic. Sci. and Technol.*, **17**, 69–96.

Goldenberg, C. and Goldhirsch, I. (2002). Force chains, microelasticity, and macroelasticity. *Phys. Rev. Lett.*, **89**(8), 084302.

Goldenberg, C. and Goldhirsch, I. (2004). Small and large scale granular statics. *Granular Matter*, **6**(2), 87–96.

Goldhirsch, I. and Goldenberg, C. (2002). On the microscopic foundations of elasticity. *Eur. Phys. J. E*, **9**(3), 245–251.

Goldhirsch, I. and Zanetti, G. (1993). Clustering instability in dissipative gases. *Phys. Rev. Lett.*, **70**(11), 1619–1622.

Goldman, D. I., Swift, J. B., and Swinney, H. L. (2004). Noise, coherent fluctuations, and the onset of order in an oscillated granular fluid. *Phys. Rev. Lett*, **92**, 174302.

Goldshtein, A. and Shapiro, M. (1995). Mechanics of collisional motion of granular materials. Part 1. General hydrodynamic equations. *J. Fluid Mech.*, **282**, 75–114.

Goldsmith, W. (1964). *IMPACT, The Theory and Physical Behavior of Colliding Solids*. Edward Arnold, London.

Goujon, C., Thomas, N., and B., Dalloz-Dubrujeaud. (2003). Monodisperse dry granular flows on inclined planes: role of roughness. *Eur. Phys. J. E*, **11**, 147–157.

Gray, J., Shearer, M., and Thornton, AR (2006). Time-dependent solutions for particle-size segregation in shallow granular avalanches. *Proc. R. Soc. Lond., Ser. A*, **462**(2067), 947–972.

Gray, J. and Thornton, A. R. (2005). A theory for particle size segregation in shallow granular free-surface flows. *Proc. R. Soc. Lond., Ser. A*, **461**(2057), 1447–1473.

Gray, J. M. N. T. and Hutter, K. (1997). Pattern formation in granular avalanches. *Continuum Mech. Thermodyn.*, **9**(1), 341–345.

Grégoire, G. and Chaté, H. (2004). Onset of Collective and Cohesive Motion. *Phys. Rev. Lett.*, **92**(2), 25702.

Grier, D. G. (2003). Fluid dynamics: Vortex rings in a constant electric field. *Nature*, **424**(6620), 267–268.

Grossman, E. L., Zhou, T., and Ben-Naim, E. (1997). Towards granular hydrodynamics in two-dimensions. *Phys. Rev. E*, **55**, 4200–4206.

Gurbatov, S. N., Saichev, A. I., and Shandarin, S. F. (1985). Model description of the development of the large-scale structure of the universe. *Akademiia Nauk SSSR, Doklady*, **285**(2), 323–326.

Haff, P. K. (1983). Grain flow as a fluid-mechanical phenomenon. *J. Fluid Mech.*, **134**, 401–430.

Halsey, T. C. and Ertaş, D. (1999). A ball in a groove. *Phys. Rev. Lett.*, **83**, 5007–5010.

Hatwalne, Y., Ramaswamy, S., Rao, M., and Simha, R. A. (2004). Rheology of active-particle suspensions. *Phys. Rev. Lett.*, **92**(11), 118101.

He, X., Meerson, B., and Doolen, G. (2002). Hydrodynamics of thermal granular convection. *Phys. Rev. E*, **65**(3), 030301.

Henein, H., Brimacombe, J. K., and Watkinson, A. P. (1983). Experimental Study of Transverse Bed Motion in Rotary Kilns. *Metall. Trans B*, **14**(2), 191–205.

Hernandez-Ortiz, J. P., Stoltz, Ch. G., and Graham, M. D. (2005). Transport and collective dynamics in suspensions of confined swimming particles. *Phys. Rev. Lett.*, **95**(20), 204501.

Hersen, P., Andersen, K. H., Elbelrhiti, H., Andreotti, B., Claudin, P., and Douady, S. (2004). Corridors of barchan dunes: Stability and size selection. *Phys. Rev. E*, **69**(1), 011304.

Hersen, P., Douady, S., and Andreotti, B. (2002). Relevant length scale of barchan dunes. *Phys. Rev. Lett.*, **89**(26), 264301.

Hill, J. M. (1997). The double-shearing velocity equations for dilatant shear-index granular materials. In *IUTAM Symposium on Mechanics of Granular and Porous Materials* (ed. N. A. Fleck and A. C. E. Cocks), pp. 251–262. Kluwer Academic Publishers, Dordrecht.

Hill, K. M., Gioia, G., and Amaravadi, D. (2004). Radial segregation patterns in rotating granular mixtures: waviness selection. *Phys. Rev. Lett.*, **93**(22), 224301.

Hill, K. M. and Kakalios, J. (1994). Reversible axial segregation of binary mixtures of granular materials. *Phys. Rev. E*, **49**(5), R3610–3613.

Hill, K. M. and Kakalios, J. (1995). Reversible axial segregation of rotating granular media. *Phys. Rev. E*, **52**(4), 4393–4400.

Hill, K. M., Khakhar, D. V., Gilchrist, J. F., McCarthy, J. J., and Ottino, J. M. (1999). Segregation-driven organization in chaotic granular flows. *Proc. Nat. Acad.*

Sci., **96**(21), 11701–11706.

Hirshfeld, D. and Rapaport, D. C. (1997). Molecular dynamics studies of grain segretation in sheared flow. *Phys. Rev. E*, **56**(2), 2012–2018.

Hong, D. C., Quinn, P. V., and Luding, S. (1999). Condensation of hard spheres under gravity. *Physica A*, **271**, 192–199.

Hong, D. C., Quinn, P. V., and Luding, S. (2001). The reverse Brazil nut problem: Competition between percolation and condensation. *Phys. Rev. Lett.*, **86**, 3423–3426.

Hornbaker, D. J., Albert, R., Albert, I., Barabasi, A.-L., and Schiffer, P. (1997). What keeps sandcastles standing? *Nature*, **387**, 765.

Howard, J. (2001). *Mechanics of Motor Proteins and the Cytoskeleton*. Sinauer, Sunderland, MA.

Howell, D., Behringer, R. P., and Veje, C. (1999). Stress fluctuations in a 2D granular Couette experiment: A continuous transition. *Phys. Rev. Lett.*, **82**(26), 5241–5244.

Hsiau, S.S. and Hunt, M.L. (1996). Granular thermal diffusion in flows of binary-sized mixtures. *Acta Mechanica*, **114**(1), 121–137.

Huan, C., Yang, X., Candela, D., Mair, RW, and Walsworth, R. L. (2004). NMR experiments on a three-dimensional vibrofluidized granular medium. *Phys. Rev. E*, **69**(4), 41302.

Huthmann, M., Aspelmeier, T., and Zippelius, A. (1999). Granular cooling of hard needles. *Phys. Rev. E*, **60**(1), 654–659.

Israelachvili, J.N. (1985). *Intermolecular and Surface Forces*. Saunders College Publishing/Harcourt Brace; 2nd edition.

Iverson, R. M. (1997). The physics of debris flows. *Rev. Geophys.*, **35**, 245–296.

Jackson, R. (2000). *The Dynamics of Fluidized Particles*. Cambridge University Press.

Jaeger, H. M., Liu, C.-H., and Nagel, S. R. (1989). Relaxation at the angle of repose. *Phys. Rev. Lett.*, **62**(1), 40–43.

Jaeger, H. M., Nagel, S. R., and Behringer, R. P. (1996). The physics of granular materials. *Physics Today*, **49**(4), 32–38.

Jenike, A. W. (1964). Storage and flow of solids, bull. no. 123. *Bull. Univ. Utah*, **53**(26), 198.

Jenike, A. W. (1987). A theory of flow of particulate solids in converging and diverging channels based on a conical yield function. *Powder Technol.*, **50**, 229–236.

Jenkins, J.T. (2007). Dense shearing flows of inelastic disks. *Phys. Fluids*, **18**, 103307.

Jenkins, J. T. and Mancini, F. (1987). Balance laws and constitutive relations for plane flows of a dense, binary mixture of smooth, nearly elastic, circular discs. *J. Appl. Mech.*, **54**, 27–34.

Jenkins, J. T. and Richman, M. W. (1985). Kinetic theory for plane shear flows of a dense gas of identical, rough, inelastic, circular disks. *Phys. of Fluids*, **28**, 3485–3494.

Jenkins, J. T. and Yoon, D. K. (2002). Segregation in binary mixtures under gravity. *Phys. Rev. Lett.*, **88**(19), 194301.

Jenny, H. (1974). *Cymatics, Vol. 2: Wave Phenomena, Vibrational Effects, Harmonic Oscillations, with their Structure, Kinetics, and Dynamics*. Basilius Press, Basel.

Jia, L. C., Lai, P.-Y., and Chan, C. K. (1999). Empty site models for heap formation in vertically vibrating grains. *Phys. Rev. Lett.*, **83**, 3832–3835.

Johnson, K. L. (1989). *Contact Mechanics*. Cambridge University Press.

Jop, P., Forterre, Y., and Pouliquen, O. (2006). A constitutive law for dense granular flows. *Nature*, **441**(7094), 727–30.

Kadanoff, L. P. (1999). Built upon sand: Theoretical ideas inspired by granular flows. *Rev. Mod. Phys.*, **71**(1), 435–444.

Kamrin, K. and Bazant, M. Z. (2007). Stochastic flow rule for granular materials. *Phys. Rev. E*, **75**(4), 041301.

Kaneko, K. (1993). *Theory and Applications of Coupled Map Lattices*. John Wiley & Sons, New York.

Keller, E. F. and Segel, L. A. (1971). Model for chemotaxis. *J. Theor. Biol.*, **30**(2), 225–34.

Khain, E. and Meerson, B. (2002). Symmetry-breaking instability in a prototypical driven granular gas. *Phys. Rev. E*, **66**(2), 21306.

Khain, E. and Meerson, B. (2003). Onset of thermal convection in a horizontal layer of granular gas. *Phys. Rev. E*, **67**, 021306.

Khain, E. and Meerson, B. (2004). Oscillatory instability in a driven granular gas. *Europhys. Lett.*, **65**(2), 193–199.

Khain, E. and Meerson, B. (2006). Shear-induced crystallization of a dense rapid granular flow: Hydrodynamics beyond the melting point. *Phys. Rev. E*, **73**, 061301.

Khain, E., Meerson, B., and Sasorov, P. V. (2004). Phase diagram of van der Waals–like phase separation in a driven granular gas. *Phys. Rev. E*, **70**(5), 51310.

Khakhar, D. V., McCarthy, J. J., and Ottino, J. M. (1997a). Radial segregation of granular mixtures in rotating cylinders. *Phys. Fluids*, **9**, 3600–3614.

Khakhar, D. V., McCarthy, J. J., Shinbrot, T., and Ottino, J. M. (1997b). Transverse flow and mixing of granular materials in a rotating cylinder. *Phys. Fluids*, **9**, 31–43.

Khakhar, D. V., Orpe, A. V., Andresén, P., and Ottino, J. M. (2001). Surface flow of granular materials: model and experiments in heap formation. *J. Fluid Mech.*, **441**, 255–264.

Khan, Z. and Morris, S. W. (2005). Subdiffusive axial transport of granular materials in a long drum mixer. *Phys. Rev. Lett.*, **94**, 048002.

Khan, Z., Tokaruk, W. A., and Morris, S. W. (2004). Oscillatory granular segregation in a long drum mixer. *Europhys. Lett.*, **66**, 212–218.

Kiyashko, S. V., Korzinov, L. N., Rabinovich, M. I., and Tsimring, L. S. (1996). Rotating spirals in faraday experiment. *Phys. Rev. E*, **54**(24), 5037–5040.

Knight, J. B., Ehrichs, E. E., Kuperman, V. Yu., Flint, J. K., Jaeger, H. M., and Nagel, S. R. (1996). Experimental study of granular convection. *Phys. Rev. E*, **54**(5), 5726–5738.

Knight, J. B., Jaeger, H. M., and Nagel, S. R. (1993). Vibration-induced size separation in granular media: The convection connection. *Phys. Rev. Lett.*, **70**(24), 3728–3731.

Koeppe, J. P., Enz, M., and Kakalios, J. (1998). Phase diagram for avalanche stratification of granular media. *Phys. Rev. E*, **58**, 4104–4107.

Kohlstedt, K., Snezhko, A., Sapozhnikov, M. V., Aranson, I. S., Olafsen, J. S., and

Ben-Naim, E. (2005). Velocity distributions of granular gases with drag and with long-range interactions. *Phys. Rev. Lett.*, **95**(6), 068001.

Kroy, K., Sauermann, G., and Herrmann, H. J. (2002). A minimal model for sand dunes. *Phys. Rev. Lett.*, **88**, 054301.

Kruse, K., Joanny, J. F., Julicher, F., Prost, J., and Sekimoto, K. (2004). Asters, vortices, and rotating spirals in active gels of polar filaments. *Phys. Rev. Lett.*, **92**(7), 078101.

Kudrolli, A. (2004). Size separation in vibrated granular matter. *Rep. Prog. Phys.*, **67**(3), 209–247.

Kudrolli, A., Lumay, G., Volfson, D., and Tsimring, L. S. (2008). Swarming and swirling in self-propelled polar granular rods. *Phys. Rev. Lett.*, **100**, 058001.

Kudrolli, A., Wolpert, M., and Gollub, J. P. (1997). Cluster formation due to collisions in granular material. *Phys. Rev. Lett.*, **78**(7), 1383–1386.

Lacombe, F., Zapperi, S., and Herrmann, H. J. (2000). Dilatancy and friction in sheared granular media. *Eur. Phys. J. E – Soft Matter*, **2**(2), 181–189.

Landau, L.D. and Lifshitz, E.M. (1980). *Statistical Physics*. Pergamon Press, N.Y.

Landau, L.D. and Lifshitz, E.M. (1987). *Fluid Mechanics*. Pergamon Press, N.Y.

Landau, L. D. and Lifshitz, E. M. (1986). *Elasticity Theory*. Pergamon Press, 3rd ed., Oxford.

Landau, L. D., Lifshitz, E. M., and Pitaevskij, L. P. (1981). *Physical kinetics*. Butterworth–Heinemann, Oxford.

Landry, J. W., Grest, G. S., Silbert, L. E., and Plimpton, S. J. (2003). Confined granular packings: structure, stress, and forces. *Phys. Rev. E*, **67**(4), 41303.

Langlois, V. and Valance, A. (2005). Formation of two-dimensional sand ripples under laminar shear flow. *Phys. Rev. Lett.*, **94**(24), 248001.

Laroche, C., Douady, S., and Fauve, S. (1989). Convective flow of granular masses under vertical vibrations. *J. Phys.*, **50**, 699–706.

Larrieu, E., Staron, L., and Hinch, E.J. (2006). Raining into shallow water as a description of the collapse of a column of grains. *J. Fluid Mech.*, **554**, 259–270.

Lebowitz, J. L. (1964). Exact solution of generalized Percus-Yevick equation for a mixture of hard spheres. *Phys. Rev.*, **133**(4A), A895–A899.

Lee, Y. H. and Kardar, M. (2001). Macroscopic equations for pattern formation in mixtures of microtubules and molecular motors. *Phys. Rev. E*, **64**(5), 056113.

Lemaître, A. (2002). Rearrangements and dilatancy for sheared dense materials. *Phys. Rev. Lett.*, **89**(19), 195503.

Lemieux, P.-A. and Durian, D. J. (2000). From avalanches to fluid flow: A continuous picture of grain dynamics down a heap. *Phys. Rev. Lett.*, **85**(20), 4273–4276.

Levin, Y. (1999). What happened to the gas-liquid transition in the system of dipolar hard spheres? *Phys. Rev. Lett.*, **83**(6), 1159–1162.

Levine, D. (1999). Axial segregation of granular materials. *Chaos*, **9**, 573–580.

Levitan, B. (1998). Long-time limit of rotational segregation of granular media. *Phys. Rev. E*, **58**(2), 2061–2064.

Li, J., Aranson, I. S., Kwok, W.-K., and Tsimring, L. S. (2003). Periodic and disordered structures in a modulated gas-driven granular layer. *Phys. Rev. Lett.*, **90**(13), 134301.

Liffman, K., Metcalfe, G., and Cleary, P. (1997). Granular convection and transport due to horizontal shaking. *Phys. Rev. Lett.*, **79**(23), 4574–4576.

Lifshitz, I. M. and Slyozov, V. V. (1958). On the kinetics of diffusion decomposition of supersaturated solid solutions. *Zh. Eksp. Teor. Fiz*, **35**(2), 479–491.

Lifshitz, I. M. and Slyozov, V. V. (1961). The kinetics of precipitation from supersaturated solid solutions. *J. Phys. Chem. Solids*, **19**, 35–50.

Lifshitz, J. M. and Kolsky, H. (1964). Some experiments on anelastic rebound. *J. Mech. Phys. Solids*, **12**, 35.

Linz, S. J. and Hänggi, P. (1995). A minimal model for avalanches in granular systems. *Phys. Rev. E*, **51**, 2538–2542.

Lioubashevski, O., Hamiel, Y., Agnon, A., Reches, Z., and Fineberg, J. (1999). Oscillons and propagating solitary waves in a vertically vibrated colloidal suspension. *Phys. Rev. Lett.*, **83**(16), 3190–3193.

Liu, C.-H., Nagel, S. R., Schecter, D. A., Coppersmith, S. N., Majumdar, S., Narayan, O., and Witten, T. A. (1995). Force fluctuations in bead packs. *Science*, **269**, 513.

Liverpool, T. B. and Marchetti, M. C. (2003). Instabilities of isotropic solutions of active polar filaments. *Phys. Rev. Lett.*, **90**(13), 138102.

Livne, E., Meerson, B., and Sasorov, P. V. (2002a). Symmetry breaking and coarsening of clusters in a prototypical driven granular gas. *Phys. Rev. E*, **66**(5), 50301.

Livne, E., Meerson, B., and Sasorov, P. V. (2002b). Symmetry-breaking instability and strongly peaked periodic clustering states in a driven granular gas. *Phys. Rev. E*, **66**, 021302.

Lobkovsky, A. E., Jensen, B., Kudrolli, A., and Rothman, D. H. (2004). Threshold phenomena in erosion driven by subsurface flow. *J. Geophys. Res.*, **109**, F04010.

Lohse, D., Bergmann, R., Mikkelsen, R., Zeilstra, Ch., van der Meer, D., Versluis, M., van der Weele, Ko, van der Hoef, M., and Kuipers, H. (2004). Impact on soft sand: Void collapse and jet formation. *Phys. Rev. Lett.*, **93**(19), 198003.

Lois, G., Lemaître, A., and Carlson, J. M. (2005). Numerical tests of constitutive laws for dense granular flows. *Phys. Rev. E*, **72**(5), 51303.

Longhi, E., Easwar, N., and Menon, N. (2002). Large force fluctuations in a flowing granular medium. *Phys. Rev. Lett.*, **89**(4), 45501.

Losert, W., Bocquet, L., Lubensky, T. C., and Gollub, J. P. (2000). Particle dynamics in sheared granular matter. *Phys. Rev. Lett.*, **85**(7), 1428–1431.

Losert, W., Cooper, D. G. W., Delour, J., Kudrolli, A., and Gollub, J. P. (1999a). Velocity statistics in vibrated granular media. *Chaos*, **9**(3), 682–690.

Losert, W., Cooper, D. G. W., and Gollub, J. P. (1999b). Propagating front in an excited granular layer. *Phys. Rev. E*, **59**, 5855–5861.

Lubachevsky, B. D. (1991). How to simulate billiards and similar systems. *J. Comp. Phys.*, **94**(2), 255–283.

Luding, S. (1997). Stress distribution in static two dimensional granular model media in the absence of friction. *Phys. Rev. E*, **55**(4), 4720–4729.

Luding, S. (2004). Micro-macro transition for anisotropic, frictional granular packings. *Int. J. Sol. Struct.*, **41**, 5821–5836.

Luding, S., Clément, E., Blumen, A., Rajchenbach, J., and Duran, J. (1994). Studies of columns of beads under external vibrations. *Phys. Rev. E*, **49**(2), 1634–1646.

Luding, S., Clément, E., Blumen, A., Rajchenbach, J., and Duran, J. (1995). Interaction laws and the detachment effect in granular media. In *Fractal Aspects of Materials* (ed. F. Family, P. Meakin, B. Sapoval, and R. Wool), Volume 367, Pittsburgh, Pennsylvania, pp. 495–500. Materials Research Society, Symposium Proceedings.

Luding, S., Clément, E., Rajchenbach, J., and Duran, J. (1996). Simulations of pattern formation in vibrated granular media. *Europhys. Lett.*, **36**(4), 247–252.

Luding, S., Lätzel, M., and Herrmann, H. J. (2001). From discrete element simulations towards a continuum description of particulate solids. In *Handbook of Conveying and Handling of Particulate Solids* (ed. A. Levy and H. Kalman), pp. 39–44. Elsevier, Amsterdam.

Makse, H. A. (1999). Continuous avalanche segregation of granular mixtures in thin rotating drums. *Phys. Rev. Lett.*, **83**(16), 3186–3189.

Makse, H. A., Ball, R. C., Stanley, H. E., and Warr, S. (1998). Dynamics of granular stratification. *Phys. Rev. E*, **58**(3), 3357–3367.

Makse, H. A., Brujic, J., and Edwards, S. F. (2004). Statistical mechanics of jammed matter. In *The Physics of Granular Media* (ed. H. H. and D. E. Wolf), pp. 45–85. Wiley-VCH, Weinheim.

Makse, H. A., Cizeau, P., and Stanley, H. E. (1997a). Possible stratification mechanism in granular mixtures. *Phys. Rev. Lett.*, **78**(17), 3298–3301.

Makse, H. A., Havlin, S., King, P. R., and Stanley, H. E. (1997b). Spontaneous stratification in granular mixtures. *Nature*, **386**, 379–381.

Makse, H. A. and Herrmann, H. J. (1998). Microscopic model for granular stratification and segregation. *Europhys. Lett.*, **43**(1), 1–6.

Makse, H. A. and Kurchan, J. (2002). Testing the thermodynamic approach to granular matter with a numerical model of a decisive experiment. *Nature*, **415**(6872), 614–617.

Malin, M. C. and Carr, M. H. (1999). Groundwater formation of martian valleys. *Nature*, **397**, 589–591.

Malloggi, F., Lanuza, J., Andreotti, B., and Clément, E. (2006). Erosion waves: Transverse instabilities and fingering. *Europhys. Lett.*, **75**(5), 825–831.

Mangeney-Castelnau, A., Bouchut, F., Vilotte, J. P., Lajeunesse, E., Aubertin, A., and Pirulli, M. (2005). On the use of Saint-Venant equations to simulate the spreading of a granular mass. *J. Geophys. Res.*, **199**, 177–215.

Mase, G. E. (1970). *Continuum Mechanics*. McGraw-Hill, New York.

Massoudi, M. and Mehrabadi, M. M. (2001). A continuum model for granular materials: Considering dilatancy and the Mohr-Coulomb criterion. *Acta Mechanica*, **152**(1), 121–138.

McNamara, S. and Falcon, E. (2005). Simulations of vibrated granular medium with impact-velocity-dependent restitution coefficient. *Phys. Rev. E*, **71**, 031302.

McNamara, S. and Herrmann, H. (2004). Measurement of indeterminacy in packings of perfectly rigid disks. *Phys. Rev. E*, **70**, 0613903.

McNamara, S. and Young, W. R. (1992). Inelastic collapse and clumping in a one-dimensional granular medium. *Phys. Fluids A*, **4**(3), 496–504.

McNamara, S. and Young, W. R. (1996). Dynamics of a freely evolving, two-dimensional granular medium. *Phys. Rev. E*, **53**(5), 5089–5100.

Medved, M. (2002). Connections between response modes in a horizontally driven granular material. *Phys. Rev. E*, **65**(2), 021305.

Meerson, B. (1996). Nonlinear dynamics of radiative condensations in optically thin plasmas. *Rev. Mod. Phys.*, **68**(1), 215–257.

Meerson, B., Pöschel, T., and Bromberg, Y. (2003). Close-packed floating clusters: granular hydrodynamics beyond the freezing point? *Phys. Rev. Lett.*, **91**, 024301.

Meerson, B. and Puglisi, A. (2005). Towards a continuum theory of clustering in a freely cooling inelastic gas. *Europhys. Lett.*, **70**(4), 478–484.

Mehrotra, A., Muzzio, F. J., and Shinbrot, T. (2007). Spontaneous separation of charged grains. *Phys. Rev. Lett.*, **99**(5), 058001.

Mehta, A. (1994). *Granular Matter: An Interdisciplinary Approach*. Springer-Verlag, New York.

Melby, P., Prevost, A., Egolf, D. A., and Urbach, J. S. (2007). The depletion force in a bi-disperse granular layer. *Phys. Rev. E*, **76**, 051307.

Melo, F., Umbanhowar, P. B., and Swinney, H. L. (1994). Transition to parametric wave patterns in a vertically oscillated granular layer. *Phys. Rev. Lett.*, **72**(1), 172–175.

Melo, F., Umbanhowar, P. B., and Swinney, H. L. (1995). Hexagons, kinks, and disorder in oscillated granular layers. *Phys. Rev. Lett.*, **75**(21), 3838–3841.

Mendelson, N. H., Bourque, A., Wilkening, K., Anderson, K. R., and Watkins, J. C. (1999). Organized cell swimming motions in Bacillus subtilis colonies: Patterns of short-lived whirls and jets. *J. Bacteriol.*, **181**(2), 600–609.

Mermin, N. D. and Wagner, H. (1966). Absence of ferromagnetism or antiferromagnetism in one- or two-dimensional isotropic Heisenberg models. *Phys. Rev. Lett.*, **17**(22), 1133–1136.

Metcalfe, G. and Shattuck, M. (1996). Pattern formation during mixing and segregation of flowing granular materials. *Physica A*, **233**, 709–719.

Metcalfe, G., Shinbrot, T., McCarthy, J. J., and Ottino, J. M. (1995). Avalanche mixing of granular solids. *Nature*, **374**, 39–42.

Metha, A. and Luck, J. M. (1990). Novel temporal behavior of a nonlinear dynamical system: The completely inelastic bouncing ball. *Phys. Rev. Lett.*, **65**(4), 393.

Mezard, M., Parisi, G., and Virasoro, M. A. (1987). *Spin Glass Theory and Beyond*. World Scientific Teaneck, NJ, USA.

MiDi, GDR (2004). On dense granular flows. *Eur. Phys. J. E*, **14**(6), 341–365.

Miller, S. and Luding, S. (2004). Event driven simulations in parallel. *J. Comp. Phys.*, **193**(1), 306–316.

Möbius, M. E., Lauderdale, B. E., Nagel, S. R., and Jaeger, H. M. (2001). Size separation of granular particles. *Nature*, **414**, 270.

Monin, A. S. and Yaglom, A. M. (1971). *Statistical Fluid Mechanics*. MIT Press Cambridge, Mass.

Moon, S. J., Shattuck, M. D., Bizon, C., Goldman, D. I., Swift, J. B., and Swinney, H. L. (2001). Phase bubbles and spatiotemporal chaos in granular patterns. *Phys. Rev. E*, **65**(1), 011301.

Moon, S. J., Swift, J. B., and Swinney, H. L. (2004). Steady-state velocity distributions of an oscillated granular gas. *Physical Review E*, **69**(1), 11301.

Moreau, J. J. (1994). Some numerical methods in multibody dynamics: Application to granular materials. *Eur. J. Mech. A*, **13**, 93.

Morse, P. M. C. and Feshbach, H. (1953). *Methods of Theoretical Physics, Part 1*. McGraw-Hill.

Mueth, D. M., Debregeas, G. F., Karczmar, G. S., Eng, P. J., Nagel, S. R., and Jaeger, H. M. (2000). Signatures of granular microstructure in dense shear flows. *Nature*, **406**, 385–389.

Mueth, D. M., Jaeger, H. M., and Nagel, S. R. (1998). Force distribution in a granular medium. *Phys. Rev. E.*, **57**(3), 3164–3169.

Mujica, N. and Melo, F. (1998). Solid–liquid transition and hydrodynamic surface waves in vibrated granular layers. *Phys. Rev. Lett.*, **80**(23), 5121–5124.

Mullin, T. (2000). Coarsening of self-organized clusters in binary mixtures of particles. *Phys. Rev. Lett.*, **84**(20), 4741–4744.

Mullin, T. (2002). Granular materials: Mixing and de-mixing. *Science*, **295**(5561), 1851.

Nakagawa, M., Altobelli, S. A., Caprihan, A., and Fukushima, E. (1993). Non-invasive measurements of granular flows by magnetic resonance imaging. In *Powders & Grains 93* (ed. C. Thornton), p. 383. Balkema, Rotterdam.

Nakagawa, M., Moss, J. L., and Altobelli, S. A. (1998). Segregation of granular particles in a nearly packed rotating cylinder: a new insight for axial segregation. *NATO ASI Series E Applied Sciences*, **350**, 703–710.

Narayan, V., Menon, N., and Ramaswamy, S. (2006). Nonequilibrium steady states in a vibrated-rod monolayer: tetratic, nematic, and smectic correlations. *J. Stat. Mech.: Theory and Exp.*, **2006**(01), P01005.

Narayan, V., Ramaswamy, S., and Menon, N. (2007). Long-lived giant number fluctuations in a nonequilibrium nematic. *Science*, **317**(5834), 105–108.

Nedderman, R. M. (1992). *Statics and kinematics of granular materials*. Cambridge University Press.

Nédélec, F., Surrey, Th., and Maggs, A. C. (2001). Dynamic concentration of motors in microtubule arrays. *Phys. Rev. Lett.*, **86**(14), 3192–3195.

Nédélec, F. J., Surrey, T., Maggs, A. C., and Leibler, S. (1997). Self-organization of microtubules and motors. *Nature*, **389**(6648), 305–308.

Newey, M., Ozik, J., van der Meer, S. M., Ott, E., and Losert, W. (2004). Band-in-band segregation of multidisperse granular mixtures. *Europhys. Lett.*, **66**(2), 205–211.

Nguyen, M. L. and Coppersmith, S. N. (1999). Properties of layer-by-layer vector stochastic models of force fluctuations in granular materials. *Phys. Rev. E*, **59**(5), 5870–5880.

Nie, X., Ben-Naim, E., and Chen, S. (2002). Dynamics of freely cooling granular gases. *Phys. Rev. Lett.*, **89**(20), 204301.

Nie, X., Ben-Naim, E., and Chen, S. Y. (2000). Dynamics of vibrated granular monolayers. *Europhys. Lett.*, **51**(6), 679–648.

Nishimori, H. and Ouchi, N. (1993). Computational models for sand ripple and sand dune formation. *Int. J. Mod. Phys. B*, **7**(9 & 10), 2025–2034.

Nixon, S. A. and Chandler, H. W. (1999). On the elasticity and plasticity of dilatant granular materials. *J. Mech. Phys. Sol.*, **47**(6), 1397–1408.

Nowak, E. R., Knight, J. B., Ben-Naim, E., Jaeger, H. M., and Nagel, S. R. (1998). Density fluctuations in vibrated granular materials. *Phys. Rev. E*, **57**(2), 1971–1982.

Nowak, S., Samadani, A., and Kudrolli, A. (2005). Maximum angle of stability of wet granular pile. *Nature Physics*, **1**(9), 50–52.

Oh, J. and Ahlers, G. (2003). Thermal-noise effect on the transition to Rayleigh-Bénard convection. *Phys. Rev. Lett.*, **91**, 094501.

O'Hern, C. S., Langer, S. A., Liu, A. J., and Nagel, S. R. (2001). Force distributions near jamming and glass transitions. *Phys. Rev. Lett.*, **86**(1), 111–114.

Olafsen, J. S. and Urbach, J. S. (1998). Clustering, order and collapse in a driven granular monolayer. *Phys. Rev. Lett.*, **81**, 4369–4372.

Olafsen, J. S. and Urbach, J. S. (1999). Velocity distributions and density fluctuations in a 2D granular gas. *Phys. Rev. E*, **60**, R2468–2471.

Olmsted, P. D. and Goldbart, P. (1992). Isotropic-nematic transition in shear flow: State selection, coexistence, phase transitions, and critical behavior. *Phys. Rev. A*, **46**, 4966–4993.

Onsager, L. (1949). The effects of shape on the interaction of colloidal particles. *Ann. NY Acad. Sci*, **51**(4), 627–659.

Orellana, C. S., Aranson, I. S., Kwok, W.-K., and Rica, S. (2005). Self-diffusion of particles in gas-driven granular layers with periodic flow modulation. *Phys. Rev. E*, **72**(4), 040301.

Orpe, A. V. and Khakhar, D. V. (2001). Scaling relations for granular flow in quasi-two-dimensional rotating cylinders. *Phys. Rev. E*, **64**, 031302.

Orza, J. A. G., Brito, R., van Noije, T. P. C., and Ernst, M. H. (1997). Patterns and long range correlations in idealized granular flows. *Int. J. Mod. Phys. C*, **8**, 953.

Ottino, J. M. and Khakhar, D. V. (2000). Mixing and segregation of granular materials. *Ann. Rev. Fluid Mech.*, **32**, 55–91.

Oyama, Y. (1939). Horizontal rotating cylinder. *Bull. Inst. Phys. Chem. Res. (Tokyo), Rep.*, **18**, 600.

Paczuski, M. and Boettcher, S. (1996). Universality in sandpiles, interface depinning, and earthquake models. *Phys. Rev. Lett.*, **77**(1), 111–114.

Pak, H. K. and Behringer, R. P. (1993). Surface waves in vertically vibrated granular materials. *Phys. Rev. Lett.*, **71**(12), 1832–1835.

Pak, H. K., van Doorn, E., and Behringer, R. P. (1995). Effects of ambient gases on granular materials under vertical vibration. *Phys. Rev. Lett.*, **74**(23), 4643–4646.

Paolotti, D., Barrat, A., Marconi, U. M. B., and Puglisi, A. (2004). Thermal convection in monodisperse and bidisperse granular gases: A simulation study. *Phys. Rev. E*, **69**(6), 061304.

Park, H.-K. and Moon, H.-T. (2002). Square to stripe transition and superlattice patterns in vertically oscillated granular layers. *Phys. Rev. E*, **65**, 051310.

Parteli, E. J. R., Duran, O., and Herrmann, H. J. (2007). Minimal size of a barchan dune. *Phys. Rev. E*, **75**(1), 011301.

Parteli, E. J. R., Schartz, V., and Herrmann, H. J. (2005). Barchan dunes on Mars and on Earth. In *Powders and Grains 2005* (ed. H. H. R. Garcia-Rojo and S. McNamara), p. 959. Balkema, Leiden, 2005.

Pedley, T. J. and Kessler, O. J. (1992). Hydrodynamic phenomena in suspensions of swimming microorganisms. *Ann. Rev. Fluid Mech*, **24**, 313–358.

Persson, B. N. J. (2000). *Sliding Friction: Physical Principles and Applications*. Springer, New York.

Pohlman, N. A., Severson, B. L., Ottino, J. M., and Lueptow, R. M. (2006). Surface roughness effects in granular matter: Influence on angle of repose and the absence of segregation. *Phys. Rev. E*, **73**(3), 031304.

Pooley, C. M. and Yeomans, J. M. (2004). Stripe formation in differentially forced binary systems. *Phys. Rev. Lett.*, **93**(11), 118001.

Pöschel, T. and Schwager, T. (2005). *Computational Granular Dynamics: Models and Algorithms*. Springer Science+ Business Media, Inc., New York.

Pouliquen, O. (1999). On the shape of granular fronts down rough inclined planes. *Phys. Fluids*, **11**, 1956.

Pouliquen, O. and Chevoir, F. (2002). Dense flows of dry granular material. *CR Physique*, **3**, 163–175.

Pouliquen, O., Delour, J., and Savage, S. B. (1997). Fingering in granular flows. *Nature*, **386**, 816.

Pouliquen, O. and Gutfraind, R. (1996). Stress fluctuations and shear zones in quasistatic granular chute flows. *Phys. Rev. E*, **53**(1), 552–561.

Prevost, A., Egolf, D. A., and Urbach, J. S. (2002). Forcing and velocity correlations in a vibrated granular monolayer. *Phys. Rev. Lett.*, **89**, 084301.

Prevost, A., Melby, P., Egolf, D. A., and Urbach, J. S. (2004). Nonequilibrium two-phase coexistence in a confined granular layer. *Phys. Rev. E*, **70**(5), 050301.

Prigozhin, L. (1999). Nonlinear dynamics of aeolian sand ripples. *Phys. Rev. E*, **60**(1), 729–733.

Pye, K. and Tsoar, H. (1990). *Aeolian Sand and Sand Dunes*. Unwin Hyman, London.

Radjai, F., Brendel, L., and Roux, S. (1996). Nonsmoothness, indeterminacy, and friction in two- dimensional arrays of rigid particle. *Phys. Rev. E*, **54**(1), 861–873.

Radjai, F. and Wolf, D. E. (1998). The origin of static pressure in dense granular media. *Granular Matter*, **1**, 3–8.

Rajchenbach, J. (1990). Flow in powders: From discrete avalanches to continuous regime. *Phys. Rev. Lett.*, **65**(18), 2221–2224.

Rajchenbach, J. (2000). Granular flows. *Adv. Phys.*, **49**(2), 229–256.

Rajchenbach, J. (2001). Dynamics of grain avalanches. *Phys. Rev. Lett.*, **88**(1), 014301.

Rajchenbach, J. (2002). Development of grain avalanches. *Phys. Rev. Lett.*, **89**(7), 074301.

Rajchenbach, J. (2003). Dense, rapid flows of inelastic grains under gravity. *Phys. Rev. Lett.*, **90**(14), 144302.

Ramirez, R., Pöschel, T., Brilliantov, N. V., and Schwager, T. (1999). Coefficient of restitution of colliding viscoelastic spheres. *Phys. Rev. E*, **60**, 4465–4472.

Ramírez, R., Risso, D., and Cordero, P. (2000). Thermal convection in fluidized granular systems. *Phys. Rev. Lett.*, **85**(6), 1230–1233.

Rapaport, D. C. (2002). Simulational studies of axial granular segretation in a rotating cylinder. *Phys. Rev. E*, **65**, 061306.

Rapaport, D. C. (2004). *The Art of Molecular Dynamics Simulation*. Cambridge University Press.

Rapaport, D. C. (2007). Simulated three-component granular segregation in a rotating drum. *Phys. Rev. E*, **76**(4), 041302.

Rayleigh, L. (1883). On the vibration of a cylindrical vessel containing liquid. *Phil. Mag*, **15**, 229–235.

Reis, P. M. and Mullin, T. (2002). Granular segregation as a critical phenomenon. *Phys. Rev. Lett.*, **89**(24), 244301.

Rericha, E. C., Bizon, C., Shattuck, M. D., and Swinney, H. L. (2001). Shocks in supersonic sand. *Phys. Rev. Lett.*, **88**(1), 14302.

Reynolds, C.W. (1987). Flocks, herds and schools: A distributed behavioral model. *ACM SIGGRAPH Computer Graphics*, **21**(4), 25–34.

Reynolds, O. (1885). On the dilatancy of media composed of rigid particles in contact. *Phil. Mag. Ser. 5*, **50-20**, 469.

Riedel, I. H., Kruse, K., and Howard, J. (2005). A self-organized vortex array of hydrodynamically entrained sperm cells. *Science*, **309**(5732), 300–303.

Risso, D., Soto, R., Godoy, S., and Cordero, P. (2005). Friction and convection in a vertically vibrated granular system. *Phys. Rev. E*, **72**(1), 011305.

Ristow, G. H. (1997). Phase diagram and scaling of granular materials under horizontal vibrations. *Phys. Rev. Lett.*, **79**(5), 833–836.

Rosato, A. D., Strandburg, K. J., Prinz, F., and Swendsen, R. H. (1987). Why the Brazil nuts are on top: Size segregation of particulate matter by shaking. *Phys. Rev. Lett.*, **58**(10), 1038–1040.

Rouyer, F. and Menon, N. (2000). Velocity fluctuations in a homogeneous 2D granular gas in steady state. *Phys. Rev. Lett.*, **85**(17), 3676.

Royer, J. R., Corwin, E. I., Flior, A., Cordero, M.-L., Rivers, M. L., Eng, P. J., and Jaeger, H. M. (2005). Formation of granular jets observed by high-speed X-ray radiography. *Nature Physics*, **1**(3), 164–167.

Samadani, A. and Kudrolli, A. (2000). Segregation transitons in wet granular matter. *Phys Rev. Lett.*, **85**(24), 5102–5105.

Samadani, A. and Kudrolli, A. (2001). Angle of repose and segregation in cohesive granular matter. *Phys. Rev. E*, **64**(5), 051301.

Samadani, A., Pradhan, A., and Kudrolli, A. (1999). Size segregation of granular matter in silo discharges. *Phys. Rev. E*, **60**(6), 7203–7209.

Sankararaman, S., Menon, G. I., and Sunil Kumar, P. B. (2004). Self-organized pattern formation in motor-microtubule mixtures. *Phys. Rev. E*, **70**(3), 031905.

Sapozhnikov, M.V., Peleg, A., Meerson, B., Aranson, I. S., and Kohlstedt, K. L. (2005). Far-from-equilibrium Ostwald ripening in electrostatically driven granular powders. *Phys. Rev. E*, **71**(1), 011307.

Sapozhnikov, M. V., Aranson, I. S., Kwok, W.-K., and Tolmachev, Y. V. (2004). Self-assembly and vortices formed by microparticles in weak electrolytes. *Phys. Rev. Lett.*, **93**(8), 084502.

Sapozhnikov, M. V., Aranson, I. S., and Olafsen, J. S. (2003a). Coarsening of granular clusters: Two types of scaling behaviors. *Phys. Rev. E*, **67**(1), 010302.

Sapozhnikov, M. V., Tolmachev, Y. V., Aranson, I. S., and Kwok, W. K. (2003b).

Dynamic self-assembly and patterns in electrostatically driven granular media. *Phys. Rev. Lett.*, **90**(11), 114301.

Savage, S. B. (1979). Gravity flow of cohesionless granular materials in chutes and channels. *J. Fluid Mech.*, **92**, 53.

Savage, S. B. (1983). Granular flows down rough inclines: Review and extension. In *Mechanics of Granular Materials: New Models and Constitutive Relations* (ed. J. T. Jenkins and M. Satake), pp. 261–282. Elsevier, Amsterdam.

Savage, S. B. (1984). The mechanics of rapid granular flows. *Adv. Appl. Mech.*, **24**, 289.

Savage, S. B. (1993). Disorder, diffusion, and structure formation in granular flows. In *Disorder and granular media* (ed. D. B. ang A. Hansen), Amsterdam, pp. 255–285. North Holland.

Savage, S. B. (1998). Analyses of slow high-concentration flows of granular materials. *J. Fluid Mech.*, **377**, 1–26.

Savage, S. B. and Hutter, K. (1989). The motion of a finite mass of granular material down a rough incline. *J. Fluid Mech.*, **199**, 177–215.

Savage, S. B. and Lun, C. K. K. (1988). Particle size segregation in inclined chute flow of dry cohesionless granular solids. *J. Fluid. Mech.*, **189**, 311–335.

Schäfer, J., Dippel, S., and Wolf, D. E. (1996). Force schemes in simulations of granular materials. *J. Phys. I France*, **6**, 5–20.

Scheffler, T. and Wolf, D. E. (2002). Collision rates in charged granular gases. *Granular Matter*, **4**(3), 103–113.

Scherer, M. A., Melo, F., and Marder, M. (1999). Sand ripples in an oscillating annular sand–water cell. *Phys. Fluids A*, **11**, 58–67.

Schorghofer, N., Jensen, B., Kudrolli, A., and Rothman, D. H. (2004). Spontaneous channelization in permeable ground: theory, experiment, and observation. *J. Fluid Mech.*, **503**, 357–374.

Schröter, M., Goldman, D. I., and Swinney, H. L. (2005). Stationary state volume fluctuations in a granular medium. *Phys. Rev. E*, **71**(3), 030301.

Schröter, M., Ulrich, S., Kreft, J., Swift, J. B., and Swinney, H. L. (2006). Mechanisms in the size segregation of a binary granular mixture. *Phys. Rev. E*, **74**(1), 011307.

Schwämmle, V. and Herrmann, H J (2003). Solitary wave behaviour of dunes. *Nature*, **426**, 619.

Shandarin, S.F. and Zeldovich, Y.B. (1989). The large-scale structure of the universe: Turbulence, intermittency, structures in a self-gravitating medium. *Rev. Mod. Phys.*, **61**(2), 185–220.

Shen, A. Q. (2002). Granular fingering patterns in horizontal rotating cylinders. *Phys. Fluids*, **14**, 462–470.

Shinbrot, T. (1997). Competition between randomizing impacts and inelastic collisions in granular pattern formation. *Nature*, **389**(6651), 574–576.

Shinbrot, T. (2004). The Brazil nut effect - in reverse. *Nature*, **429**, 352–353.

Shinbrot, T., Duong, N.-H., Kwan, L., and Alvarez, M. M. (2004). Dry granular flows can generate surface features resembling those seen in martian gullies. *Proc. Natl. Acad. Sci. USA*, **101**(23), 189–198.

Shinbrot, T. and Muzzio, F. J. (1998). Reverse buoyancy in shaken granular beds.

Phys. Rev. Lett., **81**(20), 4365–4368.

Shoichi, S. (1998). Molecular-dynamics simulations of granular axial segregation in a rotating cylinder. *Mod. Phys. Lett. B*, **12**, 115–122.

Shraiman, B. I. (2005). Mechanical feedback as a mechanism of growth control. *Proc. Natl. Acad. Sci. USA*, **102**, 3318–3323.

Silbert, L. E., Ertaş, D., Grest, G. S., Halsey, T. C., and Levine, D. (2002a). Geometry of frictionless and frictional sphere packings. *Phys. Rev. E*, **65**(3), 031304.

Silbert, L. E., Ertaş, D., Grest, G. S., Halsey, T. C., Levine, D., and Plimpton, S. J. (2001). Granular flow down an inclined plane: Bagnold scaling and rheology. *Phys. Rev. E*, **64**(5), 051302.

Silbert, L. E., Grest, G. S., and Landry, J. W. (2002b). Statistics of the contact network in frictional and frictionless granular packings. *Phys. Rev. E*, **66**(6), 061303.

Silbert, L. E., Grest, G. S., Plimpton, S. J., and Levine, D. (2002c). Boundary effects and self-organization in dense granular flows. *Phys. Fluids*, **14**(8), 2637–2646.

Silbert, L. E., Landry, J. W., and Grest, G. S. (2003). Granular flow down a rough inclined plane: Transition between thin and thick piles. *Phys. Fluids*, **15**, 1–10.

Simha, R. A. and Ramaswamy, S. (2002). Hydrodynamic fluctuations and instabilities in ordered suspensions of self-propelled particles. *Phys. Rev. Lett.*, **89**(5), 058101.

Smith, D., Ziebert, F., Humphrey, D., Duggan, C., Steinbeck, M., Zimmermann, W., and Käs, J. (2007). Molecular motor-induced instabilities and crosslinkers determine biopolymer organization. *Biophys. J.*, **93**(12), 4445–4452.

Smith, T. R. and Bretherton, F. P. (1972). Stability and the conservation of mass in drainage basin evolution. *Water Resour. Res.*, **3**, 1506–1528.

Snezhko, A., Aranson, I. S., and Kwok, W. K. (2005). Structure formation in electromagnetically driven granular media. *Phys. Rev. Lett.*, **94**(10), 108002.

Snezhko, A., Aranson, I. S., and Kwok, W. K. (2006). Surface wave assisted self-assembly of multidomain magnetic structures. *Phys. Rev. Lett.*, **96**(7), 078701.

Snoeijer, J. H., Vlugt, T. J. H., van Hecke, M., and van Saarloos, W. (2004). Force network ensemble: A new approach to static granular matter. *Phys. Rev. Lett.*, **92**(5), 54302.

Socolar, J. E. S. (1998). Average stresses and force fluctuations in non-cohesive granular materials. *Phys. Rev. E.*, **57**(3), 3204–3215.

Sokolov, A., Aranson, I. S., Kessler, J. O., and Goldstein, R. E. (2007). Concentration dependence of the collective dynamics of swimming bacteria. *Phys. Rev. Lett.*, **98**, 158102.

Song, Y., Mason, E. A., and Stratt, R. M. (1989). Why does the Carnahan-Starling equation work so well? *J. Phys. Chem.*, **93**(19), 6916–6919.

Soto, R. and Mareschal, M. (2002). Statistical mechanics of fluidized granular media: short range velocity correlations. *Phys. Rev. E*, **63**, 041303.

Srebro, Y. and Levine, D. (2003). Role of friction in compaction and segregation of granular materials. *Phys. Rev. E*, **68**(6), 061301.

Stambaugh, J., Lathrop, D. P., Ott, E., and Losert, W. (2003). Pattern formation in a monolayer of magnetic spheres. *Phys. Rev. E*, **68**(2), 026207.

Stambaugh, J., Smith, Z., Ott, E., and Losert, W. (2004). Segregation in a monolayer of magnetic spheres. *Phys. Rev. E*, **70**(3), 031304.

Staron, L., Vilotte, J. P., and Radjai, F. (2002). Preavalanche instabilities in a granular pile. *Phys. Rev. Lett.*, **89**(20), 204302.

Stegner, A. and Wesfreid, J. E. (1999). Dynamical evolution of sand ripples under water. *Phys. Rev. E*, **60**(4), 3487–3490.

Stronge, W. J. (1990). Rigid body collisions with friction. *Proc. R. Soc. Lond. A*, **431**, 169–181.

Sunthar, P. and Kumaran, V. (2001). Characterization of the stationary states of a dilute vibrofluidized granular bed. *Phys. Rev. E*, **64**(4), 041303.

Surrey, T., Nédélec, F., Leibler, S., and Karsenti, E. (2001). Physical properties determining self-organization of motors and microtubules. *Science*, **292**, 1167–1171.

Swift, J. B. and Hohenberg, P. C. (1977). Hydrodynamic fluctuations at the convective instability. *Phys. Rev. A*, **15**, 319–328.

Taberlet, N., Losert, W., and Richard, P. (2004). Understanding the dynamics of segregation bands of simulated granular material in a rotating drum. *Europhys. Lett.*, **68**(4), 522–528.

Taberlet, N. and Richard, P. (2006). Diffusion of a granular pulse in a rotating drum. *Phys. Rev. E*, **73**(4), 041301.

Taguchi, Y.-H. (1992). New origin of a convective motion: Elastically induced convection in granular materials. *Phys. Rev. Lett.*, **69**(9), 1367–1370.

Tan, M.-L. and Goldhirsch, I. (1998). Rapid granular flows as mesoscopic systems. *Phys. Rev. Lett.*, **81**(14), 3022–3025.

Tegzes, P., Vicsek, T., and Schiffer, P. (2002). Avalanche dynamics in wet granular materials. *Phys. Rev. Lett.*, **89**(9), 094301.

Tegzes, P., Vicsek, T., and Schiffer, P. (2003). Development of correlations in the dynamics of wet granular avalanches. *Phys. Rev. E*, **67**(5), 051303.

Tennakoon, S. G. K., Kondic, L., and Behringer, R. P. (1998). Onset of flow in a horizontally vibrated granular bed: convection by horizontal shearing. *Europhys. Lett.*, **99**(9), 1–6.

Thomas, C. C. and Gollub, J. P. (2004). Structures and chaotic fluctuations of granular clusters in a vibrated fluid layer. *Phys. Rev. E*, **70**(6), 061305.

Thoroddsen, S. T. and Shen, A. Q. (2001). Granular jets. *Phys. Fluids*, **13**(1), 4–6.

Toner, J. and Tu, Y. (1998). Flocks, herds, and schools: A quantitative theory of flocking. *Phys. Rev. E*, **58**(4), 4828–4858.

Toner, J., Tu, Y., and Ramaswamy, S. (2005). Hydrodynamics and phases of flocks. *Ann.Phys.*, **318**(1), 170–244.

Troian, S. M., Wu, X. L., and Safran, S. A. (1989). Fingering instability in thin wetting films. *Phys. Rev. Lett.*, **62**(13), 1496–1499.

Tsai, J.-C., Ye, F., Rodriguez, J., Gollub, J. P., and Lubensky, T. C. (2005). A chiral granular gas. *Phys. Rev. Lett.*, **94**(21), 214301.

Tsimring, L.S. and Aranson, I.S. (1997). Cellular and localized structures in a vibrated granular layer. *Phys. Rev. Lett.*, **79**, 213–216.

Tsimring, L. S., Ramaswamy, R., and Sherman, P. (1999). Dynamics of a shallow fluidized bed. *Phys. Rev. E*, **60**(6), 7126–7130.

Umbanhowar, P. B., Melo, F., and Swinney, H. L. (1996). Localized excitations in a vertically vibrated granular layer. *Nature*, **382**, 793–796.

Umbanhowar, P. B. and Swinney, H. L. (2000). Wavelength scaling and square/stripe and grain mobility transitions in vertically oscillated granular layers. *Physica A*, **288**, 344.

Underhill, P. T., Hernández-Ortiz, J. P., and Graham, M. D. (2008). Diffusion and spatial correlations in suspensions of swimming particles. *Phys. Rev. Lett.*, **100**, 248101.

Unger, T., Kertész, J., and Wolf, D. E. (2005). Force indeterminacy in the jammed state of hard disks. *Phys. Rev. Lett.*, **94**, 178001.

Urbach, J. S. and Olafsen, J. S. (2001). Experimental observations of non-equilibrium distributions and transitions in a two-dimensional granular gas. In *Granular Gases* (ed. T. Poeschel and S. Luding), p. 410. Springer-Verlag, Heidelberg, Germany.

Utter, B. and Behringer, R. P. (2004). Self-diffusion in dense granular shear flows. *Phys. Rev. E*, **69**(3), 031308.

van Duijneveldt, J. S., Heinen, A. W., and Lekkerkerker, H. N. W. (1993). Phase separation in bimodal dispersions of sterically stabilized silica particles. *Europhys. Lett.*, **21**(3), 369–374.

van Noije, T.P.C. and Ernst, M.H. (2000). Cahn-Hilliard theory for unstable granular fluids. *Phys. Rev. E*, **61**(2), 1765–1782.

van Noije, T. P. C., Ernst, M. H., and Brito, R. (1998). Ring kinetic theory for an idealized granular gas. *Physica A*, **251**, 266–283.

Vanel, L., Howell, D., Clark, D., Behringer, R. P., and Clément, E. (1999). Memories in sand: Experimental tests of construction history on stress distributions under sandpiles. *Phys. Rev. E*, **60**(5), 5040–5043.

Venkataramani, S. C. and Ott, E. (1998). Spatiotemporal bifurcation phenomena with temporal period doubling: patterns in vibrated sand. *Phys. Rev. Lett.*, **80**(1), 3495–3498.

Venkataramani, S. C. and Ott, E. (2001). Pattern selection in extended periodically forced systems: A continuum coupled map approach. *Phys. Rev. E*, **63**, 046202.

Vicsek, T., Czirók, A., Ben Jacob, E., Cohen, I., and Shochet, O. (1995). Novel type of phase transition in a system of self-driven particles. *Phys. Rev. Lett.*, **75**(6), 1226–1229.

Villarruel, F. X., Lauderdale, B. E., Mueth, D. M., and Jaeger, H. M. (2000). Compaction of rods: Relaxation and ordering in vibrated, anisotropic granular material. *Phys. Rev. E*, **61**(6), 6914–6921.

Visscher, K., Schnitzer, M. J., and Block, S. M. (1999). Single kinesin molecules studied with a molecular force clamp. *Nature*, **400**(6740), 184–189.

Volfson, D., Cookson, S., Hasty, J., and Tsimring, L. S. (2008). Biomechanical ordering in dense cell populations. *Proc. Natl. Acad. Sci. USA*, **106**. in press.

Volfson, D., Meerson, B., and Tsimring, L. S. (2006). Thermal collapse of a granular gas under gravity. *Phys. Rev. E*, **73**(6), 061305.

Volfson, D., Tsimring, L. S., and Aranson, I. S. (2003*a*). Order parameter description of stationary partially fluidized shear granular flows. *Phys. Rev. Lett.*, **90**(25), 254301.

Volfson, D., Tsimring, L. S., and Aranson, I. S. (2003*b*). Partially fluidized shear granular flows: Continuum theory and molecular dynamics simulations. *Phys. Rev.*

E, **68**(2), 021301.

Volfson, D., Tsimring, L. S., and Kudrolli, A. (2004). Anisotropy driven dynamics in vibrated granular rods. *Phys. Rev. E*, **70**, 051312.

Voth, G. A., Bigger, B., Buckley, M. R., Losert, W., Brenner, M. P., Stone, H. A., and Gollub, J. P. (2002). Ordered clusters and dynamical states of particles in a vibrated fluid. *Phys. Rev. Lett.*, **88**(23), 234301.

Wagner, C. (1961). Theory of precipitate change by redissolution. *Z. Electrochem*, **65**(7), 581–591.

Walker, J. (1982). When different powders are shaken, they seem to have lives of their own. *Sci. Am.*, **247**(3), 166.

Waller, M. D. (1938). Vibrations of free circular plates. Part 3: A study of Chladni's original figures. *Proc. Phys. Soc.*, **50**(1), 83–86.

Walsh, A. M., Holloway, K. E., Habdas, P., and de Bruyn, J. R. (2003). Morphology and scaling of impact craters in granular media. *Phys. Rev. Lett.*, **91**(10), 104301.

Walton, O. R. (1993). Numerical simulation of inelastic, frictional particle-particle interactions. In *Particulate Two-phase Flow* (ed. M. C. Roco), p. 884. Butterworth-Heinemann, Boston.

Whitham, G. B. (1974). *Linear and Nonlinear Waves*. Wiley, New York.

Wildman, R. D., Huntley, J. M., and Parker, D. J. (2001). Convection in highly fluidized three-dimensional granular beds. *Phys. Rev. Lett.*, **86**(15), 3304–3307.

Wildman, R. D., Martin, T. W., Krouskop, P. E., Talbot, J., Huntley, J. M., and Parker, D. J. (2005). Convection in vibrated annular granular beds. *Phys. Rev. E*, **71**(6), 061301.

Wildman, R. D. and Parkar, D. J. (2002). Energy nonequipartition in a vibrated granular gas mixture. *Phys. Rev. Lett.*, **88**(19), 064301.

Williams, J. C. (1963). The segregation of powders and granular materials. *Fuel Soc. J.*, **14**, 29–34.

Williams, J. C. (1976). The segregation of particulate materials: A review. *Powder Technol.*, **15**, 245–251.

Wittmer, J. P., Cates, M. E., and Claudin, P. (1997). Stress propagation and arching in static sandpiles. *J. Phys. I*, **7**, 39–80.

Wittmer, J. P., Claudin, P., Cates, M. E., and Bouchaud, J.-P. (1996). An explanation for the central stress minimum in sand piles. *Nature*, **382**, 336–338.

Wu, X. L. and Libchaber, A. (2000). Particle diffusion in a quasi-two-dimensional bacterial bath. *Phys. Rev. Lett.*, **84**(13), 3017–3020.

Yangagita, T. (1999). Three-dimensional cellular automaton model of segregation of granular materials in a rotating cylinder. *Phys. Rev. Lett.*, **82**, 3488–3491.

Yeh, S. R., Seul, M., and Shraiman, B. I. (1997). Assembly of ordered colloidal aggregrates by electric-field-induced fluid flow. *Nature*, **386**(6620), 57–59.

Zhang, W. and Viñals, J. (1997). Pattern formation in weakly damped parametric surface waves. *J. Fluid Mech.*, **336**, 301–330.

Zhou, J., Dupuy, B., Bertozzi, A. L., and Hosoi, A. E. (2005). Theory for shock dynamics in particle-laden thin films. *Phys. Rev. Lett.*, **94**(11), 117803.

Ziebert, F., Aranson, I.S., and Tsimring, L. S. (2007). Effects of crosslinks on motor-mediated filament organization. *New Journal of Physics*, **9**, 421.

Zik, O., Levine, D., Lipson, S. G., Shtirkman, S., and Stavans, J. (1994). Rotationally induced segregation of granular material in a horizontal rotating cylinder. *Phys. Rev. Lett.*, **73**(5), 644–647.

Zuriguel, I., Gray, J. M. N. T., Peixinho, J., and Mullin, T. (2006). Pattern selection by a granular wave in a rotating drum. *Phys. Rev. E*, **73**(6), 061302.

Notations

Notation	Meaning
D	diffusion coefficient
d	grain size
δ	control parameter (partial fluidization)
E	energy
e	restitution coreffcient
η	shear viscosity
$\mathcal{F}$	free energy
Fr	Froude number
f	frequency of vibration
g	gravity aceleration
Γ	driving acceleration amplitude normalized by gravity
γ	strain tensor
$\dot{\gamma}$	strain rate tensor
h	height or total thickness of a granular layer
J	mass flux
Kn	Knudsen number
$\mathbf{k}, k$	wavevector, wave number
κ	heat conductance
L	system size
λ	growth rate
m	grain mass
μ	friction coefficient
ν	density
ω	angular velocity
p	pressure
P	probability distribution
ϕ	filling fraction
$\mathbf{Q}$	orientational order parameter (tensor)
R	radius of a circular drum
Re	Reynolds number
$\mathbf{r} = [x, y, z]$	position
ρ	order parameter
Sh	Shields number
σ	stress tensor
T	granular temperature
t	time
τ	vector or quasivector of local orientation
$\mathbf{u} = [u_x, u_y, u_z]$	displacement
$\mathbf{v} = [v_x, v_y, v_z]$	velocity
Ξ	conversion rate between rolling and static layers
Z	co-ordination number
z_R	thickness of a rolling layer
z_S	thickness of a static layer
ζ	cooling rate

Index

Amplitude equations, 101, 105, 108
Angle of repose, 10, 17, 110, 133, 166, 181, 182, 192, 195, 197, 225, 236
  dynamic, 17, 161, 192, 194
  static, 17, 139, 141, 142, 161, 195
Asters, see Tubules
Avalanches, see Granular flows
Axial segregation, see Segregation

Bacteria, 23, 26, 28, 259, 260, 285, 291, 294, 296, 304
  bioconvection, 28
  nematic ordering, 26, 261
  patterns, 297, 298
  swimming, 28, 29, 259, 290, 294, 296–298, 300–302, 306, 307, 309
Bagnold relation, see Constitutive relations
BCRE model, see Granular flows
Bistability, 10, 85, 86, 89, 130, 144, 155
Boids, see Self-propelled particles
Boltzmann–Enskoq equation, see Granular gas
Brazil-nut effect, 10, 167, 174, 175
  reverse, 174, 179
Bundles, see Tubules
Burgers equation, see Clustering

Cahn–Hilliard equation, 74, 210
Cellular automata modeling, 92, 93, 108, 180, 181, 229
Cellular-automata modelling, 56
Chladni figures, 7
Chute flow, see Granular flows
Clustering, 6, 7, 59, 61, 63, 64, 66, 68, 85, 89, 249, 285, 289
  Burgers equation, 66
  electro cell, 243
  hydrodynamic description, 69
  instability, 59, 60, 63, 68, 289
  Ostwald ripening, 244
  van der Waals instability, 72
  vertical vibration, 85
Coarsening, 17, 73, 75, 89, 90, 159, 190, 191, 193, 195, 196, 199–201, 209, 212, 231, 238, 244, 246, 247, 249
  Ostwald ripening, 89, 90, 243, 247, 248
  van der Waals instability, 72, 75, 85, 90
Coaxiality principle, 53
Compactivity, see Edwards hypothesis

Constitutive relations, 35, 36, 38, 39, 42, 44, 45, 51, 55, 75, 110, 117, 135, 138, 141, 155, 159, 161, 264
  Bagnold, 54, 122–124, 137, 141, 227
  plasticity, 51, 54
Contact mechanics, see Molecular dynamics
Convection, 9, 10, 45, 78, 79, 82, 83, 92, 93, 95, 96, 113, 167, 174, 179, 250
  bioconvection, 28
  in rapid chute flows, 83
  instability, 80, 81
  mechanism for segregation, 174
Coupled map lattice, 108
Craters, 18, 19
Creep, 229
Crosslinks, see Tubules

Density inversion, 10, 78–81, 83
Diffusion, 46, 66, 109, 175, 193, 195, 198, 304
  angular, 224, 270, 272
  anisotropic, 273
Dilatancy, 10, 39, 53
Dunes, see Two-phase flows

Edwards hypothesis, 46
Einstein relation, 45
Elasticity theory, 33, 37, 51
Electrical excitation, 297
Entropy, see Edwards hypothesis
Erosion patterns, see Two-phase flows
Event-driven simulations, see Molecular dynamics

Fingering instability, 151, 164, 200, 239
Flocking, see Self-propelled particles
Flow rule, see Granular flows
Fluctuation–dissipation relation, 40, 45, 49
Force chains, 37, 51
Force distribution, 37, 39
Friction, 33, 35, 39, 49, 51, 53, 123, 126, 131, 137, 169, 174, 199, 201, 208, 218
  anisotropic, 219
  Coulomb, 57
  effective, 121, 123, 124, 133
  sliding, 32
Fronts, 12, 85, 104, 128, 141–144, 150–152, 154, 155, 195, 196, 200, 239, 284
Froude number, 80–82, 123, 124, 159, 190

Gas, granular, see Granular gas

Ginzburg–Landau theory, 54, 73, 101, 102, 104, 132, 220, 258, 275, 276
Granular flows, 10, 12, 15, 39, 44, 46, 51, 54–57, 98, 117–119, 125, 126, 128–130, 132, 134, 137, 142, 151, 153, 156, 163, 164, 181, 184, 200, 205, 220
  avalanches, 10, 12, 17, 54, 56, 117, 130, 141–144, 146–154, 156, 157, 159, 161–163, 165, 179, 184, 185, 190, 200, 236, 239
    balloon, 152
    transverse, 143
    wedge, 146, 151–153, 156
  BCRE model, 56, 117, 128–130, 132, 153, 156, 163, 181, 230
  chute, 118, 122, 124, 139, 172, 186
  dense, 36, 39, 41, 44, 46, 50, 51, 55, 98, 118, 119, 127, 128, 132, 137, 184
  flow rule, 52, 54, 123, 124
  gravity-driven, 117
  in rotating drum, 16, 159, 170, 184, 185, 191, 194, 200
  longitudinal instability, 158, 159
  rapid chute flows, 83
  rheology, 33, 118, 122, 124, 128, 137, 238
  Saint-Venant models, 56, 117, 125, 128, 137, 158
  surface, 92, 124, 128, 130, 134, 181, 189
  underwater, 12, 144, 151, 234, 239
Granular gas, 7, 40, 55, 59, 61, 66, 68, 78, 82, 177, 223, 247, 289
  clustering, 59, 61, 63, 85, 89
  instabilities, 59, 61, 72, 89
  thermal collapse, 66, 68
Granular hydrodynamics, 39, 42, 44, 51, 60, 62, 63, 66, 69, 77, 79, 80, 83, 111, 112, 133, 184, 291
Granular temperature, 9, 42, 45, 46, 59, 60, 65, 68, 83, 111, 133, 175, 177, 179, 209
Gravity-driven flows, see Granular flows

Haff's law, 59, 64, 66
Hard spheres, 79, 176, 203, 204, 254
Heaping, 7, 91, 92
Hertz law, 33
Hexagonal patterns, see Surface waves
Homogeneous cooling state, see Granular gas
Hydrodynamics, see Granular hydrodynamics
Hysteresis, see Bistability

Impact, 18, 19, 100, 230
Indeterminacy, 32, 35, 57
Inelastic collapse, 58, 63, 64, 66, 285
Inelasticity, 30, 40, 41, 57, 69, 85, 270, 283
Interfaces, 9, 96, 100, 105–107, 173, 195, 202, 246
Interstitial gas, 7, 92

Isostaticity, 35

Jamming, 39, 46, 299

Kinesin, see Molecular motors
Kinetic sieving, see Segregation
Kinetic theory, 39, 41, 43, 49, 51, 175, 269

Lattice models, 229
Leidenfrost effect, 80
Lifshitz–Slyozov–Wagner theory, 247, 248

Magnetic particles, 253
Magnetic snakes, 23
Microtubules, see Tubules
Mohr circle, see Mohr–Coulomb criterion
Mohr–Coulomb criterion, 51, 55, 128, 133
Molecular dynamics, 30, 37, 56, 58, 63, 66, 68, 75, 79, 92, 99, 100, 111, 114, 117, 132, 133, 135, 177, 179, 198, 199, 208, 212, 261, 266
  contact mechanics, 58
  event-driven, 31, 56, 57, 72, 99
  soft particle, 57, 132, 133, 208, 261
Molecular motors, 24, 210, 259, 267–270, 279, 290, 297, 304

Nematic ordering, see Ordering
Nematodynamics equations, 220, 262, 263
Normality principle, 52
Nucleation, see Coarsening

Order parameter, 39, 55, 56, 113, 118, 132–139, 141, 142, 145, 154, 200, 222
  nematic, 219, 261, 263–265
  partial fludization, 132, 138, 139
  tensor, 262
Ordering, 23, 26, 203–205, 214, 219, 222, 223, 259–262, 266, 273, 288
  crystalline, 7, 21, 26, 85, 212, 213
  instability, 273
  nematic, 21, 27, 203–205, 210, 212, 213, 219, 220, 222, 223, 259, 261–263, 266, 273, 307
  tetratic, 21, 212, 214, 223
Oscillon, 9, 96
  coupled map approach, 110
  Ginzburg–Landau theory, 102
  Swift–Hohenberg theory, 108
Ostwald ripening, see Coarsening

Parametric resonance, 98
Partial fluidization theory, 134, 135, 138, 144, 154, 159, 163, 187
Phase transition, 55, 85, 113, 132, 273, 288, 289, 296, 300, 309
Prandtl number, 79
Principle of dilatancy, see Dilatancy

Principle of normality, *see* Normality principle

q-model, *see* Force chains

Radial segregation, *see* Segregation
Rayleigh number, 79
Repose angle, *see* Angle of repose
Restitution coefficient, 30, 31, 33, 41, 45, 58, 65, 72, 77, 87, 99, 121, 125, 137, 209, 270, 291
Reynolds number, 227
Rheology, *see* Granular flows
Rice, *see* Rods
Ripples, *see* Two-phase flows
Rods, 19, 21, 27, 166, 203–210, 212, 219, 220, 263, 264, 266
   collective motion, 21, 203, 207
   Onsager theory, 203
   polar, 212, 267, 269–271, 273–275, 282, 304
   swirling, 212, 214, 222, 261, 263
Rotating drum, *see* Segregation

Saint-Venant models, *see* Granular flows
Saltation, 227, 229, 231, 239
Segregation, 10, 17, 73, 144, 167, 169, 170, 172–175, 177, 179, 181, 184, 185, 188–190, 194, 197, 199–201
   axial, 17, 73, 190, 191, 197–200, 202
      coarsening, 191, 196
      oscillatory regime, 194
      ternary, 199
   binary mixture, 167, 175, 179, 184, 201
   condensation, 175, 177, 179
   convection, 174
   kinetic sieving, 172–174, 177, 179, 184, 185, 190
   radial, 17, 184, 190, 197, 199
      avalanching, 17, 189, 225
      multi-petal, 17, 189
      rolling, 56, 179, 181, 185, 188–190
Self-organized criticality (SOC), 56, 93, 164, 165
Self-propelled particles, 222, 285, 288, 289, 291, 294, 300, 303
   flocking, 28, 29, 259, 286, 294
   Vicsek model, 285, 288, 289
Shock propagation, 65, 66, 72, 75, 111, 148, 173
Soft particle, *see* Molecular dynamics
Solid–fluid mixtures, *see* Two-phase flows
Squares, *see* Surface waves
Standing waves, 95, 96, 98
Stick–slip motion, 32, 39, 163
Strain rate, 39, 42, 52, 54, 60, 82, 118, 119, 121, 123, 134, 137, 171, 172, 263, 264, 304
Stratification, 56, 172, 179, 181, 184

Stress, 31–36, 38, 42, 44, 51–55, 60, 62, 118, 119, 121, 123, 126, 127, 132, 134, 135, 137, 142, 155, 161, 187, 235, 236, 238, 263–265, 306
Stress–strain relation, *see* Constitutive relations
Stripes, *see* Surface waves
Submonolayer, 6, 7
Surface flows, *see* Granular flows
Surface waves, 23
   dispersion relation, 96, 112
   hexagons, 9, 22, 79, 96, 100, 104, 105, 225, 226
   oscillons, 9, 96, 98, 99, 104, 108–110
   squares, 9, 40, 56, 85, 96, 98–100, 104, 108–110, 113, 115, 164, 193, 225, 278, 284, 302
   stripes, 9, 96, 99, 100, 108, 110, 185, 195, 200, 201, 225
   superoscillons, 107
Swift–Hohenberg equation, 54, 108, 113
   fluctuations, 113
   generalized, 108
   oscillons, 108
   subcritical, 115
Swimming, *see* Bacteria

Temperature, *see* Granular temperature
Tetratic ordering, *see* Ordering
Thermal collapse, 66, 68
Thermoconvection, *see* Convection
Tubules, 268, 273, 274
   asters, 27, 268, 269, 275–277, 279, 281
   bundles, 275, 281, 285
   cross-links, 24, 281–285
   cytoskeleton, 23, 267, 268, 281, 305
   vortices, 21, 27, 275, 276, 278, 281
Two-phase flows
   dunes, 34, 229, 231, 233, 239
   erosion, 13, 117, 128–130, 230, 239
   ripples, 227, 229, 233, 234, 236, 238, 239
   underwater avalanches, 12

van der Waals instability, *see* Coarsening
Vicsek model, *see* Self-propelled particles
Viscosity
   effective, 66, 74, 193
Volume fraction, 47, 49, 50, 69, 119, 123, 169, 173, 179, 225, 294
Volume function, *see* Edwards hypothesis
Vortices, 12, 15, 21, 22, 27, 83, 200, 206, 209, 211, 214, 249–251, 268, 269, 279, 308
   longitudinal, 83
Vorticity, 66, 220, 222, 252, 263, 304, 307

Yield, 51–53, 55